W9-APW-015

DEPARTMENT OF MATHEMATICS
BRONFMAN SCIENCE CENTER
WILLIAMS COLLEGE
WILLIAMSTOWN, MA 01267

Learning Discrete Mathematics
with ISETL

Nancy Baxter   Ed Dubinsky   Gary Levin

Learning Discrete Mathematics with ISETL

With 43 Illustrations

Springer-Verlag New York Berlin Heidelberg
London Paris Tokyo

Nancy Baxter
Department of Mathematical
  Sciences
Dickinson College
Carlisle, Pennsylvania 17013
U.S.A.

Ed Dubinsky
Departments of Education and
  Mathematics
Purdue University
West Lafayette, Indiana 47907
U.S.A.

Gary Levin
Department of Mathematics
  and Computer Science
Clarkson University
Potsdam, New York 13676
U.S.A.

Mathematics Subject Classification (1980): 03-01, 04-01, 05-01, 26-01, 05C20, 05B20, 68B05, 68Exx

Library of Congress Cataloging-in-Publication Data

Baxter, Nancy (Nancy H.)
  Learning discrete mathematics with ISETL.
  Includes index.
  1. Mathematics—1961– . 2. Electronic data
processing—Mathematics.   3. ISETL (Computer program
language) I. Dubinsky, Ed. II. Levin, Gary.
III. Title.
QA39.2.B39   1988   510′.78   88-29477
ISBN 0-387-96898-9 (alk. paper)

Printed on acid-free paper

Camera-ready copy provided by the authors.
Printed and bound by R.R. Donnelley and Sons, Harrisonburg, Virginia.
Printed in the United States of America.

9 8 7 6 5 4 3 2 1

ISBN 0-387-96898-9 Springer-Verlag New York Berlin Heidelberg
ISBN 3-540-96898-9 Springer-Verlag Berlin Heidelberg New York

Preface

The title of this book, *Learning Discrete Mathematics with ISETL* raises
two issues. We have chosen the word "Learning" rather than "Teaching"
because we think that what the student does in order to learn is much
more important than what the professor does in order to teach. Academia
is filled with outstanding mathematics teachers: excellent expositors, good
organizers, hard workers, men and women who have a deep understanding
of Mathematics and its applications. Yet, when it comes to ideas in Mathematics, our students do not seem to be learning. It may be that something
more is needed and we have tried to construct a book that might provide a
different kind of help to the student in acquiring some of the fundamental
concepts of Mathematics. In a number of ways we have made choices that
seem to us to be the best for learning, even if they don't always completely
agree with standard teaching practice.

A second issue concerns students' writing programs. ISETL is a programming language and by the phrase "with ISETL" in the title, we mean
that our intention is for students to write code, think about what they have
written, predict its results, and run their programs to check their predic-
tions. There is a trade-off here. On the one hand, it can be argued that
students' active involvement with constructing Mathematics for themselves
and solving problems is essential to understanding concepts. A powerful effect of computers for most students is to keep them deeply involved. This
can be very important because we believe that any time you construct
something on the computer then, with a greater or lesser degree of awareness, you will construct something in your head. On the other hand, it
can also be argued that writing computer programs forces the student to
be concerned with mathematically irrelevant issues: syntax, programming
style, debugging. Moreover, many students feel that Mathematics consists
of problem categories and a bunch of recipes for solving all problems in the
category. They believe that all that needs to be learned are the steps of
the various recipes in a category together with the ability to decide which
recipe to use for a given problem. There is a serious danger that working
with computers can reinforce this sterile view of Mathematics.

We have tried to make this text a rebuttal to the arguments against having students write programs in Mathematics courses. The language ISETL

is not only relatively free of syntax details irrelevant to Mathematics, but its syntax is very close to the usual language in which Mathematics is written. Indeed, some students who have used this book have adopted ISETL as the language which they use for taking notes in Mathematics courses. We invite mathematicians to leaf through the following pages and see how familiar are the expressions in typescript (that is how ISETL syntax appears in the book). As for seeing Mathematics as a collection of procedures, we have found that students who work on constructing ISETL implementations of mathematical concepts find these constructions far from being cut and dried. They become absorbed and are loathe to leave a problem even if they do not see right away how to do it. No matter how much we who have played the role of teachers in this approach believe that we are directing the students, the mathematical ideas that they are able to put together themselves with the help and stimulation of ISETL, these ideas our students have seen as rich, complex and applicable—and they see them as their own.

ISETL is a derivative of the programming language SETL which was developed by J.T. Schwartz and his group at the Courant Institute. SETL, a very-high-level language, is compiled and runs only on mainframes. It is very powerful and supports not only the many mathematical constructs that will be discussed in this book, but also has a number of important programming features, such as backtracking. The language and its use is fully described in the book, *Programming with Sets: an introduction to SETL* by J.T. Schwartz, R.B.K. Dewar, E. Dubinsky, and E. Schonberg (New York: Springer-Verlag, 1986).

One of us, in working with a draft of that book, came across the following phrase, written by Schwartz.

> The knowledge assumed is roughly equivalent to that which would be acquired in a freshman-level course in Discrete Mathematics.

It seemed reasonable to reverse the sense of the statement and set students to learning how to program in SETL in order to learn Discrete Mathematics. That is, in attempting to make the syntax work and in solving problems that relied on the constructs which the language supports, students learned the corresponding Mathematics, more or less automatically.

This was tried in courses at Clarkson University (Fall, 1984), Dickinson College (Fall, 1985), and University of California, Berkeley (Spring 1986). The approach seemed to be very successful, but the unwieldiness of SETL, which requires a large mainframe and is very slow, made it rather inconvenient and prevented any large-scale development. Then, in 1986 one of us developed a stripped-down version of SETL that is interactive, runs on many microcomputers, and incorporates several programming features that make the language even more useful for helping students learn Mathematics concepts. This new version, ISETL (Interactive SET Language), is fairly polished and user-friendly. In addition to mainframes such as the VAX, and intermediate machines such as the SUN, it runs on the Macintosh and IBM PC compatible machines.

There are a number of ways in which working with a language like ISETL helps students acquire Mathematics concepts. When a student constructs a process in ISETL to represent a function or calculate the greatest common divisor, then he or she is likely to be able to reflect on the dynamic aspect of such mathematical objects. We are always advising our students to THINK ISETL. It means to imagine, mentally, the actions of selecting a value in the domain, performing the operations called for in the function to obtain the result and repeating this action for other values in the domain. Or it means to think about beginning at 2 and checking every integer up to the minimum of two given positive integers to find the largest one that divides them both. We find that such reflections are helpful in thinking about the corresponding mathematical ideas.

Mathematical ideas that can be thought about as dynamic processes may sometimes have to be considered as total entities or objects. For example, in a parameterized family of functions, the functions are objects, not processes. A set may be considered as a "process" for determining which objects it contains. But if a set is to contain other sets as elements, then these sets must be objects. In ISETL, the code that implements a process can be saved in files, used later, combined with other code, and so on. Thus, it is treated as an object and tends to become so in the student's mind.

Another way in which ISETL can have a strong effect on learning is the use of procedures (called **funcs** in ISETL) to test mathematical objects for various properties. For example, in studying relations, it is possible to have a class discussion in which a large number of properties (reflexive, symmetric, transitive, and so on) are discussed in a short time, fairly vaguely. A relation (on a finite set) can be represented in ISETL as a set of ordered pairs, called a **map** and the students are given the task of writing, for each property, a **func** which accepts a relation and returns **true** or **false** depending on whether or not that relation satisfies the property. The ISETL code for doing this is usually almost identical to the mathematical definition. The student learns the concept through the action of describing it precisely for the computer. The understanding tends to be rich because the student will think about the process (which he or she constructed) that tests for the property. And the amount of energy put out by the teacher is relatively minimal. This is a very effective teaching method and we recommend its use whenever it is reasonable to construct the appropriate ISETL **func**.

Finally we would like to mention that two of the authors are engaged in research on how mathematical ideas are learned. Roughly speaking, we are trying to determine mental constructions that a student needs to make in order to understand a concept and to understand the ways in which these constructions can be made. Wherever possible, we have tried to reflect in this book the things that we are finding out about the learning process, by designing ISETL activities that tend to lead the student toward making useful mental constructions.

It is important to note that the instructor in a course that uses this book is not required to pay conscious attention to the effects of working with ISETL that we have described. The learning that takes place seems to happen somewhat spontaneously with the student as he or she is constructing mathematical concepts in ISETL. It is possible to use this text with a minimum of lecturing. We have found that students can and will read this book and they should be encouraged to do so. A great deal of the work in the course can be accomplished by assigning students tasks taken from the problems that appear at the end of each section with the teacher's contribution reduced to helping the student both to learn how to solve problems and to become aware of what he or she is doing when working on a problem. As a result, the teacher can be free for other activities such as motivation, bringing things together, providing a role model for the students and, most importantly, helping the students move through the very complex and challenging activities that are accessible to them, but not without a great deal of effort, frustration, and growth.

We have tried to incorporate several features in this book. As we just remarked, our intention is to make it a book that students are willing and able to read. A considerable amount of learning can take place by having students try to follow our explanations. These are not always written in a "spoon-feeding" style but tend to be challenging for the students. We expect our students to read actively, trying to think about things that are not instantly clear. The *Preview* section at the beginning of each chapter is offered as an "advance organizer" with the intention of guiding the student in reading through that chapter.

Material is not always presented in a linear sequence. Occasionally an idea is used before it is explained, and it appears as a "pointer" to future sections in the text. This occurs most frequently in Chapter 1 when new syntax is often introduced in context without explicit explanation. The student is expected to work out the meaning from the context, and later, when the explanation comes, coordinate her or his initial understanding with the discussion in the text. Although this is an unusual approach, we believe it is consistent with how people learn Mathematics. Learning is not always linear and even when it is, the line is not the same for everyone. Sometimes it is very helpful to struggle to understand an idea entirely on your own. This is certainly closer to the way one encounters a Mathematics concept outside of the particular class where it is "taught." It is also the case that an explanation is most useful if it comes just at the point when a student is ready for it in terms of previous knowledge and a feeling of a "need to know." Because this point comes at different times for different people, it will often seem that an explanation is in the "wrong" place.

This will cause some feelings of frustration. We believe that struggling to overcome frustration through a growth in understanding is an important part of learning, and we have tried to create such situations for students who use our text. On the other hand, this can lead to very hard times for the students and it is essential that the teacher be deeply involved in helping them through these periods. It is important to constantly point out, in very specific ways, how much they are learning and how their frustrations are shared by everyone who tries to understand Mathematics. To get students to reflect on what they are doing, the teacher can help them remember past situations (in the course and out of it) in which they struggled to understand seemingly impossible mathematical ideas which subsequently came to seem very simple.

A last feature that we want to mention is our approach to having the students learn ISETL syntax. The student's initial introduction to the language comes through what we call terminal sessions. A copy of what appears on the screen is provided in the text and the student is asked to reproduce that screen on her or his terminal. In various problems, the student is required to simply copy programs and use them. Later, the task is to modify one aspect of the given program, and finally the student reaches the point of writing complete programs on her or his own. Most of this is completed by the end of Chapter 1, at which point we expect the student to be reasonably comfortable with the language. In subsequent chapters, many things that were used in Chapter 1 are explained in detail. There appears to be quite a bit of what may look like discussion of syntax in these later chapters, for example when explaining set formers, accessing components of tuples, syntax for functions, and so on. But most of these are given because they are actually explanations of ideas behind certain mathematical terminology. It is because of the close relation between ISETL syntax and standard mathematical notation that the student may think that it is syntax that is being discussed. In fact, what is actually happening is that he or she is learning to express mathematical ideas in mathematical notation.

The study of Chapter 1, as we have noted, is designed to get the students familiar with using and interpreting ISETL. It also includes an introduction to number systems, presented in conjunction with the ISETL simple data types: `integer`, `floating-point`, `Boolean`, and `string`. This introduction includes some elementary ideas about common factors and multiples, modular arithmetic, and change of base.

Chapter 2 is about propositional calculus. We want students to become familiar with Boolean expressions, to think of them as analogous to algebraic expressions, the only difference being that the actual values involved will be `true` or `false` rather than numbers. This will provide a foundation for considering, in later chapters, functions whose domain and/or range values are Boolean. Such functions are important both for our later study of predicate calculus (Chapter 5) and for understanding mathematical induction (Chapter 7). We would like students to be able to think about Boolean expressions, calculate their values, and reason about them when given additional information. Chapter 2 also returns to the study of proofs that appeared in Chapter 1. This time we try to get the student to think about a proof as a determination of the value of a Boolean expression.

In Chapter 3 we discuss **sets** and **tuples** (finite sequences). We attempt to have the student learn to think about these concepts simultaneously as dynamic processes (formation of a set, running along a sequence) and static objects. A major topic in this chapter is the "one-liner" in which students are asked to express various mathematical statements compactly in ISETL notation. The critical point here is that in order to do this, one has to develop a global understanding of the mathematical statement. The chapter contains a large number of pointers to topics that appear later in the book.

By the time a course based on this book gets to Chapter 4, the student has spent a great deal of time with several ISETL constructs that can represent functions : **funcs**, **tuples**, **strings** and **smaps**. It is the goal of Chapter 4 for students to be strongly aware of the process involved in a function, to see these processes in situations that they are trying to make sense of and to think about the process as they use various operations on functions to make new functions. Chapter 4 also describes several applications of the function concept. The emphasis is on "function as process" because this seems to us to be very important in the beginning of a students' education in Mathematics. There is, however, lurking behind the scene, the idea of a function as an object, which will be important for students who go on to study more advanced Mathematics.

The goals of Chapter 5 are for the student to coordinate propositional calculus, sets, and functions that were studied in the previous three chapters and to construct the concept of a proposition that depends on one or more variables and is existentially or universally quantified over these variables. The discussion in the chapter takes the student from single level quantifications up through two level quantifications and on to three and higher level. The main activities that the student is engaged in include expressing situations in the language of quantification, negating a quantified statement, and reasoning about such statements. This chapter, along with Chapter 7, make up the two chapters in this book that are most influenced by two of the author's research into learning mathematical concepts.

Chapter 6 presents the student with a number of applications of **sets** and **tuples** to other ideas in Mathematics including permutations and combinations, matrices, and determinants.

The discussion of induction in Chapter 7 can move quickly and go a long way because most of the basic concepts that need to be put together in order to construct the concept of mathematical induction have already been studied. We emphasize the task of understanding a problem situation in terms of a proposition valued function of the positive integers. This initial step is often missing in the minds of students and our experience has been that conscious awareness of such a mathematical entity can come as a consequence of performing certain computer tasks. A lot of this was done already in Chapter 4 so doing it here is a helpful review or reiteration for many students. One goal of the work with `tuples` in Chapter 3 was to prepare the student for representing a statement to be proved by induction as an infinite sequence whose values must be Boolean, but are unknown before the proof. Since a `tuple` can be understood as an infinite sequence only finitely many of whose values are known at a given time, this is a particularly appropriate model that is offered to the student. We find it more effective than dominoes or stepladders. Our expectation is that as a result of the work in Chapters 2 and 4 the student can understand an implication as both a process and an object and that he or she can work with implication valued functions of the positive integers. The main work of the chapter is to put all of this together in the context of a very large number and variety of problems.

Chapter 8 is an introduction to relations and graphs. A relation (on finite sets) can be represented in ISETL as a `map` and graphs can be represented in several ways. We exploit these constructions to have the students write ISETL `funcs` that test for a variety of properties. By writing and running these programs, the student develops not only a familiarity with the formal definitions, but an intuitive feeling for what they mean.

There are a variety of courses that can be based on this book, depending on the background of the students and the intent of the instructor. It is probably the case that the first four chapters need to be covered sequentially before the rest of the book. Once that is done, there is not much interdependence among the remaining four chapters, so that any subset of them can be covered in any order.

The amount of material that can be covered in a single semester depends on the usual factors of number of class hours, student ability, instructor's style, and so on. In addition, with this book, the programming background of students makes a difference. If the students already have some programming experience (but not with ISETL), it is possible to go through most or even all of the book in one semester. At the other extreme, if students are completely new to the computer, then it is necessary to spend 2 or 3 extra weeks with Chapter 1. We have found that this is sufficient and it is possible, with students who have little or no programming experience and are not exceptionally strong in Mathematics, to cover the first four chapters and at least one chapter from the last four.

The course has been used for computer science students, mathematics majors, pre-service secondary and elementary school teachers, engineers, liberal arts students, and as an early course for future Mathematics majors. With all of these populations we have found that studying Mathematics by working with ISETL as described in this book tends to build up students' confidence in their ability to do Mathematics and increases a little bit, the level of Mathematics that it is possible for the student to understand.

This book was typeset on a Macintosh Plus using TeXtures, which is an implementation of TeX for the Macintosh, and the macro package LaTeX. It was done entirely by the authors so that we accept full responsibility for all inadequacies, both in form and content. What merit the work has would not have been possible without the help of a great number of people, some whom we can thank here.

All of the figures were done in MacDraw or MacPaint by Elizabeth White to whom we are grateful.

We would like to express our debt to Jack Schwartz for creating SETL and allowing us to modify it to obtain ISETL.

We are much influenced in our ideas about computer science by the work of David Gries. We are also influenced in our ideas about both computer science and learning by Mike Clancy whom we have known as friend, colleague, teacher, and conversationalist.

It is regrettable that we cannot explicitly name all of the students who took courses based on early versions of this book. We hope that they learned as much from us as we learned from them. There were a number of students, both graduate and undergraduate, both paid and unpaid, who assisted in the course at one time or other. We are particularly grateful to Chris Abdelmalek, Manuel Bronstein, Carol Napier, Marc Rubin, Doug Smith, and Paul Stodghill.

We thank Ann Martin for developing a complete set of notes reflecting all of the course materials and the classroom interactions the first time the course was offered at Dickinson College. These notes proved to be especially helpful in writing several of the chapters.

We owe a special thanks to John Wenner, a Dickinson College student, who as a Dana Foundation academic intern, reviewed most of the exercises and helped us with the typesetting and proofreading.

In addition to our own departments who allowed us to try something new and different, we would like to thank the Berkeley Mathematics department that allowed one of us, as a visitor, to implement a very early version of the course.

It is our pleasure to thank Richard Bayne, Arnold Lebow, Don Muench, and Mazen Shahin. These colleagues have volunteered to help us with summer courses and minicourses that we have run as a means of explaining to other mathematicians how we use this book to teach discrete mathematics.

A number of colleagues, influenced more by our enthusiasm than by anything concrete we were able to give them at the time, volunteered to try out, in their regular courses, various sets of notes on which this book was ultimately based. Their suggestions have been invaluable to us. We are very pleased to express our thanks to Julianna Dowell, Octavio Garcia, David Kramer, Arnold Lebow, and Josiah Meyer.

Finally, and sadly, we would like to pay our respects to the memory of Walter Kaufmann-Buhler who was our editor at Springer-Verlag. It is rare to find someone in the publishing industry with such sensitivity to Mathematics and Education as was possessed by Walter. We would like to thank him for the very early confidence that he expressed and support that he gave us for the project that led to this book. He was a unique individual that we were happy to have as friend and colleague. The world is a lesser place without him.

N.H.B. Spring, 1988
Ed D.
G.M.L.

Contents

Chapter 1

Numbers, Programs, and ISETL

To the Instructor

This chapter reflects, as does the entire book, our ideas on how people can learn Mathematics. The essential point is that by working to understand and use a programming language the student is, perforce, mentally active in constructing whatever concepts are required by the syntax of the language and the programs that are written in the language. When, as is the case with ISETL (interactive set language), the syntax is very close to standard mathematical language and, as in this book, the programs are carefully selected to correspond to mathematical concepts, this mental activity is automatically directed toward understanding ideas in Mathematics.

Thus, our introduction to ISETL syntax in this first chapter is designed to stimulate thinking. We do not lead the student by the hand through the syntactic details described; instead, he or she is presented with terminal screens to interpret, code to understand, and programs to write. Some explanations are offered at various points, and the full description of ISETL is available when you order ISETL.[1] In many cases, exercises will require the use of syntax and even concepts that are not explicitly discussed until later. Learning is an individual and nonlinear activity. Some students will find this manner of presentation exactly right; others will require occasional intervention by the instructor in order to gain maximum benefit from the approach in this book. Everyone will probably be moving back and forth among the various chapters.

In this chapter's discussion of integers, the computer provides both a

---

[1] This full description of ISETL is called *An Introduction to ISETL*; we will refer to it as simply the *Intro*. There is a form in the back of this book for ordering ISETL and its accompanying documentation.

motivation and an implementation for change-of-base algorithms. Working with programs corresponding to concepts such as *relatively prime, greatest common divisor*, and *modular arithmetic* is expected to help the student to think actively about these ideas. Because the calculations exist as ISETL code, assertions about them and proofs of these assertions are more real for the student—they are statements about what is or is not possible in a calculation by the computer.

Our presentation of all the various data types in terms of objects and operations on them should help the student in later studies of *abstract data types*.

The need for different kinds of numbers and the various ways of representing them is intended to prepare the student for other courses in which the standard development of the real numbers is introduced. Representation of rationals as pairs (**tuples** of length 2) of integers and writing programs to implement the arithmetic of fractions is especially helpful here, as is the presentation of ISETL equations that cannot be solved unless certain "new" kinds of numbers are constructed. Intervals of **floating-point** numbers and manipulations with them provide our first introduction to elementary operations with sets.

Much of the material in this chapter is aimed at the student's development of her or his concept of function. We exploit heavily the procedure (ISETL **func**) and the **if...then...else** clause; the procedure as a function considered to be a process that transforms objects (often, but not always, numbers) and the **if...then...else** clause to define functions differently for different portions of the domain. Recently, one of us had an interesting reaction from students who were working with this material. One student tried to formulate a question in class. He began, "Is there a name for these...these functions that are not...that are different...?" He struggled for a word, an adjective, as the class looked on encouragingly. He tried "nonmathematical" but rejected it. He offered "not defined by a formula," but the other students disagreed. He stumbled for a while, and then, he relaxed and said, quietly, "No, the question is wrong. I realize that they are all just functions." There was an almost audible sigh of relief and agreement by the class, while the instructor, in an atypical concession to decorum and physical realities, only thought about jumping in the air and clicking his heels.

At the very end of this chapter, there is some material on propositional calculus that serves as a pointer to Chapter 2.

1.1   Preview

This book is based on the principle that you can learn Mathematics by writing programs. While you are figuring out how to get the computer to do the mathematically correct thing, you are mentally confronting the mathematical ideas involved. Your mind is growing as you develop the tools for understanding these ideas. To put it differently, there is no way you can teach a computer how to do something without learning it better yourself.

This is a book on thinking—about Mathematics or learning how to think better about Mathematics—but you won't always be aware of this. Most of the time, it will feel like writing programs, working problems, or reading an explanation. However, all the time your mind is growing. We have designed exercises so that doing them will improve your ability to think mathematically—whether you know it or not. Often, students who have used this method are surprised when they look back at where they were a few months or even a few weeks ago. They are very pleased at how many difficult mathematical ideas they have learned to think about— almost without trying. But, there are some things we *don't* want to happen while you are writing programs. We don't want you to get bogged down in a lot of syntax details, and we don't want you to spend a lot of time and energy learning how to use complicated programming structures that have little to do with Mathematics.

So we are using a programming language, ISETL, whose syntax is very close to the language of Mathematics. In fact, students who have learned ISETL in Discrete Mathematics sometimes tell us that they use ISETL now for taking notes in other Mathematics courses! In addition to its syntax, ISETL's content is also very close to Mathematics. The only things about the language that may require some effort to learn are those that correspond to mathematical ideas. You may think you are only learning how to write ISETL programs. Actually, you are also learning Mathematics. You will work hard, but you will learn how to use a lot of mathematical tools, and come to understand the ideas behind them.

In this first chapter, our goal is for you to learn some preliminary things about ISETL. (If you want to read all about ISETL in one place, order and read the *Intro*.) Occasionally, as you learn things about ISETL that relate to mathematical ideas, we will digress from programming and make detours through some Mathematics. We hope the theory will be interesting because it relates to the programming tasks you are working on. This will be our philosophy throughout the whole book. Topics about ISETL will be interspersed with topics in Mathematics. Sometimes, however, the distinction between the two will be very slight. You will learn to understand many relations between mathematical objects by tracing through programs that express these relations. You will also write many programs of your own that implement abstract mathematical ideas in a concrete way. We encourage you to think about what the computer is doing when it is performing the instructions your program has given it. In writing programs and then thinking about how the computer processes the information in your programs, you will begin to form "mental images" of the ideas you are learning. By doing this, you will come to really understand these ideas.

1.2   Overview of ISETL

We begin with getting started. ISETL is a program written in the language C. It consists, essentially, of an infinite loop that keeps asking for input, over and over. Every time the user enters something, it does what the input tells it to do and then comes back asking for more input. Until you tell ISETL that you have had enough (that is, you want to quit), ISETL keeps expecting more input.

To start ISETL, you just enter the command (that is, type it and hit return)

    isetl

The ISETL program will display something like

    ISETL(1.8)
    >

The number in parentheses is just a version number and, of course, will be different as ISETL undergoes improvements. Also, the command to start ISETL may be different on different machines. The > symbol is the ISETL prompt for input. If you want to leave ISETL, enter

    > !quit

If you don't want to leave, you can start giving ISETL some input. Figures 1.1 and 1.2 show terminal sessions with some sample inputs and the results ISETL returns. Notice the semicolons that usually appear at the end of an input. Most ISETL inputs require a semicolon at the end.

In these figures, the system prompt is %. Throughout this book, anything that ISETL produces is underlined (except for the ISETL system prompt >), and what the user would enter is printed in typescript. There is no indication of the carriage returns, which, of course, the user typed at the end of every line that was entered. Try to understand what is going on and try to duplicate the terminal session on your own machine. For each particular input to ISETL, try to understand the results ISETL returns. (Better yet, try to predict what the results will be before you hit the carriage return!) You probably won't always be able to tell what is going on, but it is good for you to try to understand something on your own before getting the explanation. You should try to reason about it based on what you know and, most importantly, make some guesses. Guessing is a very important part of thinking about Mathematics. Really creative people are always guessing. You don't have to always get the right answer. It is just a first step in understanding, and if you have thought hard about guessing what something is all about, then when the explanation comes, you will be ready to assimilate it.

In Figure 1.4, you won't be able to replicate everything in a terminal session until you know a little more about using files in ISETL, but

```
% isetl
ISETL (1.8)
>  7 + 18;
25;
>  13.0 * -233.8; 4*"Hello"; 5 = 2+3;
-3.03940e+03;
"HelloHelloHelloHello";
true;
>  17
>>                      >=
>> 3*6
>> ;
false;
>  2*13.2;
2.64000e+01;
>  x := 10;
>  x;
10;
>  x + 20;
30;
>  x := x -4; x;
6;
>  newat; newat;
!1!;
!2!;
>  MaxEquals := (max(2,3)=2);
>  maxEquals;
OM;
>  MaxEquals;
false;
>  MaxEquals or ("H" in "Hello");
true;
>  !quit
```

Figure 1.1: Sample terminal session

```
% isetl
ISETL (1.8)
> 4**5; 7.0**3; 7**3.0; 2**0.5; 17.5/3.8; 17/3; 17.5/3;
1024;
```
3.43000e+02;
3.43000e+02;
1.41421e+00;
4.60526e+00;
5.66667e+00;
5.83333e+00;
```
> 3+4:
```
syntax error, line2. '!clear' or '!edit'
```
>> !edit
3+4:
```
!Enter string to replace (blank line to execute)
```
4:
```
!Enter replacement
```
4;
3+4;
```
!Enter string to replace (blank line to execute)

```
7;
> !quit
```

Figure 1.2: Sample terminal session

read carefully through it and see if you can figure out what is happening. Figure 1.5 has some more examples for you to look at.

These figures illustrate a number of ISETL features, all of which will be explained, illustrated, and applied (mostly to implement mathematical ideas) in the remainder of this book. Right now, however, we can describe some of the main features that you should be looking for as you are reading.

One way of giving input to ISETL is to type one or more expressions, each followed by a semicolon. After the carriage return, the expressions are evaluated, and their values (`OM` if you didn't provide the right preliminary information, such as values for the variables it uses) are printed on the screen. (Figure 1.1 has lots of examples of this happening.) An expression can be spread out over several lines. If the expression is not complete, then ISETL will give you the >> prompt. Evaluation does not begin until the expression is complete.

Expressions are the things that ISETL evaluates. Give ISETL an ex-

```
%  isetl
ISETL (1.8)
>  program rel_prim;
>>          rp := om;
>>          read x;          $ x must be a positive integer
>>          read y;          $ y must be a positive integer
>>          if is_integer(x) and x > 0 and
>>                     is_integer(y) and y > 0 then
>>              rp := true;
>>              for i in [2..(x .min y)] do
>>                  if (x mod i = 0) and (y mod i = 0) then
>>                      rp := false;
>>                  end if;
>>              end for;
>>          end;
>>          print x, y, rp;
>>    end;
? 8;
? 12;
8;
12;
false;
>   !quit
```

Figure 1.3: Sample terminal session

pression, and it will give you back its value. This is a lot of what functions in Mathematics are all about. You have an expression (rule); you have a specification of the kinds of preliminary values that must be provided for the calculation (domain); and you have a specification of the kinds of values that can result from the calculation (range). Then, you have a function. Indeed, if you include under "calculation" (some people use the word "algorithm" here) just about anything that you might do (such as look up the answer in a table), then you've got essentially all of the functions that come up in Mathematics—theoretical and applied.

Then, there are identifiers, which are what ISETL uses to "remember" or store information. An identifier is any sequence of letters and numbers and the underscore, _, beginning with a letter. ISETL pays attention to upper- and lowercase. Thus, the identifier MaxEquals is not the same as the identifier maxEquals, so the latter is undefined in Figure 1.1. Its value is OM, for undefined. Identifiers are the same as variables in Mathematics.

```
%  isetl
ISETL (1.8)
>  5 div 3; -5 div 3; 5 mod 3; -5 mod 3;
1;
-2;
2;
1;
>  (2<3) and ((5.2/3.1) > 0.9); (3<=3) impl (not(3=2+1));
true;
false;
>  !include relprim
!include relprim completed
? 8;
? 12;
8;
12;
false;
>  !include relprim
!include relprim completed
? 8;
? 15;
true;
>  !quit

%  isetl relprim
ISETL (1.8)
? 8;
? 12;
8;
12;
false;
>  tup := [
>> !include tup.file
!include tup.file completed
>> ];
>  tup(1); tup(2); tup(3); tup(4);
"First Component";
"Second Component";
"Third Component";
OM;
>  !quit
```

Figure 1.4: Sample terminal session

```
% isetl
ISETL (1.8)
>  even(23); even(-4); odd(23); odd(-4);
false;
true;
true;
false;
>  float(23356); fix(3.7E3);
2.33560e+04;
3700;
>  floor(1945.5334E-2); char(36); ord("P");
19;
"$";
80;
>  exp(1.5); ln(3.0); ln(1); ln("two"); sqrt(17); sqrt(3.0);
4.48169e+00;
1.09861e+00;
0.00000e+00;
OM;
4.12311e+00;
1.73205e+00;
>  s := {3=2, "set", 19.7E-1, {2..6}, {3,6..53}, {6,4..-12}};
>  s;
{1.97000e+00, false, {6, 3, 2, 4, 5}, "set",
{3, 6, 9, 12, 15, 18, 21, 24, 27, 30, 33, 36, 39, 42, 45, 48, 51},
{-8, 2, 4, 0, -10, 6, -2, -4, -12, -6}};
>  {x:   x in {1..53} | x mod 3 = 0};
{3, 6, 9, 12, 15, 18, 21, 24, 27, 30, 33, 36, 39, 42, 45, 48, 51};
>  t := [5, "tuple", char("#"), 3+5];
>  t;
[5, "tuple", OM, 8];
>  #(s); #(t);
6;
4;
>  t := t + [OM, 7 mod 5 = 2, 13.6,
>> {"TRUE", [5=3+2, 17.0E-1, 17]}];
>  t; #(t);
[5, "tuple", OM, 8, OM, true, 1.36000e+01,
 {"TRUE",[true, 1.70000e+00, 17]}];
8;
>  t(3) := 9 div 2; t;
[5, "tuple", 4, 8, OM, true, 1.36000e+01,
 {"TRUE", [true, 1.70000e+00, 17]}];
>  !quit
```

Figure 1.5: Sample terminal session

You have to give, or assign, a value to a variable. Otherwise, it is undefined and has the value OM. There are several ways that a value can be assigned to an identifier.

One is to read its value either from an external file or from the keyboard. This is how the identifiers x and y get their values in program rel_prim in Figure 1.3.[2] After executing the statement, read x;, ISETL asks you to enter a value for x by displaying the ? prompt. Whatever value you type at the keyboard is assigned to x.

Another way to give a value to an identifier is to use an assignment statement. This is what happens in Figure 1.1 when you type the line x := 10;. We read the symbol := as "gets." To find what value x gets, you first evaluate the expression on the right-hand side of the assignment statement. This value is then assigned to the identifier on the left-hand side of the statement. Once you have assigned a value to a variable, anytime this variable appears in an expression that is being evaluated, the last value that was assigned to it is used when calculating the value of the expression.

The values we are talking about have different forms, or structures, that computer scientists call data types. In ISETL, we have the simple data types: integer, floating-point, Boolean, string, atom, and undefined. We also have compound, or structured, data types: set, tuple, map, func, and file. Each data type has a collection of specific objects associated with it. For instance, associated with the type integer are all the integers that a particular implementation of ISETL can handle, or just *true* and *false* with Boolean, or even "entities without meaning" (whatever that is) with atoms. Each type also has a collection of operations (addition, union, concatenation, implies, type testers, and so on) that can be performed on its objects or groups of objects. Most of these operations have specific requirements for the type of its operands. Sometimes there is an object that is special for an operation, such as 1 for multiplication or the empty string for concatenation. A particular collection of objects along with the associated operations make up what Mathematicians call a *universal algebra*. They do really complicated things with universal algebras—but, as you can see, the idea has a very simple beginning. It's just another programming language!

Set and tuple are two important data types in ISETL. Sets, which use curly brackets, will be a fundamental theme in this entire text (ISETL stands for Interactive SET Language) and in most Mathematics that you might study in the future. Sets are illustrated in Figure 1.5. Figure 1.5 also has some examples of tuples, which are indicated by square brackets. They will be discussed at length in Chapter 3. The order of members of a tuple is fixed, but in a set it can vary. Thus, while t(3) refers to the third component of the tuple t, the expression s(3), where s is a set,

---

[2]Comments in ISETL are written using $. Everything after $ to the end of the line is ignored by ISETL. The purpose of comments is to explain parts of the program to the reader.

has no meaning, so it would take the value OM. An interesting feature of ISETL, which is illustrated by the examples in Figure 1.5, is the fact that **sets** and **tuples** can have elements of different types as members including other **sets** and/or **tuples**.

As we said, expressions are one way to input information to ISETL. Another form of input to ISETL is a program, which, like other input, may be typed on the terminal as in Figure 1.3 or included from a file. It has to have a header line as shown, and it must end with **end;**.[3] As shown in Figure 1.4, you can enter a program with the same syntax in a file, in this case, called **relprim**. The **!include** facility then treats this file exactly as if it had been typed in line for line at the terminal. The same thing can be achieved (on some systems) by typing the name of the file on the line where you call the ISETL program. You can put as many files as you like on this line. So, in order to reproduce the terminal session in Figure 1.4, you have to create the file **relprim** containing the program **rel_prim**.

When input is taken from a file, it does not have to be complete. Any additional tokens that you wish to include with it must be typed before and after as the case may be. In this way, data may be taken from a file. For example, when running the terminal session in Figure 1.4, the file **tup.file** contains

```
"First Component", "Second Component", "Third Component"
```

The sequence shown in Figure 1.4 has exactly the same result as if the following were typed at the terminal:

```
tup := ["First Component", "Second Component",
        "Third Component"];
```

Summary of Section 1.2

The purpose of this section was to help you become familiar with ISETL and your computer system. Your goals are to learn how to

- log on and off your system.

- enter ISETL, give some commands, and then quit.

- create and edit a file.

- use files for input to and output from ISETL.

- print the contents of a file.

---

[3] Many statements in ISETL end with **end**. To make it easier to read long programs, **end** may be followed by the first word of the statement. Therefore, you could end programs with **end program**. The words that may optionally follow **end** are: **program**, **if**, **for**, **while**, and **func**. Sometimes we will use the optional keyword, sometimes not.

Before beginning the exercises, find out what type of computer you will be running ISETL on and then find out how ISETL works on your particular system. (As in the text, anything that ISETL or your system produces is underlined while anything that the user is to type is in `typescript`.)

Exercises

1.2.1 Running ISETL on your system.

If you have not already done so, find out how to run ISETL on your system and then try to duplicate the lines in Figure 1.1. Pause at the end of each line you type (before you hit the carriage return) and try to predict what ISETL will output. Notice that ISETL will return four kinds of things:

1. the value(s) of the expression(s) you just gave it,

2. the ISETL prompt > asking you to provide more input,

3. the ISETL prompt >> asking you to complete an input, and

4. an error message.

If you get an error message, look carefully at the line you just typed, try to figure out what is wrong with it, and then use the ISETL line editor to make the necessary changes.

1.2.2 Creating, editing, and using the contents of a file.

a. Use your computer's editor to create a file called `rp.tst`. Enter the following lines into the file:

```
program rel_prim;
    rp := om;
    read x;          $ x must be a positive integer
    read y;          $ y must be a positive integer
    if is_integer(x) and x > 0 and
            is_integer(y) and y > 0 then
        rp := true;
        for i in [2..(x .min y)] do
            if (x mod i = 0) and (y mod i = 0) then
                rp := false;
            end if;
        end;
    end if;
    print x, y, rp;
end program;
```

b. Notice that these are the same lines that you typed if you dupli-
cated Figure 1.3. They represent a complete ISETL statement,
which is actually a **program** in this case. Re-enter ISETL, but
include the name of the file (which contains your program) on
the command line. That is, type

```
isetl rp.tst
```

When you include the name of a file on the command line,
ISETL executes all statements in the file(s) listed before it does
anything else. (This feature is not available on all systems.)

If your file, **rp.tst**, contains a syntax (typing) error, ISETL
will give you an error message. Leave ISETL, edit your file,
and try again.

If you have not made any typing errors, as ISETL executes the
code in the file it will give you the ? prompt two times. Each
time you get the prompt, type a positive integer followed by a
semicolon, to signify the end of the input. Can you figure out
what the program does?

c. Leave ISETL and re-enter without typing **rp.tst** on the command line. At the > prompt, enter

```
!include rp.tst
```

Note: Do NOT put a semicolon at the end of the line!

Again, do what seems right. When you **!include** a file, ISETL
takes whatever code you have in the file (which does not have
to be an executable, i.e., complete, statement) and includes it
as part of your current run (see Figure 1.4 for more examples).
You can **!include** a file as many times as you like. Try some
different expressions when you get the ? prompt such as

```
?.3 * (2 + 5);
? 386 mod 7;
```

Repeat the process until you have determined what the program does.

d. What is the relationship between **x** and **y** that makes the
Boolean variable **rp** have the value *true*? *false*?

e. What happens if you input a negative integer, a floating-point
number, or a character? What does the value of **rp** mean in
these cases?

1.2.3 Reading from and writing to external files. Printing the contents of
a file.

a. Sometimes, instead of entering it at the keyboard, you want
ISETL to read information from a file. To do this, you first

have to create a file whose contents are exactly what you want
ISETL to read. To continue the example from the previous
exercise, use your editor to create a file called `rp.inp` with the
contents

```
3; 5; 1245; 17890; 10 + 18; 9 div 3; 10; -10;
```

b. Next, we have to change the program in the file `rp.tst` so
that it reads the values of x and y from the input file, `rp.inp`,
instead of from the keyboard. To use a file, an ISETL program
has to open it and assign an ISETL identifier to it, for example,
`infile`, and then close the file when finished using it. To do
this, insert the line

```
infile := openr("rp.inp");
```

just after the program header and insert the line

```
close(infile);
```

just before the end of the program. Since we want to read from
`infile`, edit your **read** statements to be

```
read x from infile;
read y from infile;
```

You can get your program to run repeatedly until there is no
more information in the file by using a `while` loop and the end-
of-file function, `eof`. Insert all the statements after opening the
necessary files and before closing the files inside a `while` loop
as follows:

```
while not eof(infile) do
        List of ISETL statements
end while;
```

In the previous example, the first time through the loop the
values 3 and 5 are assigned to x and y, respectively, the next
time through, 1245 and 17890, and so on until all the values
in the file have been read. That is, ISETL will keep trying to
read new values until it reaches the end of the file.

c. Not only can ISETL read from a file, it can also write to a
file instead of writing to the screen. Change your program in
`rp.tst` to write the values of x, y, and rp to an output file
called `rp.out`.
To do this,

* Open the file for writing:
  `out:= openw("rp.out");`
* Write to the file using the **print** ... **to** ... statement:
  `print x to out;`

* When you are finished writing, close the file:
  close(out);

d. Print copies of the files rp.tst, rp.inp, and rp.out.

1.3 Integers

There are many ways to represent an integer. Humans today do it with position and the digits $0, 1, ..., 9$—probably because they have 10 fingers. ISETL represents them the same way—probably to keep the humans happy. But this base 10 representation is complicated for a machine. Base 2, using only the digits 0 and 1, is much more convenient, so conversions back and forth are always happening. How do you represent a nonnegative integer by a base 2 representation? It's done with powers of 2. You want to write the integer as a sum of powers of 2, such as

$$
\begin{aligned}
314 &= 2^8 + 2^5 + 2^4 + 2^3 + 2^1 \\
314 &= 1 \cdot 2^8 + 0 \cdot 2^7 + 0 \cdot 2^6 + 1 \cdot 2^5 + 1 \cdot 2^4 \\
&\quad + 1 \cdot 2^3 + 0 \cdot 2^2 + 1 \cdot 2^1 + 0 \cdot 2^0
\end{aligned}
$$

Then, the base 2 representation is the string of 0's and 1's representing the coefficients of the powers of 2. (Of course, the infinite sequence of leading 0's is not written.) Thus, using an obvious notation, we write

$$(314)_{10} = (100111010)_2$$

which is read "314 considered as a base 10 representation represents the same integer as 100111010 considered as a base 2 representation." Of course, this is often shortened to "314 base 10 equals 100111010 base 2."

How do we convert from base 10 to base 2 and vice versa? Converting from base 2 to base 10 is the easy case. You just add up the powers of 2 corresponding to the positions where a 1 appears. Thus,

$$
\begin{aligned}
(10010101101)_2 &= (2^{10} + 2^7 + 2^5 + 2^3 + 2^2 + 2^0)_{10} \\
&= (1024 + 128 + 32 + 8 + 4 + 1)_{10} \\
&= (1197)_{10}
\end{aligned}
$$

The idea behind converting from base 2 to base 10 directly may be easy, but for relatively large powers of 2, it is hard to do without a calculator or at least a pencil and paper! So, here is another method that can not only be done in your head but also requires far fewer multiplications than the direct method. The idea is this: multiply the first item in the string by 2, add on the next item, multiply the result by 2, add on the next item, and so on, until you reach the end of the string. Applying this method to the above example gives

$$((((((((1*2+0)*2+0)*2+1)*2+0)*2+1)*2+0)*2+1)*2+1)*2+0)*2+1$$

which needs only 10 multiplications and can easily be calculated in your head to give 1197! This is usually called Horner's method.

But what about converting from base 10 to base 2? We need to find out what the coefficients are when we write a given number, for example, 314, as a sum of powers of 2. Notice that in the above example, reading just the 1's and 0's from left to right gives the base 2 representation. Let's start with 314 and work backward by repeatedly dividing by 2 to express 314 in a similar way. If we divide by 2, we have

$$314 = 157 * 2 + 0$$

Notice that **314 div 2 = 157** and **314 mod 2 = 0**, since the operators **div** and **mod** return the quotient and the remainder from integer division when they are applied to positive integers, respectively.

If we repeat this process and divide 157 by 2, then

$$157 = 78 * 2 + 1$$

where **157 div 2 = 78** and **157 mod 2 = 1**. By substituting the right-hand side for 157 in the first equation, we get

$$314 = (78 * 2 + 1) * 2 + 0$$

Notice that we are building up the desired result from right to left. We repeat the process for 78 and so on until we reach 0. Try it and see if you get

$$314 = ((((((((1 * 2 + 0) * 2 + 0) * 2 + 1) * 2 + 1) * 2 + 1) * 2 + 0) * 2 + 1) * 2 + 0$$

which tells us that

$$314 = 2^8 + 2^5 + 2^4 + 2^3 + 2^1$$

Moreover, reading the 0's and 1's from left to right we have the base 2 representation of 314 base 10, namely,

$$(314)_{10} = (100111010)_2$$

We can use **div** and **mod** to represent the process in a table as follows.

x	x div 2	x mod 2
314	157	0
157	78	1
78	39	0
39	19	1
19	9	1
9	4	1
4	2	0
2	1	0
1	0	1

Then the base 2 representation of 314 base 10 appears in the **x mod 2** column, which we read from the bottom to the top. Why?

Here is an ISETL **func** that converts a positive integer from base 10 to base 2.

```
conv_10_2 := func(x);
        if is_integer(x) and x > 0 then
            base_2 := "";
            while x /= 0 do
                if x mod 2 = 0 then
                        base_2 := "0" + base_2;
                else
                        base_2 := "1" + base_2;
                end;
                x := x div 2;
            end;
            return base_2;
        end;
    end func;
```

What does it mean to convert an integer to a representation in, for example, base 4? Can you modify the above **func** to convert to any given base, b, where $2 < b \leq 9$? What about going from base b to base 10? Can you generalize Horner's method for base 2 to base b?

You might also enjoy thinking about a unified theory—that is, a single algorithm for converting from any given base b_1 to any specified base b_2. This can get confusing. You have to keep in mind that there are three very different objects here. There is the integer, its representation in base b_1, and its representation in base b_2. The last two are easy to get hold of—a sequence of the right kind of digits. But what is an integer? This is something psychologists and philosophers argue about. It is not the representation, because if someone says "12," you don't know exactly what is meant. For example, which of the following diagrams has 12 circles?

The answer depends on which base the symbol "12" is supposed to refer to. You might try to clarify this by writing $(12)_{10}$ or $(12)_3$ as the case may be. But what is 10? Is it the number of fingers you have (base 10) or the number of hands (base 2)? What is a number? Think about it.

Summary of Section 1.3

In this section, we showed you methods for converting from base 10 to base 2 and vice versa. The following exercises are intended to give you some practice at converting between bases and to get you to generalize the given techniques. You will also be encouraged to start thinking about how a computer processes information. For the reader who has had some previous programming experience, we have included exercises asking you to define ISETL functions that convert from base 10 to base n, base 2 to base 10, base m to base 10, and finally, base m to base n.

Exercises

1.3.1 Converting from base 10 to base 2.

   a. Convert the following integers from base 10 to base 2 by using **div** and **mod**:

$$689, \ 128, \ 3860$$

   b. On page 17, we defined an ISETL **func** that converts **integers** from base 10 to base 2. Assume you are sitting at the terminal and have typed in the **func**, conv_10_2. To use this **func**, you have to call it by name and pass it a positive **integer**. For example, to use the **func** on the **integers** in part a. you could type

   conv_10_2(689); conv_10_2(128); conv_10_2(3860);

   and ISETL would return three **strings** of 0's and 1's representing each **integer** in base 2. Now "think like ISETL." Suppose you sent conv_10_2 the value **689**, what does the **func** do with this information? That is, what are the values of x and base_2 before you enter the **while** loop? What are their values after one pass through the loop? After two passes? After the last pass? The final value of base_2 is the value that is returned by the **func**. Input **689**, "think like ISETL," and show that you get the same result as you did in part a. Repeat the process for **128** and **3860**.

1.3.2 Generalizing Exercise 1.3.1. Converting from base 10 to base n, where $2 \leq n \leq 9$.

a. Suppose you wanted to convert an integer base 10 to another base, for example, base 4. To do this you have to express the given integer as a sum of powers of 4 and then look at the coefficients of each of the powers. Notice that this time the coefficients will be $0, 1, 2$, or 3. As when converting from base 10 to base 2, using **div** and **mod** can simplify the process. Use **div** 4 and **mod** 4 to convert

$$689, \ 128, \ 3860$$

to base 4.

b. Convert 689, 128, and 3860 to base 5.

c. Explain how you would modify the ISETL **func conv_10_2** to be a **func** that converts an integer base 10 to an integer base 4.

d. Define an ISETL **func conv_from_10** that accepts a positive **integer**, **x**, and a new base between 2 and 9, **n**, and returns a **string** representing **x** in base **n**. *Hint: We need to convert the remainder that we get using* **mod** *to its corresponding character so that we can concatenate this character onto the head of the string representing* **x** *in the new base. We can do this using the predefined functions* **char** *(which returns the machine dependent character at the given position) and* **ord** *(which returns the machine dependent position of the given character). For example, we can convert 3 to "3", by writing* **char(3+ord("0"))**.

1.3.3 Converting from base 2 to base 10.

a. Convert the following from base 2 to base 10 by adding up the powers of 2 corresponding to the positions where a 1 appears. (You may use your calculator.)

$$01101101110, \ 111100000, \ 10101010101$$

b. Repeat part a. using Horner's method. (You should be able to do these in your head.)

c. Define an ISETL **func conv_2_10** that inputs a **string**, **s**, that represents an integer in base 2, and then uses Horner's method to convert the string to a base 10 integer. *Hint: We need to remove the characters from the* **string s** *one by one so that we can convert the characters to their corresponding integers (using* **ord**) *and then use these integers in the algorithm. We can do this by using the* **string** *operation:*

take x fromb s;

which removes the first character from the **string s** *and stores it in* **x**.

1.3.4 Generalizing Exercise 1.3.3.   Converting from base m to base 10, where $2 \leq m \leq 9$.

   a. Generalize Horner's method to convert the following from base 4 to base 10:

$$1232320, \ 0102013, \ 0101110$$

   b. Convert the following from base 5 to base 10:

$$01020304, \ 112233, \ 10101110$$

   c. Use a generalization of Horner's method for base 2 to define an ISETL **func conv_to_10** that accepts as input a **string s** and an **integer m**, which is the base of **s**, and outputs the **integer** that is the base 10 representation for **s**.

1.3.5 Combining the results of Exercises 1.3.2 and 1.3.4. Converting from base m to base n.

   a. Convert the following base 4 representations of integers to base 5 by first converting each one to base 2 or to base 10 and then converting the result to base 5:

$$1232320, \ 0102013, \ 0101110$$

   b. Define an ISETL **func conv** that accepts a **string s**, the current base of **s**, **m**, and the new base, **n**, and outputs the **string** representing **s** with respect to the new base. *Hint: Use* **conv_to_10** *and* **conv_from_10**.

1.4   Integer Operations

Now that the integers are firmly in your mind as objects, we can discuss the transformations that can be performed on groups of integers to make new integers or objects of other data types. In ISETL, we call these transformations operators or functions. As we said in the beginning of this chapter, expressions are what ISETL evaluates. If you think about the expression together with the values that go in and the values that come out, then you have a function or operator. Again, these can be put together to form expressions, and so on. If you think that sounds circular, look at the following example in ISETL.

```
max((213 div 7)**3, 27100);
```

Now, **213, 7, 3,** and **27100** are integers. The keyword **div** is an operator so **213 div 7** is an expression. The symbol **∗∗** stands for exponentiation and it is also an operator, so **(213 div 7)∗∗3** is an expression. Finally, **max** is a function that takes two values as input, returning the largest value. So, the whole thing is an expression. Incidentally, what is its value?

The difference between a function and an operator in ISETL is just notation and terminology. Things such as

$$6.9 - 7.8; \quad 14 > 23; \quad \{1,2,3\} + \{4,5\};$$

are expressions where -, >, and + are called operators. Since each of these operators appears in between its two operands, we say that these expressions are written in *infix* form. On the other hand, a function usually appears in front of its list of operands, with the list enclosed in parentheses. Functions, such as min, ord, and sqrt, can be used in expressions such as

$$\text{min}(238,262); \quad \text{sqrt}(152); \quad \text{ord}("S");$$

These functional expressions are said to be in *prefix* form since in each case the function appears before its associated operands.

In ISETL, we can turn operators into functions. For example, the following code makes a function out of the operator >:

```
greater := func(x,y);
           return x > y;
       end;
```

After executing this code, the expression greater(3,7) will evaluate to *false*.

Conversely, we can turn any function of two variables into an infix operator. Take, for example, the function min. The infix form is

$$238 \text{ .min } 262;$$

which, of course, has the value 238. Thus, a function (of two variables) becomes an operator by just placing a . (dot) in front of its name.

Here is another piece of notation. Suppose you have a function or operator and its name is F. Suppose also that the set of all possible input values is named D and the set of all possible resulting values is S. Then we write

$$F : D \longrightarrow S$$

which reads "F is the name of a function that may be given a value in D upon which it produces a value that is in S," or more simply, "F is a function from D to S." Thus, for example, if N^+ is the set of all positive integers that can be represented in a particular implementation of ISETL, then we can write

$$\text{even} : \quad N^+ \longrightarrow \{\text{true, false}\} \qquad \text{exp2} : \quad N^+ \longrightarrow N^+$$

for the ISETL function even, which tests the parity of an integer and the function exp2, which might have been defined by the code

```
exp2 := func(p);
              if is_integer(p) and p>0 then
                    return 2**p;
              end;
        end func;
```

If the function has two integer inputs, we use notation N × N as in

$$\text{greater} : \quad N \times N \longrightarrow \{\text{false, true}\}$$

and similarly for 3, 4, or more inputs.

Using this notation, we list in Table 1.1 the operators and functions that are already defined in the ISETL system. We use N, B, F, and S to denote, respectively, the set of all integers that can be represented in an implementation of ISETL, the set of Boolean values ({*true,false*}), the set of all **floating-point** numbers that can be represented, and the set of all character **strings**.

Most of the notation in Table 1.1 is self-explanatory, but a few comments are in order. The notation N−{0} refers to the set of all **integers** except 0. The notation N$^+$ denotes the set of positive **integers**. The notations N$^+$+{0} and F$^+$+{0} refer to the sets of all nonnegative **integers** and **floating-point** numbers, respectively. The expression char(n) evaluates to the character that is the nth one in the system's ordered list of permissible characters. Of course, its value may vary with the implementation.

If b is a positive integer, the expression a mod b is the remainder obtained when a is divided by b and a div b is the quotient obtained by that division. More precisely, it is always possible to write (uniquely)

$$a = bq + r, \quad 0 \le r < b$$

in which case, q has the value of a div b and r has the value of a mod b.

Another way to think of a mod b and a div b is to imagine the number line laid off in units of length b starting at 0 and going in both directions. The division points can be labeled $0, 1, 2, \ldots$ going to the right and $-1, -2, -3, \ldots$ going to the left. Now place a on the line according to its value. If it falls on a division point, then the value of a mod b is 0 and the value of a div b is the label of the division point. If a falls between two division points, then the value of a div b is the division point to the left of a and the value of a mod b is the amount (which will be a positive integer) by which a exceeds the division point to the left.

For example, Figure 1.6 gives some values for a mod 3 and a div 3.

Note the patterns in this table. What are all the integers x such that x mod 3 = 2? Such that x mod 3 = 0? Such that x div 3 = 1?

Some of the operations in Table 1.1 can also be defined on other domains, such as F or S. We will discuss them later. If the evaluation of an expression in ISETL leads to an operator or function being given an input

infix (operators)		
+ : $N \times N \longrightarrow N$	addition
− : $N \times N \longrightarrow N$	subtraction
* : $N \times N \longrightarrow N$	multiplication
/ : $N \times N-\{0\} \longrightarrow F$	division
** : $N \times N^+ \longrightarrow N$	exponentiation
div : $N \times N-\{0\} \longrightarrow N$	quotient after integer division
mod : $N \times N^+ \longrightarrow N^++\{0\}$	remainder after integer division (whenever both operands are positive)
= : $N \times N \longrightarrow B$	equality test
/= : $N \times N \longrightarrow B$	inequality test
< : $N \times N \longrightarrow B$	test for less than
> : $N \times N \longrightarrow B$	test for greater than
<= : $N \times N \longrightarrow B$	test for less than or equal
>= : $N \times N \longrightarrow B$	test for greater than or equal

prefix (functions)		
max : $N \times N \longrightarrow N$	maximum of two integers
min : $N \times N \longrightarrow N$	minimum of two integers
even : $N \longrightarrow B$	test for evenness
odd : $N \longrightarrow B$	test for oddness
is_integer : $N \longrightarrow B$	test for type integer
is_number : $N \longrightarrow B$	test for type integer or floating_point
sqrt : $N^++\{0\} \longrightarrow F^++\{0\}$	square root
exp : $N \longrightarrow F^+$	power of e
ln : $N^+ \longrightarrow F$	natural logarithm
log : $N^+ \longrightarrow F$	base 10 logarithm
float : $N \longrightarrow F$	convert to floating point
char : $N \longrightarrow S$	character at that position
ord : $S \longrightarrow N$	inverse of char
random : $N^+ \longrightarrow N$	random number between 0 and the given integer
sgn : $N \longrightarrow \{-1,0,1\}$	sign (-1 for negative, 0 for 0, 1 for positive)
abs : $N \longrightarrow N^++\{0\}$	absolute value

Table 1.1: Integer operations

a	-5	-4	-3	-2	-1	0	1	2	3	4	5
a mod 3	1	2	0	1	2	0	1	2	0	1	2
a div 3	-2	-2	-1	-1	-1	0	0	0	1	1	1

Figure 1.6: Some **mod** and **div** examples

value that is not in any domain over which it is defined, then the result of the evaluation will be **OM**.

It is also possible for the user to define an operator or function. Note that if you want to define a function that you can use not only during the current ISETL run but also during subsequent runs put it in a file. Then, !include the file whenever you want to use the function again.

To define a function, use the ISETL **func** facility, which has the following syntax:

```
function_name := func (list of parameters);
                  statements that implement the function
             end;
```

There are also some optional type declarations in the **func** syntax. These will be discussed in Chapter 4.

In the above code, **function_name** is an identifier. It is used to evaluate the function for some actual value(s) of its parameter(s). That is, after the above code has been performed, the identifier **function_name** followed by a list of actual values for parameters in the list given in the definition will be evaluated by performing the statements in the body of the function definition. This assignment doesn't have to be present. Without it, the code is just a piece of data whose value is a definition of a **func**. The **func** can be passed as a parameter for example but, unless you make an assignment as in this example, the **func** is not saved anywhere.

Once you have constructed a function with two parameters, you can use it to define an operator by putting a dot, ., in front of its name when you invoke it. We have already seen examples of this.

An operator that takes two parameters (such as all of the infix operators in Table 1.1 and the prefix operators, **max** and **min**) is sometimes called a *binary operator*. (What do you think we call the operator if it takes 3 or 4 parameters, or 1 parameter, or none?). Sometimes a binary operator has in its domain a special value, such as 0 for addition, that leads to a predictable result if it is the value of one of the parameters, no matter what the value of the other parameter is. Here is a table of some of the values for the operators we have mentioned. These special values are very important in Mathematics, so they usually have names.

Special value	Effect	Name
0	x+0=0+x=x	two-sided identity for addition
0	x-0=x	right identity for subtraction
1	x*1=1*x=x	two-sided identity for multiplication
0	x*0=0*x=0	two-sided absorber for multiplication
1	x div 1 = x	right identity for division
0	0 div x = 0	left absorber for division $(x \neq 0)$
1	x mod 1 = 0	no name
0	x**0 = 1	no name $(x \neq 0)$
0	0**x = 0	left absorber for exponentiation $(x \neq 0)$

Summary of Section 1.4

The purpose of this section was to present the basic features of **integer** operations in ISETL. The goals of the following exercises are to give you some practice doing the following:

- thinking some more about how ISETL evaluates an expression. (You'll be doing a lot of that in this book!)

- using an ISETL program to determine whether or not a given property holds.

- practicing with **mod**. Doing some simple proofs.

- defining simple ISETL **funcs**. Using . to convert a function with two arguments to an operator.

Exercises

1.4.1 Thinking about how ISETL performs **integer** operations. For each of the following expressions,

1. TYPE in the statement, but do not hit the carriage return.

2. PREDICT what ISETL will return. In making your prediction, think about how ISETL will process the input you have entered. RECORD your prediction.

3. HIT the carriage return. If ISETL returned something different than what you predicted or ISETL gives you an **RT error**, EXPLAIN why.

a. 2 + 3 * 2**3 - 10; 10 + 6/4;

b. 2**3**2; (2**3)**2; 2**(3**2);

c. 4/2; 4/3;

d. `12 div 4; 12 div 5; 12 div -5; -12 div 5;`
   `-12 div -4; 12 div 0;`

e. `12 mod 4; 9 mod 4; 6 mod 4; 3 mod 4; -2 mod 4;`
   `-4 mod 4; -7 mod 4;`

f. `2 = 3; (4 + 5) /= -123; (12 mod 4) >= (12 div 4);`
   `(2*4**2) < -10;`

g. `even (2**14); odd(187965*45);`

h. `max(-27, 27); min(-27, max(27, -27));`

i. `abs(min(-10, 12) - max(-10, 12));`

j. `char( ord("H")); ord( char(45));`

k. `random(6); random(6); random(6); random(6);`

l. `sqrt(25) * float(10); sgn(67); sgn(-234);`

m. `max(sgn(34)*(124 div -11)-2**5+min(20,abs(-20)),`
   `-(31 mod 32));`

1.4.2 Working with **mod**.

a. Find the value of **x mod 6** for **x = -7, -6, ..., 7**.

b. Describe all possible integer values of a for which **a mod 6 = 0**.

c. Describe all possible integer values of b for which **b mod 6 = 4**.

1.4.3 Using an ISETL program to determine whether certain properties hold.

a. Look again at the program **rel_prim** in Exercise 1.2.2. Suppose that whatever integer is read in for **x**, the value of the integer that is given for **y** has twice the value of **x**. In this case, can you always be sure of what the value of the Boolean **rp** is? If so, what is **rp** and how can you be sure? If not, why not? (Refer to the program when justifying your response.)

b. List some relationships between the values of the variables **x** and **y** for which you can always be sure of the resulting value of **rp**.

c. If **rp** is *true*, then we say that **x** and **y** are *relatively prime*. If **x** and **y** are relatively prime, and **y** and **z** are relatively prime, then is it always the case that **x** and **z** are relatively prime? Explain why it is or give a counterexample.

1.4.4 ISETL **funcs**: using . to convert a binary **func** to an operator.

a. Type the following ISETL **funcs** into a file:

```
F := func(x);
        if is_integer(x) then
                return 2 * x**4 + 6 * x - 18;
        end;
    end func;

times := func(x,y);
                return x * y;
        end;
```

b. Enter ISETL and type the name of the file that includes your funcs on the command line or include them using !include. (Now they are both available for you to use as often as you want.) Call the two functions with a variety of inputs (both valid and invalid). Use the . notation when using the func times as an operator. Why can't you use . to convert F from functional form to operator form?

c. Write an ISETL func that accepts two positive integers or two negative integers, n and m, and returns the square root of their product.

   i) Run your func on a variety of inputs, using both functional and operator forms.

   ii) What would happen if you input one positive and one negative integer?

   iii) What happens if one of the inputs is 0?

   Define an ISETL func discr that accepts three numbers (integer or floating-point), a, b, and c, and returns their *discriminant*, namely, $b^2 - 4ac$.

d. i) Run it a few times.

   ii) Where have you seen this formula before?

   iii) Can you use . to express this func in operator form?

   iv) For what values of a, b, and c does the func return 0? A positive number? A negative number? OM?

1.4.5 Re-express the following in as simple a way as you can, using x, max, min, and abs, where x is of type integer.

a. x .max −x

b. min(x, −x)

c. max(x, 0) + min(x, 0)

d. (x .max 0) + (−x .max 0)

1.4.6 Simplify the following expressions to x, y, max(x,y), or min(x,y).

a. max(x, x)

b. min(y, y)

c. max(y, min(x, y))

d. min(max(x, y), x)

e. max(x, max (x, y))

f. min(min(x, y), y)

1.5   Special Binary Integer Operators

1.5.1   Extending Binary Operators

We have discussed converting binary functions to infix operators. Sometimes it is useful to extend a binary function to a function having 3, 4, or even more arguments. The simplest example of this is the binary operator addition. The summation

$$1 + 2 + 3 + 4 + 5 + 6 + 7 + 8 + 9 + 10 = 55$$

can, of course, be written as

$$1 + 2 + \cdots + 10 = 55$$

which allows us to write longer ones, such as

$$7 + 8 + \cdots + 50 = 1254$$

or even more complicated ones, such as

$$3 + 6 + \cdots + 27 = 135$$

All of these can be thought of as applying the binary operator + to the first two terms and then applying it to the result of that addition together with the third term and so on. In other words,

$$(\cdots ((((1 + 2) + 3) + 4) + 5) + \cdots + 10) = 55$$

In Mathematics, we use the symbol $\sum$ for this kind of expression. We will discuss this in some detail when we get to Chapter 4, Section 3.6. In ISETL, we use the following notation for these expressions:

```
%+[1..10];
%+[7..50];
%+[3,6..27];
```

where ISETL evaluates expressions such as %+[1..10] exactly the way we did above when we inserted all the parentheses.

The compound operator % will be discussed more fully in Chapter 3, Section 7. It can be used with any predefined binary operator, for example, +, or any binary function, such as max. (It does not work with a user-defined binary operation formed by a . (dot).) For example, the expression

%max[6, -12, 14, 3, 8];

in ISETL has the value 14. Try to work through, in your mind, the details of how this computation is made.

A less simple but useful way of thinking about max applied to a set of integers is that it returns the smallest of all integers that are greater than or equal to every integer in the set. It is not so easy to think about expressing max this way. Similar situations, however, come up rather often in Mathematics, and in other, more complicated situations, it can be quite useful. So try to think about this definition of max where the situation is simple. (In fact, thinking along these lines, how could you describe the min applied to a set of integers?)

Any two integers will have a maximum. This is a basic property of the ordering of integers. It is also true that any finite set of integers has a maximum. The situation changes completely if we change the meaning of "order." Suppose, for instance, we said that a positive integer a *comes before* b if a divides b with no remainder. If we denoted the maximum of two integers according to this rule by d_max, then, for example, $d\_max(3, 12)$ is 12 because 3 divides 12, and $d\_max(8, 4)$ is 8 because 4 divides 8, whereas $d\_max(5, 12)$ is OM since neither 5 divides 12 nor 12 divides 5.

To play with this idea a little, let's make d_max an ISETL binary operator. To keep everybody honest, we should make sure the parameters are positive integers. We'll use this test a lot so let's encapsulate it as follows.

```
is_two_pos_int := func(a,b);
                    return is_integer(a) and a>0 and
                           is_integer(b) and b>0;
               end;
```

Now, any time the expression is_two_pos_int(x,y) appears, it will be evaluated to *true* or *false* depending on whether or not x and y are positive integers. Then, we can write

```
d_max := func(a,b);
             if not is_two_pos_int(a,b) then return; end;
             if (a mod b = 0) then return a; end;
             if (b mod a = 0) then return b; end;
         end func;
```

With this func, d_max(5,12) will be evaluated to OM, while d_max(2,12) and d_max(12,2) both return 12. Now, an interesting question arises. What are the values of the following two expressions?

```
%d_max[1, 2, 3, 4, 5, 6];
%d_max[1, 2, 3, 6];
```

In both cases, ISETL will evaluate the expression to OM. Do you agree? Did you think the right answer to the second one was 6? What did you think the answer to the first one was? Does the word "associativity" have any place in this discussion? Convince yourself that the value of both expressions actually is OM by thinking carefully about how ISETL evaluates each of them (note, for example, that d_max(5,OM) is OM). What about the usual way of expressing the meaning of "max applied to a set" that we mentioned previously. If we used that interpretation with d_max, what would happen? Finally, try to predict the value of

```
%d_max[6,1,2,3];
```

In reading other texts, you should be aware that mathematicians often use exactly the same symbol for the extended binary operator as for the original and leave the reader to determine the meaning from the context. Thus, the above expression could be written in mathematical notation as simply

$$d\_max[6, 1, 2, 3]$$

1.5.2   Greatest Common Divisor

Now, we consider another binary operator, the *greatest common divisor* (gcd) of two positive integers. This is the largest positive integer that divides both of them. We can state the definition in ISETL code.

```
gcd := func(a,b);
         if not is_two_pos_int(a,b) then return; end;
         return % max {n:   n in [1..min(a,b)] |
                     (a mod n = 0) and (b mod n = 0)};
       end;
```

With this definition the following equalities are true:

$$
\begin{array}{llll}
\gcd(8, 12) & = & \max\{1, 2, 4\} & = & 4 \\
\gcd(120, 90) & = & \max\{1, 2, 3, 5, 6, 10, 15, 30\} & = & 30
\end{array}
$$

Such a computation is easy for a computer, but gets rather tedious for humans unless the numbers are fairly small. A method that is sometimes better for humans, but hard to program, is to express both integers as a product of primes. Then, the gcd is the product of all the primes that appear in both factorizations, each prime counted once for each time it appears in both. For example, writing 8 and 12 as a product of primes, we have

$$8 = 2 \cdot 2 \cdot 2$$
$$12 = 2 \cdot 2 \cdot 3$$

Then, since the prime 2 appears in both factorizations twice,

$$\gcd(8, 12) = 2 \cdot 2 = 4$$

About 3000 years ago, Euclid found an algorithm for computing $\gcd(a, b)$ that is pretty good for both humans and machines. What he did was to divide a by b to get a remainder r (that is, he computed $a \bmod b$). Then, he replaced a with b and b with r and repeated the division. He continued this way until the remainder was 0. In that case, the last nonzero remainder was the answer. Here is what the calculations look like:

$$
\begin{array}{llll}
a & = & q_1 b + r_1, & 0 < r_1 < b \\
b & = & q_2 r_1 + r_2, & 0 < r_2 < r_1 \\
r_1 & = & q_3 r_2 + r_3, & 0 < r_3 < r_2
\end{array}
$$

$$
\begin{array}{llll}
\vdots & & & \\
r_{n-2} & = & q_n r_{n-1} + r_n, & 0 < r_n < r_{n-1} \\
r_{n-1} & = & q_{n+1} r_n + r_{n+1}, & r_{n+1} = 0
\end{array}
$$

and the greatest common divisor is r_n.

We can express Euclid's algorithm a little more compactly in ISETL code.

```
gcd_euc := func(a,b);
            if not is_two_pos_int(a,b) then return; end;
            while b /= 0 do
                [a,b] := [b, a mod b];
            end while;
            return a;
          end func;
```

The statement [a,b] := [b, a mod b]; has the effect of first evaluating the expressions b and a mod b and then assigning them to a and b respectively.

There are several observations we can make about the Euclidean algorithm. First of all, using the Euclidean algorithm does not require any more computations than following the definition, which makes the main computation once for each of $1, 2, ..., \min(a, b)$ for a total of $\min(a, b)$ times. In the Euclidean algorithm, each iteration of the loop is for $r_{n+1} < r_n$, so the drop is at least one each time. But, it could be more. For example,

$$120 = 1 * 90 + 30 \qquad 8 = 0 * 12 + 8$$
$$90 = 3 * 30 + 0 \qquad 12 = 1 * 8 + 4$$
$$8 = 2 * 4 + 0$$

So,

$$\gcd(120, 90) = 30 \qquad \text{and} \qquad \gcd(8, 12) = 4$$

Thus, instead of doing the main computation 90 times or 8 times, with the Euclidean algorithm it is only done 2 times or 3 times, respectively.

The Euclidean algorithm is also a convenient tool for considering theoretical questions. First of all, it is clear that since the remainder is reduced by at least 1 each time and the procedure halts when the remainder reaches 0 it will always finish, so we do get an answer (provided that a and b are positive integers, otherwise the result is OM). But, how do we know it is the right answer? This must be proved.

Proof that gcd_euc(a,b) *is the greatest common divisor.*

> Since we are sure that gcd_euc(a,b) returns a positive integer, let's call it d. Now, in the notation of our first description of the algorithm $d = r_n$. The last equation in that description tells us that d divides r_{n-1}. Moving up one equation, we see that r_n divides both r_n and r_{n-1}, so it divides r_{n-2}. The next equation up looks like
>
> $$r_{n-3} = q_{n-1}r_{n-2} + r_{n-1}$$
>
> and since d divides both r_{n-2} and r_{n-1}, it divides r_{n-3}. Continuing all the way up, we obtain from the second equation by the same argument that d divides b and from the first equation that d divides a.
>
> Now that we know d divides both a and b, we must prove that it is the largest integer with this property—that is, *divides a and b*. Suppose that some integer, e, divides both a and b. We want to show that $e \leq d$. This time we start at the top with the first equation and work our way down.
>
> The first equation tells us that since e divides both a and b, it divides r_1. We conclude from the second equation that e divides r_2 and so on. The next to the last equation tells us that e divides r_n. But, $r_n = d$, so e divides d. Therefore, $e \leq d$. Since e was any common divisor of a and b, we have shown that the common divisor d is greater than or equal to any common divisor. That is, d is the greatest common divisor of a and b.

End of proof.

The Euclidean algorithm also tells us that the greatest common divisor of two positive integers is unique. This is because `gcd_euc(a,b)` is a definite finite computational procedure that will always produce the same answer for the same a and b. The above proof shows that any number that satisfies the definition of gcd is equal to the result of this computation. Hence, there can be only one such number.

You can reason to this same uniqueness conclusion another way. Suppose that d_1 and d_2 were both greatest common divisors of a and b. Then, d_2 is a common divisor, and d_1 is greater than or equal to any common divisor. Hence, $d_1 \geq d_2$. Symmetrically, d_1 is a common divisor, and d_2 is greater than or equal to any common divisor, so $d_2 \geq d_1$. Hence, $d_1 = d_2$.

Incidentally, gcd is related to the **program rel_prim** that we discussed at the beginning of this chapter. Do you see the connection? Everything that was done for gcd can be turned upside down to get a concept denoted lcm—the *least common multiple*. We will pursue this in the exercises.

1.5.3 Modular Arithmetic

Let us turn now to another kind of binary operation, *modular arithmetic.* This is again a topic that is very important in Mathematics and will be a major theme in this book. Here is the ISETL code for addition and multiplication, *modulo 6:*

```
add_mod6 := func(a,b);
            if is_integer(a) and is_integer(b) and
                    a in {0..5} and b in {0..5} then
                return (a+b) mod 6;
            end;
         end;


mult_mod6 := func(a,b);
            if is_integer(a) and is_integer(b) and
                    a in {0..5} and b in {0..5} then
                return (a*b) mod 6;
            end;
         end;
```

Notice that these operators will only give a result if the parameters have one of the six values, 0, 1, 2, 3, 4, or 5. The domain is a finite set. Notice also that the two functions are exactly the same except for the names and a single character (which one?). In computing with these, you might want to use shorter names, so you could write

```
a6 := add_mod6;          m6 := mult_mod6;
```

Then, the expressions

$$3 \text{ .a6 } 4; \qquad 2 \text{ .a6 } 4; \qquad 3 \text{ .m6 } 5; \qquad 3.\text{m6 } 4;$$

will have the values **1**, **0**, **3**, and **0**, respectively. These operators also have special values like the predefined operators. For example, no matter what (legal) value **x** has, we have

$$\text{x .a6 } 0 = 0 \text{ .a6 } \text{x} = \text{x}$$
$$\text{x .m6 } 1 = 1 \text{ .m6 } \text{x} = \text{x}$$

Thus, 0 and 1 are two-sided identities for **add_mod6** and **mult_mod6**, respectively. Notice also that you can always "go back" to 0 with **add_mod6**. That is, for example, if we start with 2, then

$$\text{x .a6 } 2 = 0$$

will be true if **x** = 4. If we started with 1, then **x** = 5 would work, etc. But you can't always get back to 1 with **mult_mod6**. What value of **x** will make the following equation true?

$$\text{x .m6 } 2 = 1$$

There isn't any!

Summary of Section 1.5

In this section, we introduced the powerful operator %, which extends a binary operator to a **set** or **tuple** of operands. We also used ISETL to study the greatest common divisor of two integers and to prove a few simple facts about this concept. Finally, we introduced modular arithmetic, which will be of continuing importance throughout the book.

Exercises

1.5.1 The compound operator %. Explain how ISETL evaluates each of the following inputs:

    a. %min {1,2,3,4};

    b. %min [4,3,2,1];

    c. %max {-3,3,-4,2,-6,1};

    d. %d_max [5,10,30,120];

    e. %d_max [120,10,30,5];

    f. %d_max [5,10,15,20,25,30];

1.5.2 The gcd.

  a. Calculate the gcd of the following pairs of integers using three different methods, namely,

    i) the definition: trace through the ISETL **func gcd** (see page 30), which finds the max of the set of common factors.

    ii) factorization: write each integer as a product of primes and then find the product of the common primes, if there are any.

    iii) Euclid's algorithm: trace through the ISETL **func gcd_euc** (see page 31).

$$(67, 98), \ (1024, 256), \ (768, 176)$$

How does each method compare in terms of efficiency? Note: You might want to type in the two **funcs, gcd** and **gcd_euc**, and then compare how long it takes each **func** to calculate the result for each of the inputs.

  b. Trace through the body of the **func gcd_euc** to show that $\gcd(a, b) = a$ whenever b is a multiple of a.

  c. Compare the results of the **program rel_prim** (see Exercises 1.2.2 and 1.4.3) and the **func gcd_euc** when they are both given the same input.

  d. If **rel_prim** returns true, what does **gcd_euc** return?

  e. When you know what the greatest common divisor of two particular integers is, how can you be sure that the two integers are relatively prime?

  f. Give another definition for relatively prime in terms of the greatest common divisor.

1.5.3 The lcm.

  a. The greatest common divisor of two positive integers is the largest integer that divides them both. The least common multiple, on the other hand, is the smallest integer that is a multiple of both. For example,

$$\operatorname{lcm}(12, 20) = 60 \ \text{and} \ \operatorname{lcm}(8, 15) = 120$$

If a and b are two positive integers, then we know that $a \cdot b$ is always a multiple of both a and b. What we need to find out is if there are any smaller integers that are also multiples and then take the smallest one of these.

Recall what happens in the **func gcd**. We look at all the integers that are candidates for common divisors, namely, all the integers between 1 and `min(a,b)`. If a candidate is a divisor, then we put it in a set. Finally, we take the maximum of this set (of divisors). We can find the lcm in a similar manner, only this time we are looking for candidates for multiples between 1 and $a \cdot b$. If a candidate is a multiple of both a and b, we put it in a set. Then, we take the minimum of the set. With this definition,

$$\mathrm{lcm}(12, 20) = \min \{60, 120, 180, 240\} = 60$$

and

$$\mathrm{lcm}(8, 15) = \min \{120\} = 120$$

Write an ISETL `func` `lcm` (using the same format as the `func` `gcd`) that accepts two positive integers and returns their least common multiple. Test your function on a number of different inputs.

b. Trace through your `func` (that is, "think like ISETL") to find the lcm of the following pairs of integers:

$$(2, 12), \ (6, 10), \ (3, 5), \ (4, 9)$$

c. Our second method for finding $\gcd(a, b)$ consisted of looking at the prime factors of a and b. There is a similar method for finding $\mathrm{lcm}(a, b)$. Consider the examples in part b.

   i) In each case, express a, b, and $\mathrm{lcm}(a, b)$ as products of primes.

   ii) Compare the three products. What appears to be happening here?

   iii) What is the general rule for using the prime factors of a and b to find the lcm?

d. If b is a multiple of a, what is $\mathrm{lcm}(a, b)$?

e. Compare the results of the program `rel_prim` and your `func` `lcm` when they are both given the same input.

f. If a and b are relatively prime, what is $\mathrm{lcm}(a, b)$?

g. If you are given the least common multiple of two integers, how can you be sure that they are relatively prime?

h. Give another definition for relatively prime in terms of the least common multiple.

i. Explain why $\gcd(a, b) \cdot \mathrm{lcm}(a, b) = a \cdot b$ for any positive integers a and b.

j. What is $\gcd(n, n + 1)$?

k. What is lcm($n, n + 1$)?

l. If a and b are positive integers and s and t are integers, what can we say about gcd(a, b) if we know

    i) $as + bt = 2$.

    ii) $as + bt = 3$.

    iii) $as + bt = 4$.

    iv) $as + bt = 6$.

m. Prove that if a and b are relatively prime, then gcd($a-b, a+b$) = 1 or 2.

1.5.4 Max and d_max.

a. As we discussed in the text, you can think of the maximum of a set of integers as being the smallest integer that is greater than or equal to every item in the set. Mathematicians sometimes call this the *least upper bound* or *lub* of the set. Give a similar definition for the minimum of a set of integers. How do you think mathematicians refer to the minimum?

b. Give a definition of d_min analogous to the definition of d_max given in the text.

c. Write an ISETL **func** that will evaluate d_min(x, y).

d. Evaluate

    i) d_min($3, 12$).

    ii) d_min($5, 12$).

    iii) % d_min[$5, 10, 15, 20, 25, 30$].

    iv) % d_min[$30, 25, 20, 15, 10, 5$].

e. Give a relationship between d_max(a, b) and d_min(a, b)

f. Explain under what conditions on a, b the following statements are true or false:

    i) [a, b] = [d_max(a, b),d_min(a, b)].

    ii) {a,b} = {d_max(a, b),d_min(a, b)}.

1.5.5 Modular arithmetic.

a. In the table below, for each value of x, write a value of y between 0 and 6, such that $(x + y) \bmod 7 = 0$. When this equation is satisfied, we say that y is the *additive inverse* of x *modulo 7*.

x	0	1	2	3	4	5	6
y							

b. For each value of x, write the value of y between 0 and 6, such that $(x \cdot y)$ mod $7 = 1$. When this equation is satisfied, we say that y is the *multiplicative inverse* of x modulo 7.

x	0	1	2	3	4	5	6
y							

c. Write ISETL **funcs** for addition and multiplication mod 7, **add_mod7** and **mult_mod7**, that accept two integers and return the appropriate result. Use your **funcs** to show that your entries in the tables in parts a. and b. are correct.

d. What if you tried to answer the questions in parts a. and b. for modulo 8? What about a multiplicative inverse of x modulo 8? Make a guess about what seems to be happening here. What "type" of number (e.g., even, odd, prime, divisible by a certain number, etc.) does n have to be in order to guarantee that every integer between 1 and $n - 1$ has both an additive and a multiplicative inverse modulo n?

e. Suppose that $r = a$ mod n. Show that r is unique. *Hint: Assume that $r_1 = a$ mod n and $r_2 = a$ mod n. Use the definition of mod to show that $r_1 = r_2$.*

f. Use the definition of mod to show that if a mod $n = 0$ and b mod $n = 0$ then $(a + b)$ mod $n = 0$.

g. Use part f. to show that the sum of two even integers is even.

h. If a mod $b = 0$ and b mod $c = 0$, does it follow that a mod $c = 0$? If it does, explain why; otherwise, give a counterexample.

i. Is it true that $((a \bmod n) + (b \bmod n))$ mod $n = (a+b)$ mod n, for all values of a, b, and n? If it is, explain why; otherwise, give a counterexample.

j. Show that if n and k are positive integers then

$$n \times (n+1) \times \cdots \times (n+k-1) \bmod k = 0.$$

k. Show that the sum of two odd integers is even.

1.6   Random Integers

"Pick a card at random." "The air molecules in this room are distributed randomly." "A random student signed up for this course." The word *random* is used all the time in ordinary conversation—as if we all knew what it means. But, did you ever ask anyone for a precise definition? Most people define it by a synonym, such as "equally likely," or give an example, such as flipping coins. You could ask a professional statistician, who will give you an answer, but you better be ready to listen for a very long time.

One thing you probably won't get is a satisfactory explanation of what the weather report means when it says, "the probability of rain today is 60%"—especially if you hear this report on your car radio while driving through a thunderstorm. Was the prediction right or wrong? What does the statement even mean? It certainly doesn't mean that if today comes 1000 times and you pick some of these times "at random," then it will rain on about 6 out of every 10 of your days.

It is very hard to say what random means. This is a philosophical question. It is also hard to describe an algorithm to generate a set of *random numbers* and even harder to justify the algorithm. These are mathematical questions that turn out to be related to something called *Galois fields*.

We won't answer these questions in this book. We will remain in a state of not knowing what random numbers are, how they might be produced, or why a method works. Nevertheless, we have an ISETL function called *random*. If n has a positive **integer** value, then the value of

random(n);

will be an **integer** between 0 and n. If you knew what random meant, you could try to check whether the integer you get is "random" (although that's not even the right form of the question). If this ISETL function worked, then you could use it in very complicated programs that simulated important human activities, such as shuffling cards, spinning a roulette wheel, or predicting the weather. In any case, the next time someone says to you "pick a number at random" ask her or him what range they want it in, run to your nearest ISETL implementation, and use the function **random** to get them one.

1.7 Floating-Point, Rational, and Real Numbers

Working with integers, it is possible to create situations that call for numbers that are not integers. For instance, suppose you wanted to find values of x that would make the following ISETL expression evaluate to *true*.

$$4 * x = 1; \tag{1.1}$$

You can't do this with **integers**—you need a new kind of object. For this example, the number 0.25 will make the equation *true*. This is an example of a **floating-point** number.

A **floating-point** number in ISETL is a sequence of at least one digit, followed by a period, . (sometimes called a *decimal point*). For example,

63.86, 0.012, 0.53, -5.0, 675000.0

are all ISETL **floating-point** numbers. Notice that there must be a digit to the left of the decimal point. The following expressions

.012, -.63456, 63.

are not valid representations of **floating-point** numbers in ISETL, and their use will constitute a syntax error.

You can think of a number that can be represented by a **floating-point** number as an integer multiplied by a power of 10. If the power is positive, the number is also an integer. If it is negative, you get a number that is called a *decimal*. In ISETL, the symbol **e***b* can be added at the end of a **floating-point** number, where *b* is an integer. It means that the number is to be multiplied by 10^b. This is sometimes called *scientific notation*. Thus, the following expressions all represent the same number in ISETL.

-63.86, -6.386e1, -6386.0e-2, -0.006386e4

and in scientific notation, this number is written

$$-6.386 \times 10^1 \qquad \text{or} \qquad -0.6386 \times 10^2$$

But, **floating-point** numbers are not enough to satisfy all equations of the form of equation (1.1). For example, no explicit **floating-point** value for **x** will make the following ISETL expressions evaluate to *true*.

$$3 * x = 1; \tag{1.2}$$

$$14 * x = 3; \tag{1.3}$$

$$-9 * x = 17; \tag{1.4}$$

For this, you need *fractions* or *rational numbers*. It is easy to see that, for example, no **floating-point** value of **x** will make equation (1.2) true. Suppose there were such a value. Since 3 and 1 are positive, this value would have to be positive as well. Now, a **floating-point** number is an integer, let's call it **n**, multiplied by a (positive or negative) power of 10, for example, **b**. Then, if Eq. (1.2) were satisfied, this integer would be positive and we would have

$$3 \cdot n \cdot 10^b = 1$$

If $b > 0$, then the left-hand side is too big to be equal to 1, and if $b < 0$, multiply both sides of this equation by 10.0^{-b} to get

$$3 \cdot n = 10^{-b}$$

Now, the left-hand side is an integer divisible by 3, but the right-hand side is a power of 10, so it is not divisible by 3. Hence, they can't be equal.

So, in order to solve equations such as those in (1.2), (1.3), and (1.4), we need rational numbers. But, how can you represent rational numbers in ISETL? You can't use decimals because solutions of equations such as those in (1.2), (1.3), and (1.4) have *infinite decimal representations*, such as

0.333...   or   0.214285714285714285714285...

which satisfy equations (1.2) and (1.3), respectively. One way to represent rational numbers is to go back to fractions and represent the fraction n/d as an ordered pair, $[n, d]$, of integers. Thus, the solutions of equations (1.2), (1.3), and (1.4) would be represented as

$$[1, 3], \quad [3, 14], \quad [17, -9]$$

respectively. Then, you could write procedures for the arithmetic operations such as addition, multiplication, and even equality. For example, you could do this for multiplication in ISETL with a **func** as follows:

```
rat_mult := func(x,y);
            if(is_tuple(x) and is_tuple(y) and
                is_integer(x(1)) and is_integer(x(2)) and
                is_integer(y(1)) and is_integer(y(2)) and
                (x(2)/=0) and (y(2)/=0)) then
            return [x(1)*y(1), x(2)*y(2)];
            end;
        end;
```

There are lots of things you would still have to do before you can solve equations with these fractions. For instance, how would you represent the number 3, or how do you determine that $[2, 3]$ and $[8, 12]$ represent the same fraction? But, it can all be done, and this is actually the way mathematicians do it. You will have a chance to work on this in the exercises.

There are equations for which even the rationals are insufficient for the solutions. For example,

$$x ** 2 = 3; \tag{1.5}$$

$$x ** 2 = -1; \tag{1.6}$$

or you could go beyond equations and describe numbers geometrically. For example, what is the ratio of the circumference of a circle to its diameter? What is the length of the hypotenuse of a right triangle, both of whose legs have length 1? The answers to these questions require discussions of *real* numbers and *complex* numbers, but the study of such objects goes beyond the scope of this book.

Finally, we discuss intervals of **floating-point** numbers. If a and b are **floating-point** numbers, then the following four expressions in mathematical notation (not ISETL!) represent sets of **floating-point** numbers:

$$[a, b], \quad (a, b), \quad (a, b], \quad [a, b)$$

The first set consists of all those **floating-point** numbers greater than or equal to a and less than or equal to b. That is, all **floating-point** numbers between a and b including the *endpoints* a and b. The second is the same except that the endpoints are excluded. The third excludes a and includes b, and the last includes a but excludes b.

Be careful not to confuse this with our notation for `tuples` or fractions. For example, $[-1.0, 3.0]$ could mean the set of all `floating-point` numbers from -1.0 to 3.0 or it could mean the `tuple` with two components, `-1.0` and `3.0`. You have to look at the context to tell. We live with this ambiguity because the notation for intervals is solidly established in common practice and the notation for `tuples` is stored in many computers.

Notice also that there is no way to represent an interval of `floating-point` numbers completely in ISETL. There are too many of them. But, you can think of all the `floating-point` numbers in a particular interval as being on a number line. For example, the following number lines represent the intervals $[-1.5, 1.25]$ and $(0.0, 2.7]$:

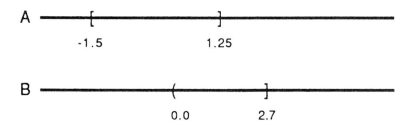

You can use pictures like this to figure out certain operations on intervals. For instance, the *intersection* of the two intervals shown (that is, all `floating-point` numbers that are in both intervals) is written

$$
\begin{array}{ccccc}
\text{Math} & & \text{ISETL} & & \text{Math} \\
A \cap B & = & \texttt{A * B} & = & (0.0, 1.25]
\end{array}
$$

Here, we are using two different notations to represent the operation of intersection—standard mathematical ($\cap$) and ISETL (*).[4] The *union* of the two intervals (the set of `floating-point` numbers in one or the other or both) is written

$$
\begin{array}{ccccc}
\text{Math} & & \text{ISETL} & & \text{Math} \\
A \cup B & = & \texttt{A + B} & = & [-1.5, 2.7]
\end{array}
$$

where again we have used the mathematical notation ($\cup$) and the ISETL notation (+). There is also the *difference* (mathematical and ISETL nota-

---

[4]You may also use `inter` and `union` in place of * and +, respectively.

tions agree), consisting of those numbers in A but not in B:

$$A - B = [-1.5, 0.0]$$

and those numbers in B but not in A:

$$B - A = (1.25, 2.7]$$

Then, using the definitions of union, intersection, and difference, we can define the *symmetric difference*, consisting of those **floating-point** numbers in one but not both of A and B:

Math	ISETL	Math
$(A \cup B) - (A \cap B)$ =	(A + B) - (A * B) =	$[-1.5, 0.0] \cup (1.25, 2.7]$

Summary of Section 1.7

This section in the text was concerned with representation of **floating-point** and rational numbers in ISETL. The need for something more than just integers was motivated by looking at the types of equations we can't solve with just integers, which leads us to **floating-point** numbers. But, there are still equations that can't be solved, so we look at the rationals. Even the rationals don't give us everything we need to solve a general equation, but real numbers and complex numbers lie beyond the scope of the book, so we stop with the rationals! Finally, we looked at intervals of **floating-point** numbers and their associated operations—union, intersection, and difference.

Exercises

1.7.1 Representations of **floating-point** numbers in ISETL.

    a. Each of the following representations of the **floating-point** number −678900.0 is INVALID in ISETL. Explain why ISETL will return a syntax error for each one.

```
-6789E2,  -6789.0E2.0,  .6789E6,
-678900e0.0,  -6789000*10**-1
```

    b. All the following are VALID representations in ISETL. Group together the representations that stand for the same **floating-point** number:

```
-678.9E02,  -0.006789E7,  -0.6789*10**5,
-67890.0*10.0**-5,  -67890.0,  -67890000.0E-3,
-678.900E-3,  -0.0006789E+3,  -0.6789*10.0**0,
-67890.0e0,  -67890*10**0,  -0.67890,
```

1.7.2 Solutions to equations.

    a. Which of the following equations do not have **floating-point** solutions?

$$
\begin{aligned}
\texttt{x**2} &= 2 \\
\texttt{x**2} &= 4 \\
\texttt{x**2 - 2.25} &= 0 \\
\texttt{3*x + 9} &= 12 \\
\texttt{3*x**2 + 5*x + 2} &= 0 \\
\texttt{7*x} &= \texttt{sqrt(x)}
\end{aligned}
$$

    b. Prove that the following equations do not have **floating-point** solutions:

      i)   `11*x` `=` `2`
      ii) `-9*x` `=` `17`

1.7.3 Representing rational numbers in ISETL:

    a. Define an ISETL **func** that accepts two fractions (each written as a **2-tuple**) and returns their sum. Test your **func** on a variety of inputs, including ones with 0 in the denominator. When testing your **func**, use both functional and operator (.) notation.

    b. Define an ISETL **func** that accepts two fractions and returns *true* if they are equivalent and *false* otherwise. (Notice that 1/2 and −2/−4 are equivalent, while 1/2 and 4/−7 are not.) Again, run your **func** with different inputs, using both functional and operator notation.

    c. Define an ISETL **func** that accepts a fraction and returns the fraction reduced to its lowest terms. For example, if you passed your **func** `[3,12]`, it would return `[1,4]`.

    Hint: Use gcd. *(ISETL can call one* **func** *from inside another* **func**. *They do not have to be in the same file, but a* **func** *must be defined before it is used.)*

1.7.4 Using interval notation.

Suppose A, B, C, and D are the intervals $(-10.5, 2.3)$, $[-2.3, 10.5]$, $[0, 10.5)$, and $(10.5, 15.0]$, respectively. We say that A is an *open* interval since A does not contain either of its endpoints, B is *closed* since it contains both its endpoints, and C and D are *half-open* (or *half-closed*) since each one contains exactly one of its endpoints.

    a. Give a verbal description of the set of all **floating-point** numbers contained in each of the intervals A, B, C, and D. For example, "A is the collection of all **floating-point** numbers that are greater than 10.5 and are less than 2.3."

b. Draw the number line representation for each interval.

c. Assume that $\mathcal{F}$ is the set of all **floating-point** numbers.

    i) How do you represent $\mathcal{F}$ using a number line?

    ii) Do you think $\mathcal{F}$ is open? Closed? Neither? Both? *Hint: $\mathcal{F}$ does not have any endpoints!*

d. Simplify each of the following to $\mathcal{F}$, a specific interval, or the union of two or more intervals that do not intersect. Notice that we are using ISETL notation +, *, and – for union, intersection, and difference, respectively, and F for $\mathcal{F}$.

i)	A + B	viii)	A * B	xv)	A + B + C
ii)	B + C	ix)	B * C	xvi)	A + B + C + D
iii)	B - C	x)	A * D	xvii)	A * B * C
iv)	C + D	xi)	C * D	xviii)	(A * B) - (B * C)
v)	B + D	xii)	F * B	xix)	F - (F - B)
vi)	F + A	xiii)	B - D	xx)	B - (B - A)
vii)	A - B	xiv)	F - A	xxi)	(B - C) * A

e. Determine which of the following expressions is true for all intervals I, J, K. If the expression is false, give a counterexample.

    i) $I \cup J = J \cup I$

    ii) $I - J = J$

    iii) $(\mathcal{F} - I) \cup I = \mathcal{F}$

    iv) $I \cup (J \cap K) = (I \cup J) \cap K$

    v) $I - (I - J) = I \cap J$

    vi) $I \cap I = I$

    vii) $(\mathcal{F} - I) \cup (\mathcal{F} - J) = \mathcal{F} - (I \cup J)$

    viii) $I \cup I = I$

    ix) $\mathcal{F} - (I \cap J) = (\mathcal{F} - J) \cup (\mathcal{F} - I)$

1.8   Floating-Point Operations

The **floating-point** operations that are predefined in ISETL are almost the same as the ones defined for integers shown in Table 1.1. For completeness, we list them all in Table 1.2. The notation is the same as in Table 1.1 (see page 23).

Of course, there are *precedence rules* governing the use of these and all operations in ISETL. Precedence rules determine the order in which the various parts of the expression are evaluated. For example, in the expression

```
a+b*c;
```

infix (operators)

+	:	$F \times F \longrightarrow F$	addition
-	:	$F \times F \longrightarrow F$	subtraction
*	:	$F \times F \longrightarrow F$	multiplication
/	:	$F \times F\text{-}\{0\} \longrightarrow F$	division
**	:	$F^{+}+\{0\} \times F \longrightarrow F$	exponentiation
=	:	$F \times F \longrightarrow B$	equality test
/=	:	$F \times F \longrightarrow B$	inequality test
<	:	$F \times F \longrightarrow B$	test for less than
>	:	$F \times F \longrightarrow B$	test for greater than
<=	:	$F \times F \longrightarrow B$	test for less than or equal
>=	:	$F \times F \longrightarrow B$	test for greater than or equal

prefix (functions)

max	:	$F \times F \longrightarrow F$	maximum of two floating-point numbers
min	:	$F \times F \longrightarrow F$	minimum of two floating-point numbers
is_floating	:	$F \longrightarrow B$	test for type floating_point
is_number	:	$F \longrightarrow B$	test for type integer or floating_point
-	:	$F \longrightarrow F$	unary minus
sqrt	:	$F^{+}+\{0\} \longrightarrow F^{+}+\{0\}$	square root
exp	:	$F \longrightarrow F^{+}$	power of e
ln	:	$F^{+} \longrightarrow F$	natural logarithm
fix	:	$F \longrightarrow N$	truncate to integer
ceil	:	$F \longrightarrow N$	smallest integer not less than
floor	:	$F \longrightarrow N$	largest integer not greater than
random	:	$F \longrightarrow F$	random number between 0 and the given number
sgn	:	$F \longrightarrow \{-1,0,1\}$	sign (-1 for negative, 0 for zero, 1 for positive)
abs	:	$F \longrightarrow F^{+}+\{0\}$	absolute value

Table 1.2: Floating-point operations

- Operators are listed from highest priority (tightest binding) to lowest priority.

- Operators are left associative unless otherwise indicated.

- "nonassociative" means that you cannot use two operators on that line without parentheses.

`CALL`	anything that is a call to a function — func, tuple, string, map, etc.
`# - +`	unary operators
`?`	nonassociative
`%`	nonassociative
`**`	right associative
`* / mod div`	
`+ - with less union inter`	
`.ID`	infix use of binary function
`in notin subset`	
`< <= = /= > >=`	nonassociative
`not`	unary
`and`	
`or`	
`impl`	
`iff`	
`exists forall`	
`where`	

Table 1.3: Precedence of operators in ISETL

the evaluation of the multiplication is done before the addition, so it is the same as

$$a+(b*c);$$

A precedence rule can always be overridden by the insertion of parentheses. The full list of ISETL precedence rules (which are exactly the same as the mathematical precedence rules) is given in Table 1.3.

As with integers, it is possible for the user to define her or his own functions that will last as long as the particular run of ISETL in which they are defined. Again, they can be preserved by putting them in a file and using !include to include them.

ISETL control structures can be used to define functions that are a little more complicated than simple expressions. For example, suppose you wanted a function that described the temperature of a rod that was made up of two materials separated by a heat source. Suppose at the source itself, the temperature of the rod is exceptionally high, for example, 500 degrees. But, as soon as you move away from the source, the temperature plunges immediately to 300 degrees. Furthermore, as you move to the right, the temperature continues to drop by twice the distance from the source, and as you move to the left, the temperature drops gradually by 20 times the square root of the distance from the source. Thus, to determine the temperature of the rod at any given point, you need three separate equations. To do this, you might write the following ISETL code.

```
bar_temp := func(x);
            if x = 0.0 then          $ x is at the source
                return 500.0;
            elseif x > 0.0 then      $ x is to the right
                return 300.0 - 2.0 * x;
            else                     $ x is to the left
                return 300.0 - 20.0 * sqrt(abs(x));
            end if;
        end func;
```

A graph of this function would look like

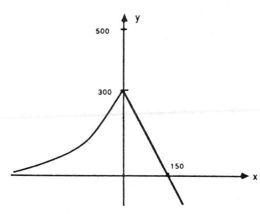

Graph of bar_temp

It can get much more complicated. Suppose for example, you divided all of the **floating-point** numbers into infinitely many intervals, for example, at the integers, and you wanted to have a function whose graph in each of these intervals looked like the graph of the function x^2 in the unit interval $[0, 1]$. That is, in each interval, you wanted your function to look like the following:

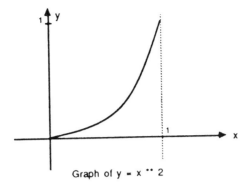

Graph of y = x ** 2

Now, you can change the shape of this graph just by multiplying the expression x^2 by a constant coefficient. If you multiply it by a coefficient greater than 1.0, say 3.5, then you stretch the graph:

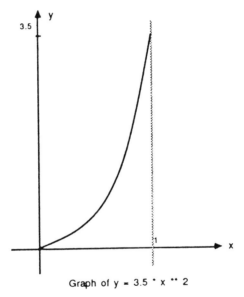

Graph of y = 3.5 * x ** 2

or, if you multiply by a positive number smaller than 1.0, such as 0.3, you squeeze it:

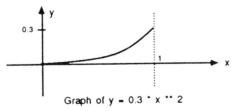

Graph of y = 0.3 * x ** 2

Finally, by adding a number, you can raise it (positive number) or lower it (negative number):

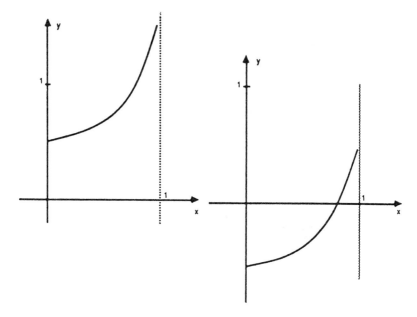

Graph of y = x ** 2 + 1/2   and   y = x ** 2 - 1/2

Now, here is the rub. Suppose you wanted to do this differently in each interval. For instance, suppose you wanted to define a function so that, in the interval $[n, n + 1)$, it looked like x^2, but with a coefficient of n and raised by $1/n$. Thus, in the interval $[n, n + 1)$, it would look like

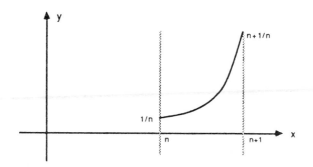

Then, the graph of the entire function (on the positive side) would look something like

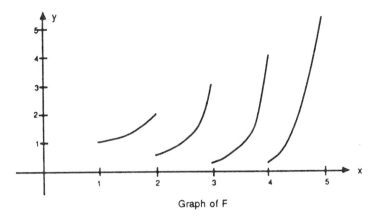

Graph of F

Of course, our definition of the function does not work in the interval [0,1), so the function is not defined there, and there is no graph in that portion of the plane. The ISETL code to implement this function might look like this:

```
f := func(x); local n;
        if is_floating(x) and (x > 1) then
            n := floor(x);
            return n*(x-n)**2 + 1/n;
        end if;
    end func;
```

Here is a very important point. We have given three different explanations of this function: verbal description, graph, and ISETL code. Both the verbal description and the graph are vague about what happens at the endpoints. The ISETL code cannot be. It can be wrong, but it can't be vague. If you read the code, you can tell that the value of the function when x is a whole number (other than 0) is $1/x$.

The usual way that this kind of definition is denoted in Mathematics is the following.

$$f(x) = n(x - n)^2 + \frac{1}{n}, \qquad \text{for} \qquad n \leq x < n+1, \quad n \neq 0$$

What we have been doing is to define a function *in parts*. Another way to do this is to use recursion, although it can get fairly complicated. Suppose, for example, we define a function on the interval $[0, 1)$ by some simple formula, say $3 - 2x$, and then we define the function on $[1, 2)$ by using the same expression with a little adjustment. Thus, if there were no adjustment, the value on $[1, 2)$ would be given by $3 - 2(x - 1)$, and if we continued this on subsequent intervals, the function would be periodic.

Let us introduce an adjustment, however, by subtracting x and multiplying the whole expression by 0.5. Thus, the value of the function on the interval $[1, 2)$ would be $\frac{1}{2}(5 - 3x)$. Continuing in this manner, we obtain

Interval	Expression for value of function
$[0, 1)$	$3 - 2x$
$[1, 2)$	$\frac{1}{2}(5 - 3x)$
$[2, 3)$	$\frac{1}{4}(8 - 5x)$
$[3, 4)$	$\frac{1}{8}(13 - 9x)$
$[4, 5)$	$\frac{1}{16}(22 - 17x)$

To finish it off, we define the function for negative values by making it symmetric about the vertical axis, that is, $F(-x) = F(x)$. The ISETL code for this function reads as follows:

```
F := func(x);
        if not is_number(x) then return; end;
        if x<0 then return F(-x); end;
        if x<1 then return 3-2*x; end;
        return 0.5*(F(x-1)-x);
     end;
```

and its graph looks like

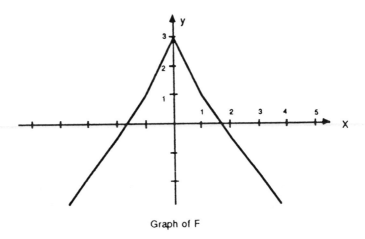

Graph of F

Summary of Section 1.8

The topics covered in this section were **floating-point** operations and defining, graphing, and reasoning with user-defined functions.

Exercises

1.8.1 Think about how ISETL performs floating-point operations.

For each of the following expressions,

1. TYPE in the statement, but do not hit the carriage return.

2. PREDICT what ISETL will output. In making your prediction, think about how ISETL will process the input you have entered. RECORD your prediction.

3. HIT the carriage return. If ISETL returned something different than what you predicted or ISETL gives you an RT error, EXPLAIN why.

a. 2.0 * (0.5 + 6.5) ** 2 / 5.0 - 10.2;

b. -3.0 ** 4.0; 4.0 ** 0.5;   -4.0 ** 0.5;

c. sgn(48.678) = (ceil(48.678) - floor(48.678));

d. floor(-45.67) <= -fix(45.67);

e. abs(-67.0E-4) >= 0.998;

f. random(5.6);   random(-95.67);   random(1007.0E-2);

g. random(22.4);   random(22.4);   random(22.4);

h. max(min(3.5,-9.87),-9.87E0);

i. min(sqrt(0.16),0.16);

j. 23.4 .max 67.2 .min -23.4 .min -abs(-67.2);

k. exp(ln(-4.5));

l. exp(-567.894) > 0.0; ln(-0.8E-3) <= 1.0;

1.8.2 Determine whether each of the following statements is *true*, *false*, or undefined (i.e., returns OM) for all values of x, where x is a floating-point number. If the statement is *true*, explain why; otherwise, give a counterexample, in which ISETL returns *false*.

a. abs(float(x)) = float(abs(x));

b. fix(float(x)) = float(fix(x));

c. floor(x) <= fix(x);

d. ceil(x) >= fix(x);

e. ceil(x) - floor(x) = 1;

f. exp(ln(x)) = x;

g. ln(exp(x)) = x;

h. sgn(abs(x)) = 1;

i. `sqrt(x) <= x`; whenever `x` is nonnegative.

1.8.3 Let b and c be positive integers and t be a positive **floating-point** number, where 0<b<c and b*t<=c. Then, the expression

$$\{n * t : n in [0,b..c]\};$$

evaluates to a **set** of **floating-point** numbers, starting at 0.0, increasing by a fixed increment and continuing as long as they do not exceed a certain upper bound.

Make an ISETL program out of this expression using **read** to obtain values for t, b, and c; **print** to return the **set**; and **!include** to run the program. Then, answer the following questions by using a combination of experimentation and reasoning. (Note: [2,4..10] = [2,4,6,8,10], [3,6..13] = [3,6,9,12], and so on.)

   a. Run your program for
      i) t = 0.001, b = 5, and c = 500
      ii) t = 2.0, b = 7, and c = 300

      What is the fixed increment in each case? What is the maximum item in the set? What is the minimum item?

   b. Pick other values for b, c, and t, and try the program for them. Answer the same questions as you did in part a. about the increment and about the upper and lower bounds for each new collection of inputs.

   c. Find a formula (in terms of b, c, and t) for the fixed increment.

   d. Find a formula for the least upper bound (maximum) of the set. Find one for the greatest lower bound (minimum). See Exercise 1.5.4.

   e. What happens if we don't require things such as 0 < b < c, t > 0, and b * t <= c?

1.8.4 Use ISETL to define functions in parts.

   a. Put the following lines in a file:

```
F := func(x);
        if not is_floating(x) then
            return;
        elseif x < 0.0 then
            return x - 1.0;
        else
            return x + 1.0;
        end if;
    end func;
```

i) After doing a !include on your file, evaluate F for various **floating-point** numbers. What happens if you try to evaluate F at something other than a **floating-point** number, for example, at a **character** or an **integer**?

ii) Give a verbal description of the action of the function.

iii) Draw a graph of the function.

iv) Is the function continuous?

v) If the function is not continuous, how can you change the definition of the function so that it is continuous?

b. Put the following lines in a file:

```
G := func(x);
        if not is_floating(x) or x <= 0.0 then
            return om;
        elseif x <= 1.0 then
            return sqrt (x);
        else
            return G(x-1.0) + 1.0;
        end if;
    end func;
```

Answer the same questions for G as you did for F in part a.

c. Consider the following graph:

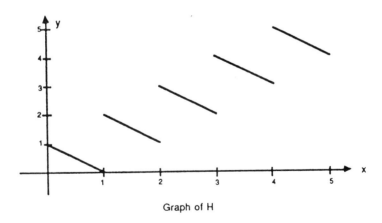

Graph of H

i) Give a verbal description of the action of a function, H, which this graph might represent.

ii) Write an ISETL **func** that accepts a **floating-point** number, **x**, and returns the value of H at **x**. Run your **func** on a variety of inputs.

d. Consider a function, such that for each integer n, its graph looks like n/x for x negative and nx for x nonnegative, where $n \leq x < n + 1$.

   i) Draw a graph of the function.

   ii) Is the function continuous?

   iii) Write an ISETL **func** that implements this function.

1.9 Other Simple Data Types

We list for completeness four other simple data types that are discussed in the remainder of the book: character **strings**, **atoms**, **undefined**, and **file**.

A character **string** is a sequence of 0 or more characters not separated by a carriage return and enclosed in double quotes, ". The list of characters used by ISETL is given in Table 1.4. A string of 0 characters, called the *null string*, is written as two successive double quotes, "". The operations on **strings** include *concatenation* (+), *replication* (*), *length* (#), tests for *equality* and *inequality* (=, /=), tests for *lexicographic order* (<, >, <=, >=), a test for *substring* (**in, notin**), and a type test (**is_string**). It is also possible to access a single character in a **string** by its position. This uses function notation. We can illustrate these operations and substrings with the terminal session in Figure 1.7.

The null **string** is a two-sided identity for concatenation, that is, for any **string s**, the following expressions will evaluate to *true*:

```
s + "" = "" + s;
"" + s = s;
```

These **string** operations lead to an *algebra of strings* that is very important in the study of automata. We will briefly examine automata in Chapter 4.

Atoms are objects in ISETL. They are created by the operation **newat** (which takes no parameters). This operation creates an object that is different from every other ISETL object in existence at the time. Once created, all an **atom** does is exist. The only operations relative to **atoms** are **newat**, **is_atom**, and tests for equality and inequality.

Surprisingly, it is possible to do a lot with **atoms** even though they have so little going for them. Like any ISETL object, an **atom** can be assigned as the value of an identifier or made an element of a **set** or a component of a **tuple**. Hence, **atoms** can be in the domains and images of **maps**. We will use them in setting up data bases (as examples of functions) in Chapter 4, where we discuss graphs and trees. Here is an example of a whimsical application.

There is an old science fiction story in which a hardened criminal commits an ugly crime, and in the ensuing police chase, his vehicle is smashed

Any of the printable characters on the keyboard except for the double quote may be typed directly into a string. The backslash convention may be used to enter special characters. When pretty-printing, these conventions are used for output. In the case of formated output, the special characters are printed. (See the *Intro* for information about formated output.)

\b	backspace
\f	formfeed (new page)
\n	newline (prints as CR-LF)
\q	double quote
\r	carriage return (CR)
\t	tab
octal	character represented by *octal*
	Refer to an ASCII chart for meaning.
other	*other* — may be any character
	not listed above.

In particular, "\\" is a single backslash. You may type, "\"" for double quote, but the pretty printer will print as "\q". ASCII values are limited to '\001' to '\377'.

```
>        %+ [char(i): i in [1..127]];
"\001\002\003\004\005\006\007\b\t\n\013\f"
+"\r\016\017\020\021\022\023\024\025\026"
+"\027\030\031\032\033\034\035\036\037 !"
+"\q#$%&'()*+,-./0123456789:;<=>?@ABCDEF"
+"GHIJKLMNOPQRSTUVWXYZ[\\]^_`abcdefghijk"
+"lmnopqrstuvwxyz{|}~\177";
```

Table 1.4: ISETL's character set

```
% isetl
ISETL(1.8)
>  s := "AB"; t := "ab";
>  s+t;
"ABab";
>  3*s;
"ABABAB";
>  s=t; s/=t;
false;
true;
>  "B" in s+t; "Ab" in s+t; "Ba" in s+t;
true;
false;
true;
>  "book" < "box"; "box" <= "book";
true;
false;

>  s(1); t(3); (s+t)(3);
"A";
OM;
"a";
>  !quit
```

Figure 1.7: Sample terminal session

and his body is totally wrecked. But, still he retains a spark of life. Since
this is the future, it is possible to completely rebuild him. Organ transplants,
skin grafting, new blood, even a new brain are provided. His memory is
gone (both conscious and subconscious), so he receives a total psychological
rehabilitation. It takes a long time, but, eventually, he is completely cured.

   Then, of course, he must go on trial. His lawyer uses as a defense the
claim that this is not the same person who committed the crime. This is
a philosophical question, but there is also the practical point that there is,
indeed, no way to identify this person as the hardened criminal. Every facet
of him, physical, emotional, chemical, is different. So there is a problem.

   If, however, newat had been invoked and somehow identified with this
man when he was born, then he would have a unique identity. This would
resolve the practical issue but not the philosophical issue. How would you
deal with that?

   OM is the value that an identifier has when it has not been given any
other value. It means that the variable is *undefined*. It is also often the

value of ill-defined expressions. You can check whether or not an ISETL object is undefined by using the type tester is_om.

File is the data type used to read information from and to print to an external file in your operating system. The technical details of its use are explained in the *Intro*. See page 14 for an example.

Summary of Section 1.9

In this section, we discussed character strings, atoms, files, and OM. You already know how to read and write to a file, so the initial exercises will give you some practice working with strings, atoms, and OM and ask you to think about some tricky assignment statements while reviewing the other types that were discussed in this chapter. The later exercises consist of computer experiences that serve as pointers to things to come.

Exercises

1.9.1 Manipulating strings—thinking about how ISETL performs string operations.

Type in each of the following lines. Before you hit the carriage return, predict what ISETL will return. Compare your prediction to ISETL's actual output. If the two differ, explain why.

a. s := "abra cadabra"; t := "fee-fi-foo-fum";

b. s; t; w;

c. #s; #t;

d. s(4); s(5); t(16);

e. s + t; s + "dabra"; "abra " + s; s + w;

f. 2*s; 3*s(36); s;

g. s < t; t > t; "aa" >= s; s <= 2*s;

h. "rac" in s; "Abra" in s; "rafe" in s + t; " " in 4*s;

i. t(4) + t(7) + t(11); (s + t)(20); (s + t)(30);

j. (t + s) in (2*t + 3*s + s(5));
   "a" = (s + " " + s)(17);

1.9.2 Assigning values to variables in ISETL.

Type in the following lines and think ISETL.

a. n := 58; x := 1.5; s1 := "help "; s2 := "ME!";

b. y := n*s2; y; y := char(n); y; y := ord(y); y; Y;

c. n := n mod 3 + n div 3; n; n*r + n;

d. print "please enter an integer"; read m;

e. m; n; m := m*n + n**2 - 10; m; min(n,max(n,m));

f. x := sqrt(2.0*x + 3.0 + float(n + 9 mod 4)); x;

g. z := newat; z; w := newat; w; z = w; z = z; w /= w;

h. t := s1 + s2; t; s2 := s1; s2; s1 := s2; s1;

i. print "Tell me your two favorite"
        + " floating-point numbers";

j. read a, b, c;

k. a; b; c; d;

1.9.3 Taking a first look at the Boolean connectors, and, or, not, =, and
impl.

Each of the following expressions is a Boolean expression, that is,
it has the value *true* or the value *false*. Type in the following lines
and see if you can GUESS whether ISETL will output *true* or *false*.
(Don't worry if you don't guess correctly—we will be talking a lot
about these things in the next chapter.)

a. is_floating (1.456) and is_floating (-78.0);

b. is_floating (98.0E4) and is_floating (198);

c. is_integer (-1020.0) and is_floating (1020);

d. is_string ("ISETL") or is_integer (10**2);

e. is_string ("ISETL") or is_floating (10**2);

f. is_integer (10.0) or is_integer (2.0 * 4.0);

g. not is_floating (23.6); not is_floating (236);

h. is_string ("hi there") = is_integer (0);

i. is_floating (1.2) = is_string (1.2);

j. is_integer (9) impl is_floating (2.3);

k. is_floating (9.88) impl is_string (234);

l. is_integer (9.9) impl is_string("abc");

m. is_atom (3) impl is_atom (9);

n. true; false;

o. true and true; true and false;

p. `false and true; false and false;`

q. `true or true; true or false;`

r. `false or true; false or false;`

s. `not true; not false;`

t. `true = true; true = false;`

u. `true impl true; true impl false;`

v. `false impl true; false impl false;`

w. `a := 56.897; p := is_floating (a); a; p;`

x. `z := newat; q := is_integer (z); z; q;`

y. `p and q; p or q; p impl q; not p; not q; p = q;`

Chapter 2

Propositional Calculus

To the Instructor

In this chapter, we are setting a number of goals for the cognitive development of the student. The most fundamental construction that we are after is that the student should have a mental model of *proposition*. This should take the form of a representation of *Boolean valued variables* that can be combined by logical connectives to form Boolean expressions corresponding to logical statements in English. When the variables are replaced by specific Boolean values (*true* or *false*), then the expression has a Boolean value. We will try to stimulate this construction by having the student experience the interaction between a logical statement in English, an ISETL (or mathematical) expression that represents the statement, and the computer activities involved in storing and evaluating the expression. Translation back and forth between English and ISETL will be an important activity that will also contribute to the student's ability to work with formal notation.

Once the concept is firmly in mind, it should be possible for the student to begin to reason about propositions. We start with negation, and, here again, the key to understanding will be for the student to always try to think about how the computer evaluates an expression and its negation. Reasoning includes answering questions about a Boolean expression, such as "What do you know about the value of the expression if you know that this variable (or this clause) is definitely true?"

Thinking about Boolean expressions, and in particular, implication, leads naturally into the notion of *proof.* We intend that the student will come to understand something of the need for proof, the difference between a hypothesis and a conclusion, and the transition from the former to the latter, as well as the variety of methods of proof that might be in one's repertoire. We will make these abstract ideas concrete by having the student work with truth tables, validity arguments, and a review of some of the proofs about integers that were discussed in Chapter 1.

Finally, we want to introduce two goals that will be of increasing importance as the book unfolds. The first is that in addition to thinking about a proposition as a process, the student should begin to think of a proposition as an object—an object that can be an element of a set or on which a function can act. To this end, we will have the student work with programs that manipulate **strings** representing propositions. The second goal is a pointer to the existential and universal quantifiers of predicate calculus, which will be the main topic of Chapter 5. The student's introduction to this (with credit to David Gries) will be via disjunction and conjunction of finitely many Boolean expressions. These can be negated, iterated, and otherwise manipulated to form very complex statements that, because they are concrete, will be accessible to the student. The notation of **exists** and **forall** will be introduced—as ISETL syntax—in this relatively simple context. The corresponding mathematical notations $\exists$ and $\forall$ will also be used.

2.1 Preview

A proposition is a statement that can be either true or false. Sometimes this statement will have variables and their value will determine whether it is true or false. The purpose of this chapter is to help you learn to work with such statements, which are sometimes called *Boolean expressions*. You will gain experience with expressing propositions in formal language, which can be either ISETL or standard mathematical notation. You will learn to combine propositions with operations called *Boolean operators*. You will work on learning how to reason about Boolean expressions, to determine if they are true, false, or you can't tell.

The topic of this chapter, propositional calculus, is essential to your mathematical development. The reason is that so much of Mathematics is about deciding which statements about mathematical objects and their relations are true and which are false. Determining that a statement is true is, in Mathematics, what is called a proof. Determining that it is not always true amounts to producing a *counterexample*, a particular case in which the statement is false. You will study some general methods for making proofs and practice with a number of different kinds of situations.

Finally, at the end of the chapter we give two pointers to future material. There is a brief introduction to quantification, which is fully treated in Chapter 5, and, in order to be able to study functions in which the variables take on values that are propositions (essential for Chapters 4, 5, and 7), we provide some exercises that will start you thinking about propositions as objects, just like numbers.

2.2 Boolean Variables and Expressions

Sometimes you put **integers** or **strings** into an expression and what comes out is *true* or *false:*

```
3 > 5; "s" in "trash basket"; odd(17 mod 3);
```

Such Boolean expressions are important in writing programs that branch in different directions depending on the value of certain quantities. For example, here is an ISETL **func** that calculates the solution of the general quadratic equation $ax^2 + bx + c = 0$:

```
Bin_root := func(a,b,c);
                if a=0.0 and b=0.0 then
                    if c=0.0 then
                        print "No equation";
                    else
                        print "No solution";
                    end if;
                elseif a=0.0 then
                    print "One solution:   ", -c/b;
                elseif b**2=4*a*c then
                    print "One multiple solution:   ",
                        -b/(2*a);
                elseif b**2<4*a*c then
                    print "No real solutions";
                else
                    print "Two solutions:   ",
                        (-b+sqrt(b**2-4*a*c))/(2*a),
                        (-b-sqrt(b**2-4.0*a*c))/(2*a);
                end if;
            end func;
```

In this program, the Boolean expression after **if** is evaluated. If its value is *true*, then the clause after **then** is evaluated; otherwise, it goes on and does the same thing with **elseif**. If no expression evaluated to *true*, then the clause after **else** is evaluated. Thus, for example, Bin_root(0.0,0.0,1.0) would print the value **No solution** and Bin_root(1.0,0.0,-1.0) would print the two solutions **1.0** and **-1.0**.

Sometimes the variables themselves are Boolean. For example, suppose you wanted to think about a sentence such as the following.

> If I finish my homework before dinner and it does not rain,
> then I will go to the ball game.

You might decide that a good way to begin analyzing it is to express it in ISETL syntax. Observe first that the following three statements are in the sentence.

I finish my homework before dinner.
It does rain.
I will go to the ball game.

Each of these three may be considered to be a statement about reality. It could be true or false. More importantly, each of these statements could be, true in one situation and false in another. What we are saying is that they are variables whose possible values are one of the two elements of the set {*true, false*}. We say that these statements are *Boolean variables*, and that the *domain* of these variables is the set {*true, false*}.

We could name them and represent them with identifiers in ISETL, for example, finish, rain, and go for the three statements, respectively. Then, the original sentence can be expressed in ISETL as

$$(\text{finish and not rain}) \text{ impl go} \qquad (2.1)$$

Notice that *and* and *not* were essentially translated, but the *if. . . then* clause was replaced by the single connective, impl, which we read, "implies."

Now the sentence is in a form to be analyzed, but before getting into that, let's say a little more about various uses of such expressions. Look at the function we defined on page 51. We used a Boolean expression to restrict the domain of this function to the floating-point numbers not in the interval [0,1). Similarly on page 52 we used expressions whose values were Boolean to define the function.

Another important use of Boolean values occurs in programming with while loops. For example, look again at the ISETL func we defined on page 17, which converts a positive integer from base 10 to base 2. The statements inside the while loop are repeatedly executed as long as ("while") the Boolean expression x /= 0 is *true*. The truth value of the Boolean expression is determined before entering the loop. If the value is *true*, then the statements inside the loop are executed, and the Boolean expression is tested again to determine whether or not the loop should be re-entered. Notice that if the Boolean expression that controls a while loop were never to become *false* the loop would execute forever, and you would have an *infinite loop*.

Now let's go back to the expression (2.1). Evaluating such an expression (for a particular situation) means to figure out the truth value of the whole expression once you know the truth value of each of the variables finish, rain, and go. For example, suppose you had a situation in which you finished your homework, it did not rain, but you didn't go to the ball game. This would mean you had

$$(true \wedge \neg false) \implies false$$
$$(true \wedge true) \implies false$$
$$true \implies false$$
$$false$$

(where we are using the mathematical symbols $\land$ for **and**, $\lor$ for **or**, $\neg$ for **not**, and $\Longrightarrow$ for **impl**). So, the value of the expression in this situation is *false*. You could get the same result by writing the following ISETL code:

```
>   finish := true;
>   rain := false;
>   go := false;
>   (finish and not rain) impl go;
false;
```

If you wanted to know the result for all possible values of the three Boolean variables (how many possibilities are there?), then you would have to be a little more systematic and talk about variables, connectives, and their truth tables.

We have already discussed Boolean variables. The standard Boolean operations are **and**, **or**, **impl**, **=**, and **not**. The first four connect two Boolean expressions to form a Boolean expression, and the last applies to a single expression. The meaning of these connectives is almost exactly what you might expect, but there are a few fine points, so let's describe them completely. The best way is to specify the value of the expression for all possible values of the variables that appear. For the first four, since there are two variables, each of which has two possible values, there are a total of four possibilities. For the last, there is only one variable, so there are only two possibilities. We can arrange these facts in a *truth table*, Table 2.1, where we assume that P and Q are two Boolean variables, T denotes *true*, and F denotes *false*. There might be a couple of values in Table 2.1 that

P	Q	$P \land Q$	$P \lor Q$	$P \Longrightarrow Q$	$P = Q$	$\neg P$
T	T	T	T	T	T	F
T	F	F	T	F	F	F
F	T	F	T	T	F	T
F	F	F	F	T	T	T

Table 2.1: Definitions of Boolean connectives

you find puzzling, perhaps because they differ from English usage. For example, in English, the word "or" often means one or the other, but not both. In formal language (which is what we are talking about), the English word "or" means *inclusive or*.

What is perhaps more troubling for beginners is the fact that if P has the value *false* then $P \Longrightarrow Q$ has the value *true* no matter what value Q has. An intuitive way to think of this is that if a false premise is accepted, then any conclusion will follow from it. For instance, if someone says, "If you get above 95 on the final, then you will receive an A in the course," they are still telling the truth when you get an 85 and receive a B and when you get an 89 and receive an A. More formally, you can perhaps see from the truth tables that the following equality is always true:

$$(P \Longrightarrow Q) \quad = \quad ((\neg P) \vee Q)$$

To convince yourself that the two expressions are equal, evaluate them for all four possible values of P and Q and see that you always get the same truth value. Once this equation is accepted, then it is clear that if P is *false*, then $\neg P$ is *true* and the result is *true* no matter what the value of Q is.

We already mentioned exclusive or, which is symbolized *xor*. There are some other interesting operators that can be defined.

$$
\begin{aligned}
(P \ xor \ Q) &= ((P \vee Q) \wedge (\neg(P \wedge Q))) \\
(P \ nor \ Q) &= ((\neg P) \wedge (\neg Q)) \\
(P \ nand \ Q) &= ((\neg P) \vee (\neg Q)) \\
(P \ ximpl \ Q) &= ((\neg P) \ xor \ Q)
\end{aligned}
$$

You might enjoy trying to think up "real-life" meanings for these. In the exercises, you will have a chance to think about their truth tables. Be careful, though. The standard connectives and, or, impl, and not are implemented in ISETL, but *xor*, *nor*, *nand*, and *ximpl* are not. Of course, you can define them with a func if you need them.

It's time to admit that we have been using the English language a little sloppily. We have spoken about Boolean expressions and also about the value of a Boolean expression. Actually, a Boolean expression standing alone does not have a value. Look at

```
(x or y) impl (y and z);
```

This is a Boolean expression. It is a string of symbols that uses logical connectives and, in fact, amounts to correct ISETL syntax. But it doesn't have a value. If you ran it, you would get OM. What you can do, however, is assign Boolean values to x, y, and z first, and then use ISETL to evaluate the expression. Evaluating the expression still doesn't mean that the expression itself has a value. It simply means that it can be used together with other things (such as assignments to the identifiers that appear in it) to obtain values. Thus, the following code does produce a value:

```
>   x := true;
>   y := false;
>   z := false;
>   (x or y) impl (y and z);
false;
```

You can also do it with a func as follows:

```
F := func(x,y,z);
        return (x or y) impl (y and z);
    end;
```

After defining this **func**, if x, y, and z are not assigned values, then

```
(x or y) impl (y and z);
F(x,y,z);
```

will both return **OM**, but

```
F(true, false, false);
```

will return **false**. What has happened here is that a function (which we call a *Boolean function of 3 variables*) has been created using a Boolean expression with three identifiers (or *literals*). Then, we evaluated the function with x = *true*, y = *false*, and z = *false*.

This distinction between Boolean expressions and Boolean functions can be very important, but in this book we will usually continue our informal way of mixing a Boolean expression and the function that can be created using it.

Summary of Section 2.2

This section introduced you to the basic tools used in propositional calculus: Boolean variables, operators (connectives), and expressions. We discussed translating English sentences (statements whose values are either *true* or *false*) into Boolean expressions in ISETL and vice versa.

We defined the truth values of the Boolean connectives by giving their truth tables. Finally, we discussed how to evaluate an expression by assigning to each of the identifiers that occur in the expression a specific truth value.

The following exercises give you some practice using the Boolean operators, translating sentences back and forth between English and ISETL, and evaluating expressions. In the final exercise, you are asked to write ISETL **funcs** for the operators *xor*, *nand*, *nor*, and *ximpl*.

Exercises

2.2.1 Translating from ISETL code to English. Evaluating an expression.

a. Let the identifier **rain** represent the statement "It is raining," **grump** represent "James is grumpy," and **cancel** represent "The picnic is canceled."

Write the following expressions in simple sentence form. Note that usually the connective = can be translated to "if and only if."

   i) `rain and grump and cancel`

   ii) `rain impl (cancel and not grump)`

   iii) `(not grump) = (not rain)`

   iv) `((not rain) or (not cancel)) impl (not grump)`

   v) `not ((not rain) and (not grump))`

   vi) `rain or (grump and (not cancel))`

   vii) `grump = (rain and cancel)`

   viii) `(rain impl (not (not grump))) or (not cancel)`

b. Enter ISETL. Assign `rain` and `cancel` the value *true*, and `grump` the value *false*, by typing

$$\texttt{rain := true;}$$
$$\texttt{cancel := true;}$$
$$\texttt{grump := false;}$$

* TYPE in each of the expressions in part (a), followed by a semicolon.

* BEFORE you hit the carriage return, PREDICT what ISETL will return by evaluating each part of the expression using the definitions of the connectives. (Hint: Evaluate the parenthesized portions of the expression first just as you would with an algebraic expression.)

* RECORD your prediction.

* HIT the carriage return, and COMPARE ISETL's response with your prediction. If the two differ, try to explain why.

2.2.2 Translating from English to ISETL. Evaluating an expression for all possible combinations of the values of its variables.

a. For each of the following statements, choose names for the variables it depends on, and then write the statement as a Boolean expression in ISETL.

   i) If I don't go to the party, then I don't study Discrete Mathematics or I don't go to bed late.

   ii) When $x <= y$ then $y > z$, but when $x > y$ then $y <= z$.

   iii) If the utility costs go up or the request for initial funding is denied, then a new computer will be purchased if and only if we can show that the current computing facilities are not adequate.

   iv) If apples cost more than cranberries, then we will make cranberry sauce, but (and) if they cost the same or less than cranberries, then we will make applesauce.

   v) At least one of the integers $B_1, B_2, B_3,$ or B_4 equals 6.

   vi) Every one of the integers C_1, C_2, C_3 is greater than -2.

b. Now that we have a Boolean expression representing each statement in part a., we can determine the truth or falsity of the statement for every possible combination of values for the variables on which the statement depends. One way to do this is to give the truth table for the expression just like we gave the truth tables for the operators and, or, impl, =, and not. For instance, in the example we gave in the text:

> If I finish my homework before dinner and it does not rain, then I will go to the ball game.

we chose variables finish, rain, and go, and then expressed the statement in ISETL code as follows:

```
(finish and (not rain)) impl go;
```

To evaluate this expression for all possible values of finish, rain, and go, our truth table will have eight lines (why?), and might look like Table 2.2 (where we have written f for finish, r for rain, g for go, T for *true*, and F for *false*). Notice how this

f	r	g	not r	f and (not r)	(f and (not r)) impl g
T	T	T	F	F	T
T	T	F	F	F	T
T	F	T	T	T	T
T	F	F	T	T	F
F	T	T	F	F	T
F	T	F	F	F	T
F	F	T	T	F	T
F	F	F	T	F	T

Table 2.2: (finish and (not rain)) impl go

table has one column for each of the variables, and then one column for each parenthesized part of the expression, starting with the innermost portion of the expression and working our way out. Give the truth table for each of the expressions in part a.

c. Developing truth tables is tedious work. There has to be an easier way to do this! Another way is to write an ISETL program that will determine the truth value of an expression in each case. Our program has to iterate through all the possible combinations of values for the identifiers that appear in the statement. We can do this with an ISETL for loop. For instance, looking again at the example in part b., our program might look like program HighHopes.

```
program HighHopes;
    for finish, rain, go in {true, false} do
        print [finish, rain, go];
        print (finish and (not rain)) impl go;
    end;
end;
```

The **for** loop will iterate through all 8 possible combinations of values for the variables, and for each combination it will print out the values of **finish**, **rain**, and **go** and the corresponding value of the expression. One pass through the **for** loop corresponds exactly to one line in the associated truth table, and vice versa.

Write an ISETL program for each of the statements in part a. In each expression, try to visualize ISETL iterating through all the combinations of values of the variables and, for each combination, try to predict what your program will print.

2.2.3 Write an ISETL program that will print out the various ranges of values of x for which the expression

$$\frac{(x-1)(x-2)}{(x-3)(x-4)}$$

is positive, negative, zero, or undefined. Try to minimize the number of **if** clauses by using Boolean expressions.

2.2.4 Translating from English to mathematical notation. Take each of the six statements in Exercise 2.2.2, part a., and express it in mathematical notation.

2.2.5 It is possible to negate an expression in ISETL by inserting the operator **not** in front of the expression. It is more informative however, to expand such an expression, "absorbing" the **not** within the expression. For example, the negation of the statement

> If I finish my homework before dinner and it does not rain, then I will go to the ball game.

can be written as

> I finish my homework before dinner and it does not rain, but (and) I don't go to the ball game.

Now, consider the statement

> If I don't go to the party, then I don't study Discrete Mathematics or I don't go to bed late.

Which of the following four statements is the negation of this statement?

1. I go to the party and I study Discrete Mathematics and I go to bed late.

2. I go to the party and I don't study Discrete Mathematics and I go to bed late.

3. I don't go to the party and I don't study Discrete Mathematics, but I do go to bed late.

4. I don't go to the party and I study Discrete Mathematics and I go to bed late.

2.2.6 Consider the expression

$$(p \Longrightarrow q) = ((\neg p) \lor q)$$

The expression on the right-hand side and the one on the left are equivalent if and only if the equality expression evaluates to *true* for all combinations of values of the variables p and q.

a. Prove the the two expressions are equivalent by giving the truth table for the equality.

b. Prove that the two expressions are equivalent by writing an ISETL program that evaluates the equality.

2.2.7 How much work does it take to determine the value of an expression for all possible combinations of values for its variables? Suppose you have a Boolean expression with n variables, say $p_1, p_2, \ldots, p_n$.

a. If you used a truth table, how many lines would there be in your truth table?

b. If you used an ISETL program, how many times would ISETL execute the statement list in the body of the following loop?

```
for p1, p2, ..., pn in {true, false} do
      statement list
end;
```

2.2.8 The user-defined operators *xor*, *nand*, *nor*, and *ximpl*.

a. Give the truth table for each operator.

b. Use truth tables to show that

i) $(p \text{ } nor \text{ } q) = (\text{not}(p \text{ or } q))$

ii) $(p \text{ } nand \text{ } q) = (\text{not}(p \text{ and } q))$

iii) $(p \text{ } xor \text{ } q) = (\text{not}(p = q))$

iv) $(p \ ximpl \ q) = (p = q)$

c. To be able to use these operators in ISETL, we must define ISETL **funcs** corresponding to each one. If we put all these **funcs** in one file and then **!include** the file when we enter ISETL, we will be able to use the operators along with the predefined operators **and, or, impl, =,** and **not.**

For example, we can define *xor* as follows:

```
xor := func(p,q);
          if is_boolean(p) and is_boolean(q) then
              return (p or q) and (not (p and q));
          end;
      end;
```

We can then call the function **xor** by assigning values to **p** and **q** and writing

```
                     xor (p,q);
```

or we can use the . notation to make **xor** an operator:

```
                      p .xor q;
```

We can also pass the values directly, for example

```
                  true .xor false;
```

i) What happens if you call **xor** without first assigning values to **xor**'s arguments?

ii) Put the **func** for **xor** into a file and then write ISETL **funcs** for the other user-defined operators. Test your operators for all possible combinations of inputs.

2.3   Reasoning and Boolean Expressions

"Is the value of a particular Boolean expression *false?*" This question should call to mind a picture. Perhaps it is a picture of a specific example of a Boolean expression or of a general expression with a bunch of Boolean operators and literals or maybe it is a piece of ISETL code. In any case, there are variables—or identifiers—somewhere in the picture, and before the question is answered, values must be assigned to them. Then, a process takes place, and the result is the value that was asked for. This process might be performed by the computer if the expression were a piece of ISETL code, it might be written out with pencil and paper if it were a homework assignment, or you might just work it out in your head. It is all the same thing. So if you are ever stuck with figuring out an expression, ask yourself what ISETL would do with it.

The question can get more complicated. Suppose that, instead of thinking about the value of the expression for one set of values for the variable, the idea was to see if it was true for *all* values. If this is the case, then the expression is called a *tautology*. Here are some examples.

```
(X and (Y or Z)) = ((X and Y) or (X and Z));
(not (X and Y)) = (not X or not Y);
```

You can check that each of these expressions always evaluates to *true* by forming a truth table, just like we did in the previous section. Table 2.3 gives the truth table for the first expression, where we have used the following shorthand notation:

Let L be the left-hand side of the equation, namely X and (Y or Z)

Let R be the right-hand side, (X and Y) or (X and Z)

X	Y	Z	Y or Z	X and Y	X and Z	L	R	L = R
T	T	T	T	T	T	T	T	T
T	T	F	T	T	F	T	T	T
T	F	T	T	F	T	T	T	T
T	F	F	F	F	F	F	F	T
F	T	T	T	F	F	F	F	T
F	T	F	T	F	F	F	F	T
F	F	T	T	F	F	F	F	T
F	F	F	F	F	F	F	F	T

Table 2.3: (X and (Y or Z)) = ((X and Y) or (X and Z)) is a tautology

Another kind of expression that is always *true* is what is called a *valid argument*. This is a Boolean expression that represents an implication between two expressions. These are often presented as *word problems*. Consider the following argument:

I run to class or it is not late.
If my alarm rang, then it is late.
Therefore, it is late or my alarm did not ring or I don't run to class.

Is this a valid argument? First, let's clarify what is meant by the technical phrase *valid argument*. It doesn't mean that any (or all) of the three statements is true. It means that the implication between the conjunction of the first two and the third is always true, no matter whether the individual statements or any parts of them are true or false.

If ISETL does return a *false*, or if one of the lines in the truth table for the implication does evaluate to *false*, then the implication is not a tautology, and the argument associated with it is not valid. A specific set of values that makes the implication *false* is called a *counterexample*. You only need to find one such set of values for the argument not to be valid.

This gets clearer if we express it in ISETL syntax. First, let's pick out and assign symbols to the variables, that is, the parts of the statement that are simple factual assertions. We will write

Run to represent "I run to class"
Ring for "My alarm rings"
Late for "It is late."

Then, the entire argument can be symbolized with the first two statements joined by and and this conjunction implying the third statement.

((Run or not Late) and (Ring impl Late)) impl
(Late or not Ring or not Run);

To say that this is a valid argument means that the implication, that is, the entire statement, is *true* no matter what the values of Run, Ring, and Late are. There are several ways to determine this. One is to make a truth table as we did before. Another is to write ISETL code that will do it. Here is one simple version:

```
for Run, Ring, Late in {true, false} do
    print ((Run or not Late) and (Ring impl Late)) impl
                    (Late or not Ring or not Run);
end;
```

Then, you just look at the output and make sure that nothing but *true* comes out.

There is a shorter way to do this kind of problem that lets you think instead of compute. Notice that since an implication will always be true whenever at least one of the hypotheses is false (why?) we will be sure that the argument is valid if we can show that the conclusion is true when all the hypotheses are true. This method is called a *direct proof.*

There is another way of thinking called *indirect proof.* Observe that an implication can only be false if the hypothesis is true and the conclusion is false. In this problem, the only possibility for the conclusion to be false is if Late is *false* and both Ring and Run are *true*. But in this case, it follows that Ring impl Late is *false*, and so the hypothesis cannot be true. Therefore, the statement is always true.

To summarize, we have seen three ways to deal with such a problem: a truth table, ISETL code, and reasoning. They are not so different. The ISETL code is nothing more than a computer implementation of the truth table. Instead of listing the eight possibilities in rows, it runs through them with a for loop. As for the reasoning, it would really be very hard to do if you do not have in mind a solid understanding of the computer looping through the possibilities so that you could realize that you didn't need to think about all of them.

At this point, you shouldn't have much trouble understanding what is meant by the negation of a Boolean expression. The negation is a second expression that uses exactly the same variables as the first and has the property that for each set of values of the variables if the original expression is true then the new one is false and vice versa. Thus, the negation of

$$A \wedge (\neg B \vee C)$$

is

$$\neg A \vee (B \wedge \neg C)$$

which you can check by a truth table, an ISETL program, or straight reasoning.

Look at Table 2.4. This gives the negations of the four standard operators. You can choose your method for checking them. The first two of

Expression	Negation
$S \wedge T$	$\neg S \vee \neg T$
$S \vee T$	$\neg S \wedge \neg T$
$S \implies T$	$S \wedge \neg T$
$\neg S$	S

Table 2.4: Negation of standard operators

these are called *De Morgan's laws* and they should be fairly clear to you by now. If you have any difficulty with the third, then recall the following tautology:

$$(S \implies T) = ((\neg S) \vee T)$$

Clearly, if two Boolean expressions are equal, then so are their negations, so the third line of the table follows from the second and fourth lines. That is, we have

$$\begin{aligned} (\neg(S \implies T)) &= (\neg((\neg S) \vee T)) \\ &= (\neg(\neg S) \wedge (\neg T)) \\ &= (S \wedge (\neg T)) \end{aligned}$$

This kind of calculation provides a useful tool for analyzing a statement given in English. Suppose you are trying to negate the following English sentence:

> If a student gets an A and the course work was mechanical or the teacher made a mistake, then the student didn't learn anything or the teacher did not make a mistake.

The first thing is to name the variables and express the statement in ISETL syntax. Thus, we might let

A represent "the student gets an A"
M represent "the course work was mechanical"
T represent "the teacher made a mistake"
L represent "the student learned something."

Then, the statement in ISETL reads

(A and (M or T)) impl (not L or not T)

To negate it, we first understand it as being of the form

P impl Q

where

P = (A and (M or T)) and Q = (not L or not T)

By our table, the negation of this is

P and not Q

which translates back to

(A and (M or T)) and not(not L or not T)

Now, the last part, not Q, is of the form, not(X or Y), which is (not X and not Y), so the expression translates to

A and (M or T) and L and T

We can then put this back into English to obtain the negation of the original statement.

The student gets an A; the work was mechanical or the teacher made a mistake; the student learned something; and the teacher made a mistake.

Moreover, with a little thought, you can see that the "or" clause can be dropped without changing the truth value of the statement.

Finally, a good way to check that you are following all this and a useful mental exercise is contained in the following two examples.

Suppose you knew that *James will be grumpy tomorrow*. What can you say about the truth value of the following statement? Is it a tautology?

If it is not raining and the picnic is not canceled, then James will not be grumpy tomorrow.

Well, after an obvious renaming of variables, we can call the statement that we know as true by the letter G, and the statement in question becomes

```
(not R and not C) impl (not G)
```

Then, we can argue as follows: If R or(and) C is *true*, then the hypothesis of our statement is false, so the statement must be true. On the other hand, if both R and C are *false*, then the hypothesis is true but the conclusion is false. Thus, our statement is not true in every case, so it is not a tautology.

For a second example, suppose that the following statement is false:

If it is not raining, then the picnic is canceled.

That is, $(\neg R \implies C)$ is *false*. What can be said about the following statement?

The picnic is canceled iff it is raining and James is grumpy.

(Note that "iff" stands for "if and only if," which is a way of saying that the truth values of the two statements are identical in all cases.)

This statement translates as

$$C = (R \wedge G)$$

Now, we argue as follows. Given that $(\neg R \implies C)$ is *false*, we know that $(\neg R)$ must be *true* (and, hence, R must be *false*) and that C must be *false*. But, then, $R \wedge G$ is *false* because R is *false*. Thus, both sides of the equality have the value *false*, so under the given condition, the statement is true for every possible choice of values for its variables.

Summary of Section 2.3

In this section, you were introduced to reasoning and Boolean expressions. The following exercises involve proving (by writing truth tables and by writing ISETL programs) the *laws of equivalences*, and they involve determining whether or not an argument is valid. You will also negate Boolean expressions and see if a statement's value can be determined when some information concerning the values of its variables is specified.

Exercises

2.3.1 Practicing with the laws of equivalences.

The following is a list of equivalent expressions along with their "names." You are already familiar with some of these based on the examples in the text and based on your own intuition. They can be used in a number of ways, such as to simplify a complicated expression, to show that two expressions are equivalent, or to express the negation of a given expression without having a leading *not*. For the sake of completeness, we list them all here. In order to give you practice with both notational systems, we alternate expressing the laws in ISETL and mathematical notation.

Commutative laws

$$\begin{aligned} (\text{p and q}) &= (\text{q and p}) \\ (\text{p or q}) &= (\text{q or p}) \\ (p = q) &= (q = p) \end{aligned}$$

Associative laws

$$\begin{aligned} ((p \wedge q) \wedge r) &= (p \wedge (q \wedge r)) \\ ((p \vee q) \vee r) &= (p \vee (q \vee r)) \\ ((p = q) = r) &= (p = (q = r)) \end{aligned}$$

Distributive laws

$$\begin{aligned} (\text{p and (q or r)}) &= ((\text{p and q}) \text{ or } (\text{p and r})) \\ (\text{p or (q and r)}) &= ((\text{p or q}) \text{ and } (\text{p or r})) \end{aligned}$$

DeMorgan's laws

$$\begin{aligned} (\neg(p \wedge q)) &= ((\neg p) \vee (\neg q)) \\ (\neg(p \vee q)) &= ((\neg p) \wedge (\neg q)) \end{aligned}$$

Law of negation

$$(\text{not (not p)}) = \text{p}$$

Law of the excluded middle

$$(p \vee (\neg p)) = true$$

Law of contradiction

$$(\text{p and (not p)}) = \text{false}$$

Law of implication

$$(p \Longrightarrow q) = ((\neg p) \vee q)$$

Law of equality

$$(\text{p} = \text{q}) = ((\text{p impl q}) \text{ and } (\text{q impl p}))$$

or-**simplification laws**

$$\begin{aligned} (p \vee p) &= p \\ (p \vee true) &= true \\ (p \vee false) &= p \end{aligned}$$

and-simplification laws

$$
\begin{aligned}
\text{(p and p)} &= \text{p} \\
\text{(p and true)} &= \text{p} \\
\text{(p and false)} &= \text{false}
\end{aligned}
$$

Of course, each of the laws can be proved by reasoning directly using the definitions of the Boolean operators, by giving the truth table, or by writing an ISETL program. In any case, we must show that each expression is a tautology—that is, that the expression is always true for all possible combinations of values for its variables.

    a. Use a truth table to prove the law of equality.

    b. Write an ISETL program to prove the distributive laws.

    c. Give verbal explanations for why the laws of negation and contradiction hold.

2.3.2 Simplifying Boolean expressions.

Use the laws of equivalences to simplify the given expressions to one of the following: *true*, *false*, p, q, p and q, or p or q.

    a. (p and q) or (p and not q)

    b. (p or (q or p)) or not q

    c. p impl (q impl (p and q))

    d. (not p and q) or p

    e. (p or q) or not p

    f. not p impl (p and q)

    g. (not p or q) and (q or p)

    h. not (not (not (not p or not q)))

2.3.3 Determining whether or not a given expression is a tautology.

For each of the following expressions, write an ISETL program to determine whether or not it is a tautology. If it is, then prove it using the laws of equivalences—that is, starting with either the left-hand side of the expression or the right-hand side, apply the laws to get the other side. If it is not a tautology, then give specific values for each of the expression's variables that make the expression *false*.

    a. $(\neg p \implies \neg q) = (p \implies q)$

    b. $(\neg(p \implies (q \wedge r))) = (p \wedge (\neg q \vee \neg r))$

    c. $(p \wedge (q \vee r)) = ((p \wedge q) \vee r)$

    d. $(p \implies q) = (\neg(q \implies p))$

Å

e. $((p \Longrightarrow q) \land (q \Longrightarrow r)) = (p \Longrightarrow r)$

2.3.4 Negating expressions.

Give the negation of each of the following statements by performing the following steps: translate the sentence to ISETL (see Exercise 2.2.2), negate the ISETL expression, simplify using the laws, and translate the expression back to English.

   a. If I don't go to the party, then I don't study Discrete Mathematics or I don't go to bed late.

   b. When $x <= y$ then $y > z$, but when $x > y$ then $y <= z$.

   c. If the utility costs go up or the request for initial funding is denied, then a new computer will be purchased if and only if we can show that the current computing facilities are not adequate.

   d. If apples cost more than cranberries, then we will make cranberry sauce, but if they cost the same or less then we will make applesauce.

   e. One of the integers B_1, B_2, B_3, or B_4 equals 0.

   f. Every one of the integers C_1, C_2, and C_3 is greater than -2.

2.3.5 Reasoning with expressions.

   a. Given that the value of (p impl q) is *false*, determine the value of the following expression.

   (not p or not q) impl q

   b. If $(p \Longrightarrow q)$ is *true*, can you determine the value of

   $$\neg p \lor (p = q)$$

   c. If you do go to the party (that is, *I go to the party* is *true*), can you determine the value of

   If I don't go to the party, then I will study Discrete Mathematics and I will not stay up late?

   d. Look again at the expression in part c. Can you find its value if you don't go to the party?

   e. If p_1 is *true*, while the values of p_2, p_3, p_4, and p_5 are unknown, can you determine the values of

   i) $\neg(p_1 \lor p_2 \lor p_3 \lor p_4 \lor p_5)$
   ii) $\neg(p_1 \land p_2 \land p_3 \land p_4 \land p_5)$

   f. If $(p \Longrightarrow r)$ is *true*, can you determine the value of

   $$(p \Longrightarrow q) \land (q \Longrightarrow r)$$

g. If (p impl r) is *false*, can you determine the value of

(p impl q) or (not (q and r))

2.3.6 Reasoning about valid arguments.   Proofs by enumeration.

Consider the following arguments:

1. If it rains, then the picnic will be canceled and James will be grumpy. The picnic is not canceled. Therefore, James is not grumpy.

2. If the labor market is perfect, then the wages of persons with the same job will be equal. But, it is always the case that wages for such persons are not equal. Therefore, the labor market is not perfect.

3. If it is before two o'clock, Joe can go visit Julie and take the bus home afterward. If it isn't earlier than two, then Joe doesn't have time to visit Julie and he must take the train. Joe visits Julie. Therefore, it is before two o'clock.

4. If GM located its new plant in Smalltown, PA, then property values in Smalltown would be rising and Smalltown's population would be increasing. Smalltown's property values are rising, but its population is decreasing. Therefore, GM did not locate its new plant in Smalltown.

5. If investments fail to increase each period, then income will decline several periods later. But, investments are higher this period than last. Therefore, no new decline is indicated.

6. If a country is poor, then it cannot devote much of its time to technological development. Moreover, if a country cannot devote much of its time to technological development, then its income will not change. Therefore, if a country is poor, then its income will not change.

7. If Dad praises me, then I can be proud of myself. Either I do well in sports or I cannot be proud of myself. If I study hard, then I cannot do well in sports. Therefore, if Dad praises me, then I do not study hard.

Recalling that an argument is *valid* if the implication determined by the argument is true for all combinations of values, and is a *fallacy* otherwise, determine the validity of each of the arguments above. If the argument is a fallacy, give specific values for the argument's variables that make the argument false.

First, express the argument as an implication and then

> a. write an ISETL program that prints the value of the implication for each combination of values of the variables that occur in the argument.
>
> b. give the truth table for the implication.

2.4 Methods of Proof

Ambrose Bierce, in his *Devil's Dictionary*, defined a *proof* to be a "more or less plausible argument, agreed upon by at least two people." DeMillo, Lipton, and Perlis pursued this notion more seriously in suggesting that a proof was a "social activity." In any case, it seems hard to think about the concept of proof without some idea of a dialectic between a presentation of an argument and an acceptance of its validity.

We are using "validity of an argument" in the same sense as we did in the previous section. There is an implication. The implication, of course, has a hypothesis and a conclusion. A proof consists of anything that establishes (by whatever means) that for every set of values of the variables, the implication is true.

In the last section, we discussed one method of proof that might be called *proof by enumeration*, or maybe more appropriately proof by *exhaustion*. We described how you can use truth tables or an ISETL program to determine the validity of an argument. Using this method, we looked at all possible situations, that is, every combination of values for its variables, and checked that the desired implication was true for each combination.

Based on the definition of implication and on its properties, however, there are shorter ways of showing that it is true without looking (exhaustively) at every case. In this section, we would like to discuss some of these other methods with you.

Notice that an implication is always true whenever its hypothesis is false, no matter what the value of its conclusion. So, to show that an implication is always true, all we really need to do is to show that it is true whenever its hypothesis is true. Thus, it suffices to consider only those combinations of values of variables for which the hypothesis is true and show that in each case the conclusion (and, hence, the implication itself) is true. This is called a *direct proof.*

Let's look again at the argument

$$((Run \lor \neg Late) \land (Ring \implies Late)) \implies (\neg Ring \lor Late \lor \neg Run)$$

which we proved valid by exhaustion in the last section (see page 76). We could use a truth table to prove this directly by checking out each line in the table where the hypothesis is *true* and making sure that the conclusion is *true*.

An alternate proof is to run ISETL code, such as the following:

```
for Run, Ring, Late in {true, false} do
    if ((Run or not Late) and (Ring impl Late)) = true then
        print not Ring or Late or not Run;
    end;
end;
```

If the argument is valid, then ISETL will print all *trues*. Otherwise, it will print at least one *false*.

Using direct proofs to prove theorems in Mathematics usually involves doing the proof by reasoning. You assume that all the hypotheses hold (are true), and then you show that the conclusion holds. You have already worked with some proofs of this type in this book. In Section 1.5.2, we discussed the proof that the **func gcd_euc** computes a number that is equal to the greatest common divisor of the two positive integers that are given as parameters to **gcd_euc**. In the proof, we began by assuming that the hypothesis was true. That is, we assumed that for any two positive integers a number is computed by **gcd_euc**. Then, we showed that the conclusion held, namely, that this number is the greatest common divisor of the two positive integers. We did this by running through the program with an arbitrary pair of positive integers in mind and argued *directly* that the result of **gcd_euc** divides both of the given integers and that it is divisible by any integer that divides both of the integers. So, the result must be the greatest common divisor.

Another method of proof is the *indirect proof*. The idea behind this method comes from the fact that $(P \implies Q)$ and $(\neg Q \implies \neg P)$ are equivalent statements. In this case, in order to prove that $(\neg Q \implies \neg P)$ is always *true*, we assume $\neg Q$ is *true* (that is, the conclusion is false) and then show that $\neg P$ is *true* (that is, the hypothesis is false).

Of course, when using an indirect proof to determine whether or not an argument is valid, we may use the appropriate lines in a truth table, an ISETL program, or reasoning. Let's look at the ISETL code for an indirect proof of the previous example:

```
for Run, Ring, Late in {true, false} do
    if (not Ring or Late or not Run) = false then
        print (Run or not Late) and (Ring impl Late);
    end;
end;
```

Since the argument is valid, ISETL will return a list of (how many?) *falses*.

Mathematicians use indirect proofs to prove theorems all the time. We used an indirect proof in Section 1.7 to show that no floating point value for x will make the equation 3 * x = 1 true. We supposed that there was such a value and then showed that this assumption leads to an impossible situation, namely, an equality that can't possibly hold.

Another widely used method of proof is *proof by contradiction*. This is a cross between a direct and an indirect proof. When using this method you assume that the hypothesis holds and that the conclusion is false. You then show that this is impossible—that is, you get a "contradiction"—so the conclusion must be true. This is what we did when we showed that the greatest common divisor is unique. We assumed that a greatest common divisor existed, but that it was not unique, that is, that there were two distinct greatest common divisors of the given positive integers. Then, we showed that they were equal, which was a contradiction. Hence, it must have been wrong to think that there were two different ones!

Summary of Section 2.4

The following exercises provide you with some practice proving arguments and theorems in a variety of ways. In Section 2.3, we discussed how you can use enumeration to show that an argument is valid. Using that method involves showing that an implication, $P \implies Q$, is *true* for all values of the variables in P and Q. However, because of the properties of the logical connective $\implies$, there are alternate methods for showing that an argument is valid that involve a lot less work than the method of enumeration. The first method we discussed in this section is called a direct proof. This consists of showing that the conclusion is true whenever all the hypotheses are true. Another method is to show that at least one of the hypotheses is false (and hence the entire hypothesis is false) whenever the conclusion is false. This is called an indirect proof.

Proofs that arguments are valid can be done in either one of these methods via appropriate truth tables, ISETL programs, or by reasoning. Proofs that theorems are valid can be done using reasoning.

Exercises

2.4.1 Showing that arguments are valid using a direct proof.

Look again at the arguments that are given in Exercise 2.3.6 and at your responses to the questions asked in the problem. You already know whether or not the arguments are valid, but let's examine each one again using a direct proof. Recall that when using a direct proof to show that an argument is valid you need to show that the conclusion is true whenever all the hypotheses are true.

a. Write an ISETL program that prints the value of the conclusion for each combination of the values of the variables for which all the hypotheses are *true*. Based on your output, determine the validity of each argument.

      b. Consider the enumeration truth tables that you developed in Exercise 2.3.6, part b. By considering just the lines where the hypotheses are all true, develop direct proofs for each of the arguments.

      c. For each valid argument, use reasoning to show directly that the argument holds.

2.4.2 Showing that arguments are valid using an indirect proof.

Recall that, in general, when using an indirect proof to show that an argument is valid, you must show that at least one of the hypotheses (and, hence, the entire hypothesis) is false, whenever the conclusion is false.

      a. Write an ISETL program that prints the value of the hypothesis for each combination of values that makes the conclusion false. Based on your output, determine the validity of each argument in Exercise 2.3.6.

      b. Consider the enumeration truth tables that you developed in Exercise 2.3.6, part c. By considering just the lines where the conclusion is *false*, develop indirect proofs for each of the arguments.

      c. For each valid argument, use reasoning to show indirectly that the argument holds.

2.4.3 Use reasoning to prove each of the following theorems, if it is true, otherwise, provide a counterexample. For each statement, be sure to state clearly what the hypotheses are and what the conclusion is and, when it is to be proved, give the method of proof (direct or indirect) that you are going to use.

      a. The sum of two odd integers is an even integer.

      b. The sum of two primes is never a prime.

      c. The sum of four consecutive integers is divisible by 4.

      d. If x and y are **integers**, then

$$\texttt{abs(x*y) = abs(x)*abs(y);}$$

will always return the value *true*.

      e. If n is a positive integer, then $n^4 - n^2$ is divisible by 3.

      f. If n is a positive integer, then $n^2 - 2$ is not divisible by 3.

      g. If $x^2 - 5x + 6 = 0$, then either $x = 2$ or $x = 3$.

      h. **(x-y) mod n = 0** if and only if x and y have the same remainder when divided by n.

   i. If the product of two positive real numbers is a, then it is not the case that both of them are greater than $\sqrt{a}$.

   j. If the department offers 10 courses and has 9 meeting times at its disposal, then 2 courses must meet at the same time.

   k. If x is a real number and $|x| > 3$, then $x^2 > 9$.

2.5   Predicate Calculus: First Pass

Suppose we want to consider the following Boolean statement:

One of the integers B_1, B_2, B_3, or B_4 is less than 0.

Using our method of translating into ISETL syntax, we would write

```
(B1 < 0) or (B2 < 0) or (B3 < 0) or (B4 < 0)
```

Then, the negation of the statement in ISETL would look like

```
(B1 >= 0) and (B2 >= 0) and (B3 >= 0) and (B4 >= 0)
```

This would then translate back to English as

All of B_1, B_2, B_3, and B_4 are greater than or equal to 0.

This could get rather tedious if there were a lot more than four integers. It could even get complicated if we just mix the two types of statements. (Did you realize that there are two types?) The first, which says that something is the case for at least one of the integers is called an *existential quantifier*. It can be symbolized in ISETL as

```
exists x in {B1, B2, B3, B4} | x < 0;
```

We read this ISETL expression as

There exists an x in the set $\{B_1, B_2, B_3, B_4\}$ such that x is less than 0.

If we assign integer values to B_1, B_2, B_3, and B_4, then the expression has the value *true* or *false*. If you find this a little vague, try to evaluate the expression in ISETL. (Don't forget to assign values to B_1, B_2, B_3, and B_4.)

    The second statement says that something is true about every one of the integers. This is called a *universal quantifier* and is implemented in ISETL as follows:

```
forall x in {B1, B2, B3, B4} | x >= 0;
```

In standard mathematical notation, these two kinds of quantifiers are written

$$\exists x \in \{B_1, B_2, B_3, B_4\} \ni x < 0$$

$$\forall x \in \{B_1, B_2, B_3, B_4\}, \ x \geq 0$$

Now let's take a look at a mixed statement.

> Every one of the integers $x_1, x_2, x_3, x_4, x_5, x_6, x_7$, and x_8 that is greater than 10 is divisible by at least one of x_1, x_3, or x_5.

You could spend a long time writing this out just using the operators **and** and **or**. Then, if you tried to negate it, you might have a hard time keeping everything straight. But, look how easy it goes if you use the quantifiers and the following two rules:

1. The negation of a statement of the form

   ```
   forall x in S | P(x)
   ```

   is the statement

   ```
   exists x in S | not P(x)
   ```

   where P(x) is a Boolean expression containing the variable x.

2. The negation of a statement of the form

   ```
   exists y in T | Q(y)
   ```

   is the statement

   ```
   forall y in T | not Q(y)
   ```

These are really just common sense. The first says that if your statement asserts that something is true for *all* values, then the negation is that it is false for *at least* one of them. On the other hand, if your statement says that something is true for *at least* one of them, then the negation is the assertion that it is false for *all* of them. We can also prove that they are true using the definitions of **forall** and **exists** along with DeMorgan's law.

Now, returning to our "mixed statement," we can code it in ISETL as follows:

```
forall x in {x1, x2, x3, x4, x5, x6, x7, x8} |
       (x>10 impl (exists y in {x1, x3, x5} | x mod y = 0));
```

Then, if we wanted to negate it, we first think of it in the following form:

$$\texttt{forall x in S | P(x)}$$

where `P(x)` is the second line in our statement. Then, by the first rule, the negation is

$$\texttt{exists x in S | not P(x)}$$

which means we must negate the second line. This has the form

$$\texttt{x>10 impl (exists y in T | Q(x,y))}$$

and its negation is

$$\texttt{x>10 and not (exists y in T | Q(x,y))}$$

or, by the second rule,

$$\texttt{x>10 and (forall y in T | not Q(x,y))}$$

Putting it all together, we obtain

```
exists x in {x1, x2, x3, x4, x5, x6, x7, x8} |
     (x>10 and (forall y in {x1, x3, x5} | x mod y /= 0))
```

Finally, we can translate this back to English to obtain for the negation of the original statement

> At least one of x_1, x_2, x_3, x_4, x_5, x_6, x_7, and x_8 is greater than 10 and not divisible by any of x_1, x_3, or x_5.

A parting question: If you know that x_2, x_4, x_6, x_7, and x_8 are all less than 10, can you conclude anything about whether the original statement or its negation is true?

If you found the material in this section a little hard to follow, don't worry, it was just an introduction. Chapter 3 should make things a lot clearer for you, and we will return to these ideas in a big way when we come to Chapter 5.

Summary of Section 2.5

In this section, you met **exists** ($\exists$) and **forall** ($\forall$) for the first time. We will use these ideas a lot in the rest of the book. One purpose of the following exercises is to give you some practice expressing a quantified statement in both ISETL and mathematical notation, negating and simplifying the expression, and then translating the negation back to English again. Another purpose of the exercises is to have you think about the relationships between **exists**, **forall**, and **not**. We should mention that the idea for the problems in Exercise 2.5.4 comes from David Gries' book, *The Science of Programming*, Springer-Verlag, 1981.

Exercises

2.5.1 The definitions of **exists** and **forall** for a sequence of Boolean expressions. Assume P(k), P(k+1), ..., P(n) is a sequence of Boolean expressions. Then, if $k \le n$,

```
(exists i in [k..n] | P(i)) =
            (P(k) or P(k+1) or ... or P(n))
```

and

```
(forall i in [k..n] | P(i)) =
            (P(k) and P(k+1) and ... and P(n))
```

   a. Express each of the following as a conjunction or disjunction of P(i)'s. Simplify if possible.

     i) `exists i in [1..5] | P(i)`
     ii) `exists i in [5..8] | P(i)`
     iii) `forall i in [2..6] | P(i)`
     iv) `forall i in [4..4] | P(i)`
     v) `exists i in [1..9] | not P(i)`
     vi) `forall i in [1..3] | not P(i)`
     vii) `not (forall i in [4..9] | not P(i))`
     viii) `forall i in [1..8] | (P(i) impl P(i+1))`
     ix) `exists i in [2..7] | (P(i) and not P(i+1))`

   b. Given the list

$$[2, -7, -20, 10, 82, 56, 9, 0, 30]$$

suppose P(i) is *true* if and only if the ith item in the list of integers is positive and divisible by 10. Find the truth value of the quantified expressions in part a.

   c. What are the values of

```
exists i in [k..n] | P(i)
```

and

```
forall i in [k..n] | P(i)
```

when $k > n$?

2.5.2 Relationships between **exists**, **forall**, and **not**. Assume P(k), P(k+1), ...,P(n) is a sequence of Boolean expressions.

   a. Use the definitions in Exercise 2.5.1 and the equivalence laws to prove the following rules:

    i) not (exists i in [k..n] | P(i))
       = (forall i in [k..n] | not P(i))
   ii) not (forall i in [k..n] | P(i))
       = (exists i in [k..n] | not P(i))
  iii) $(\exists i \in [k..n] \ni P(i)) = \neg(\forall i \in [k..n], \neg P(i))$
  iv) $(\forall i \in [k..n], P(i)) = \neg(\exists i \in [k..n] \ni \neg P(i))$

b. Use these rules to negate and simplify each of the statements in Exercise 2.5.1, part a.

2.5.3 Numerical quantification. Assume P(k), P(k+1), ..., P(n) is a sequence of Boolean expressions. Then,

$$\#\{i : \quad i \text{ in } [k..n] \mid P(i)\}$$

is the number of i between k and n for which P(i) is *true*. This is called *numerical quantification*.

For example, if P(1), P(3), and P(5) are *true* while all the other P(i) are *false*, where $1 \le i \le 10$, then

$$\#\{i : \quad i \text{ in } [1..10] \mid P(i)\} = 3$$

a. Given the list $[2, 7, 33, 20, 7, 6, 14, 10, 27]$, suppose P(i) is *true* if and only if the ith item in the list is i times a prime number. Find

  i) $\#\{i : \quad i \text{ in } [1..9] \mid P(i)\}$
  ii) $\#\{i : \quad i \text{ in } [1..8] \mid P(i) \text{ or } P(i+1)\}$

b. Use the numerical quantifier to define

  i) the existential quantifier.
  ii) the universal quantifier.

2.5.4 Translation, negation, and simplification. Assume $b(k..n)$ is an arbitrary list of integers, $b_k, b_{k+1}, \ldots, b_n$, where the list is assumed to be empty whenever $k > n$. Consider the following statements:

1. Every element in $b(k..n)$ is zero.

2. At least one element in $b(2..7)$ is even.

3. The items in $b(j..k)$ are in descending order.

4. Some items in $b(1..9)$ are in $b(11..15)$.

5. None of the elements in $b(1..100)$ is equal to 0.

6. Each item in $b(k..n)$ is in $b(n + 1..m)$.

7. All the elements in $b(1..n)$ are in $b(p..q)$.

8. Those items in $b(1..n)$ that are not in $b(j..k)$ are in $b(p..q)$.

9. Every positive item in $b(1..n)$ is a multiple of 6.

10. For each x in $b(1..n)$ that is even, there is an item in $b(n+1..m)$ that is divisible by $x + 2$.

11. There is an element in $b(4..p)$ that is positive and that does not divide any other element in $b(4..p)$ evenly.

12. There is an item x in $b(i..j)$ that is a multiple of 7 and whose square is not in $b(n..m)$.

Perform the following tasks.

a. Express each statement as a Boolean expression in either ISETL or mathematical notation.

b. Negate the expression and simplify.

c. Express the negation in English.

Note: This is what your answers might look like for statement 1. If for each i in $[k..n]$ we say that P(i) is *true* if and only if the ith item in the list, namely, b_i or b(i), is equal to 0, then the sentence

Every element in $b(k..n)$ is zero.

can be written as

```
forall i in [k..n] | P(i)
```

or we can omit the reference to P(i) and write the statement directly as

```
forall i in [k..n] | b(i) = 0
```

Then, the negation is

```
not (forall i in [k..n] | b(i) = 0)
        = exists i in [k..n] | not (b(i) = 0)
        = exists i in [k..n] | b(i) /= 0
```

and we can express the negation in English as

There exists an item in $b(k..n)$ that is not zero.

2.5.5 Use the rules to negate and simplify the following statement:

```
forall j in [1..n] | (exists t in [j+1..m] |
        (forall k in [1..m] | k + t is not prime))
```

2.6   Propositions as Objects

This last section of the chapter is a pointer to future material. In this chapter, we have tried hard to get you to think about a proposition as a process that is dynamic. That is, whenever you see a statement like the ones we have been discussing, we want you to have some sort of picture in your mind of a bunch of variables all interconnected to each other with `ands`, `ors`, parentheses, and so on. However, the picture is more useful if it is in motion. We want you to be thinking about someone going around and tagging each of these variables with a value of *true* or *false*. Then, all these *trues* and *falses* are churned around according to the definitions of the operators and a single *true* or *false* value comes out. This process is a *Boolean function of Boolean variables*, and it is often what one needs to do with a Boolean expression. Having a strong mental image of what is going on will help you in figuring out a lot of Mathematics.

On the other hand, there are some situations in which we want you to keep your picture static. Continue thinking of the proposition, with its identifiers and operators, as a process that may be in motion. But, all this is encapsulated in some sort of container and becomes a single, stationary object. The hard part is to be able to keep both the dynamic picture and the static one in your head at the same time and to be able to decide which to concentrate on in a particular situation. It is not so very different from what you do with atoms (the ones in Physics, not in ISETL). You can think of an atom as consisting of a bunch of electrons all flying around a neutron; but, you can also think of it as an object, sitting still and waiting to combine with other atoms to form a molecule.

Summary of Section 2.6

Our goal is to help you think of a proposition as a process and as an object. To help you with the second point of view, we have one rather long exercise that deals with propositions as `strings` (which are surely objects) and another that is a setup for the future when we will talk about sets of propositions.

Exercises

2.6.1 In Figure 2.1, there is an assignment to `nec_symp` of a `set` of `strings` of the form `"D impl S"`, where D is the name of a disease and S is a symptom that must be present if one has the disease. Place in a file an ISETL program that will

  1. initialize `nec_symp`.

2. input two **strings** and assign them to identifiers, **first** and **second**.

3. use **string** concatenation to construct a single **string** of the form "D impl S" where D and S are the values of **first** and **second**, respectively.

4. print **true** or **false** indicating if the **string** is in nec_symp.

Run your program with various data.

```
nec_symp := {"polio impl high fever", "polio impl backache",
    "polio impl glands swollen", "polio impl pink eyes",
    "polio impl nausea", "polio impl anemia",
    "polio impl coughing","polio impl insomnia",
    "measles impl low temperature","measles impl backache",
    "measles impl fainting spells", "measles impl nausea",
    "measles impl anemia", "measles impl diarrhea",
    "measles impl insomnia", "flu impl running nose",
    "measles impl trembling",
    "flu impl glands swollen", "flu impl trembling",
    "flu impl migraine", "flu impl nausea",
    "flu impl coughing",
    "flu impl diarrhea", "flu impl insomnia",
    "diphtheria impl sore throat",
    "diphtheria impl pink eyes",
    "diphtheria impl fainting spells",
    "diphtheria impl anemia", "diphtheria impl coughing",
    "diphtheria impl migraine", "diphtheria impl diarrhea",
    "diphtheria impl insomnia"};
```

Figure 2.1: Definition of the set nec_symp

2.6.2 This is a problem that gives you an opportunity to apply Boolean expressions to answering questions about a manufacturing process. The statement of the problem is much longer than the work you are asked to do with it now, but we will be returning to it later in the book.

First, the statement of the problem: A manufacturer produces a certain chemical that is made by combining 3 ingredients, an acid, a base, and a catalyst. For each ingredient, there are several alternative materials that can be used, and the manufacturer is about to switch over to a new set of materials. For example, there could be the possibility of making the base ingredient from chalk, animal bones, and so on.

In the proposed changeover, there are 4 possible materials for the acid, 5 for the base, and 4 for the catalyst. This gives 80 possible choices that could be made for the new manufacturing procedure. Some of them however, are ruled out by certain conditions. There are some pairs of materials that cannot go together, there are some limitations on the changeover cost that will be exceeded by certain combinations of materials, there is a balance between the cost of the new materials and the cost of the new operation, and there are certain quality standards that must be met.

To describe these conditions in detail, we will use **strings "A1"**, **"A2"**, **"A3"**, and **"A4"** to denote the four materials for making the acid, **"B1"**,...,**"B5"** for the base, and **"C1"**,...,**"C4"** for the catalyst.

The specific conditions that must be met are as follows

1. Certain pairs of materials are not permissible. They are listed in Table 2.5.

2. For at least one of the three materials chosen, both of the following conditions must hold

   a) The changeover cost must be less than $100,000. Table 2.6 gives the changeover costs for various materials in the column ChngOvr Cost.

   b) If the new operating cost is more than 110% of the old, then the new material cost must be not more than 115% of the old material cost. These percentages are listed for each material in Table 2.6 under the columns Op Percent and Mat Percent, respectively.

3. At least two materials must both meet quality standard 1 and both must also meet one of quality standards 2 or 3. The materials that meet these standards are listed in Table 2.7 under the columns Qual 1, Qual 2, and Qual 3.

For this problem, you are to write a program that reads in three **strings** considered to be a choice for the acid, base, and catalyst and does the following:

1. Checks that it is valid input. That is, it must be a **string** of two characters, the first **A**, **B**, or **C** and the second a digit in the right range.

2. Determines whether it satisfies the three conditions.

Run your program with various inputs. Also, try a few by hand. Which is the more convenient way to do it?

Acid-Base	Acid-Catalyst	Base-Catalyst
A1, B1	A1, C1	B1, C1
A1, B2	A1, C2	B1, C2
A2, B1	A1, C3	B2, C1
A3, B2	A2, C1	B5, C3
A3, B3	A3, C1	
A3, B4	A4, C1	
A3, B5	A4, C4	

Table 2.5: Inadmissible pairs

Material	ChngOvr Cost	Op Percent	Mat Percent
A1	$ 81,000	89%	102%
A2	70,000	74	111
A3	105,000	109	99
A4	93,000	123	87
B1	101,000	119	119
B2	100,000	111	144
B3	93,000	106	98
B4	82,000	104	104
B5	98,000	100	114
C1	101,000	119	106
C2	72,000	109	108
C3	89,000	79	104
C4	121,000	121	101

Table 2.6: Changeover costs and operating and material cost ratios

Qual 1	Qual 2	Qual 3
A1	A1	A1
A2	A3	A2
A3	A4	A3
A4	B3	A4
B2	B5	B1
B3	C1	B2
B4	C3	B4
B5	C4	B5
C2		
C3		

Table 2.7: Quality control conditions satisfied by materials

Chapter 3

Sets and Tuples

To the Instructor

The goal of this chapter is to motivate students to understand **sets** and **tuples** (sequences with finitely many non-**OM** components) as objects to which certain operations can be applied. By working with a **set** as a collection of objects and realizing that a **set** can be a member of another **set**, the student begins to think of a **set** itself as an object. Along these same lines, we try to strengthen the notion of a proposition as an object by giving many examples in which a proposition is an element of a **set**. We also attempt to objectify **sets** and **tuples** by having the student define ISETL **funcs** that return a **set** or a **tuple**.

To help the student develop dynamic mental images associated with the **set** and **tuple** operations and with the formation of new **sets** and **tuples**, we encourage the student to think about how the computer might process the given information. It has been our experience that "thinking ISETL"—that is, thinking loop-test-evaluate—really helps the students to understand the meaning of complicated **set** expressions and to construct their own.

With the help of ISETL, the students learn to do hard problems involving abstract **set** and **tuple** formation. They learn to write compact expressions for very complicated statements. These **set** and **set** former expressions are referred to as *one-liners* since they can be expressed using one line of ISETL code. From a strictly programming point of view, the one-liner may not be the best way to think about a complicated expression. The goal for learning Mathematics, however, is to develop the ability to understand the full logical structure of such expressions and writing ISETL one-liners can help.

This chapter contains many pointers to important things to come. Students will become familiar with sets of ordered pairs and variables and domains (which are both important with respect to functions and relations), sets of propositions and indexed collections of sets (important respectively for proposition valued functions and set valued functions with integer domains), **tuples** of propositions (important for induction), and counting problems (important for permutations and combinations). Furthermore, we include several pointers to predicate calculus. ISETL supports the quantifiers **exists** and **forall**, and it has been our experience that using them comes very naturally to the students when they are forming complicated **set** expressions, such as the set of all prime numbers between 2 and 100. So, when discussing one-liners and primes in Section 3.6, we introduce the ISETL syntax for **exists** and **forall** in a supportive setting. As a further pointer to quantification, we give examples of the conjunction and disjunction of sets of propositions in Section 3.7.

3.1   Preview

A *set* is a "bunch" of objects. But, what types of things can be a part of the bunch? Actually, it's possible for a set to contain nothing at all (in which case we say it is *empty*) or it can contain anything: maybe another set, a proposition, or perhaps even both. For example, it makes sense to talk about the set of all items in your room, the set of all dollar bills currently in your pocket (is this set empty?), the set of all foods you dislike, the set consisting of everything in your universe, the set of all integers, or even the set of all sets of integers. Because we'll be working a lot with ISETL, we'll restrict most of our examples to *finite* sets, that is, sets that have a finite number of objects in them. But most of what we'll be discussing carries over to "big" sets, such as the set of all real numbers.

A set is an object to which certain operations can be applied. Just as you can apply certain operations to integers, such as addition, subtraction, multiplication, and division, you can apply certain operations to sets—union, intersection, difference, and complement. So, just as adding two integers gives you an integer, finding the intersection of two sets gives you a set. For example, if you have the set containing all your friends and the set containing everyone in your family, then the intersection of these two sets is the set consisting of everyone who is a family member and is your friend. (Does this new set include your family dog?) It turns out that the set operators are closely related to the Boolean operators. For instance, set intersection and the Boolean operator **and** are related. Can you see why?

In addition to being able to form new sets from old sets by applying the set operations, you can also form new sets by iterating through the members of a given set (say the set of all socks in your drawer), testing whether or not the item satisfies a specified condition (does it have more than one hole?), and if it passes the test, putting it into the new set (the set of all socks in your drawer that need darning). Using this process to construct a set is called *set formation*, and it allows you to express some very complicated statements in very compact form. In fact, all you need is one line of ISETL code, which looks just like mathematical set notation, to construct sets, such as the set of all integers between 2 and 100 that are prime or the set of all integer right triangles whose sides are less than or equal to 30. We call these expressions *one-liners*.

A one-liner is a very powerful construct. We'll encourage you to think about how ISETL uses a loop-test-evaluate process to evaluate one-line **set** formers, and we'll encourage you to think ISETL as you design and test your own **set** formers. Our goal is to get you to develop a mental image associated with the construction process and thinking ISETL will help you do this. That's a promise.

We also discuss sequences, which can be represented in ISETL by using **tuples**. As far as **sets** and **tuples** are concerned, the objects in a **tuple** are ordered (there is a first item, a second item, and so on), while the objects in a **set** are not. Furthermore, the same object can be a member of a **tuple** more than once (in different positions), while an object can be in a **set** at most once. **Tuples** behave a lot like character **strings**, except that an item in a **tuple** can be of any valid ISETL type (including **OM**), whereas the items in a **string** must be characters. As with **sets** you can form new **tuples** from old using the **tuple** operations or you can construct a new **tuple** using a **tuple** former.

A footnote: Most of the material in this book was designed in accordance with how you, the student, learn things. We've held lots of interviews, asked students before you lots of probing questions, and tried to determine how you put things together and what are your cognitive difficulties. We hope that this book is based on how you think about the ideas being presented and not just on how we as mathematicians see them. What follows are (totally irrelevant) excerpts from a conversation one of the authors (**N**) had with her seventeen-year-old daughter (**K**) concerning some of the interviews for this chapter:

> **K**: What are you going to be doing today, Mom?
> **N**: I'm going to spend the day interviewing my students about sets.
> **K** (hearing the word "sex" instead of "sets"): What do you want to do that for?
> **N**: Each student has a different level of understanding, and I want to find out what level they are at.
> **K**: How are you going to do that?
> **N**: I'm going to ask them lots of probing questions and tape record our sessions.
> **K**: Won't they be embarrassed?
> **N** (thinking that K is referring to the students being embarrassed about having the session taped): I know these students real well. They may be a little embarrassed at first, but once

we get into it, things will go just fine.

K: But, WHY do you have to ask them about this?

N: So I'll know how to teach them about it in a way that makes sense to them.

K: But why you? I learned everything I know from my brother ...

N (a little surprised that her kids have spent time sitting around talking Mathematics): I'm not talking about the simple stuff, but the really sophisticated stuff, like set formation.

K: What????

What follows is an introduction to "sets education." Have fun.

3.2 Introduction to Sets

A *set* is a collection of objects. If a given object is contained in a set, we say that the object is a *member* or an *element* of the set or simply that it is *in* the set.

A set is often defined by simply listing all the objects in the set separated by commas. The usual practice (in both Mathematics and ISETL) is to enclose this list between a pair of curly brackets. For instance, the following **sets** are defined by listing their elements:

> {1, 34, -789, 22}
> {-2.12, 6, sqrt(5), "ABC"}
> {{"Mark", "Karin"}, {"Malgosia", "Ewa", "Hania"}}
> {*true*}

Sometimes a set does not contain any items at all, in which case we say that the set is *empty* and write {} in ISETL or $\emptyset$ in Mathematics to indicate that we are talking about the empty set.

When thinking of a set as a list, however, there are two important things to remember. The first is that the list representing a set does not imply that the items in a set are ordered. A set is a collection of objects. It does not make any sense to talk about the "first" item in a set because any item in the set can appear at the head of the list, and you will still have exactly the same set! To help remember that this is the case, think about a particular set. Then pretend that you dump all the objects in your set into a bag of some sort. Now, reach into your bag, pull out an object, and put it into a list representing your set. Continue adding items from your bag to the list until your bag is empty. Dump the objects back in the bag and repeat the process. The two resulting lists look different (unless you were extremely lucky), but the collection of objects that the lists represent are identical. So we say that the lists represent exactly the same set. This means, for instance, that the ISETL expressions

$$\{"a", "b", "c"\};$$

and

$$\{"b", "c", "a"\};$$

have the same value, since they both contain the same collection of objects, namely, the characters "a", "b", and "c". In fact, if you enter ISETL and type

$$\{"a", "b", "c"\} = \{"b", "c", "a"\};$$

ISETL will return *true*. Whenever every item in one **set** is in the other and vice versa, the two given **sets** are said to be *equal*.

The second important thing to notice about a set is that a set does not have repeated items, even though a particular item may appear more than once in the list. A given object can be in a particular set at most once. Think again about a set as a bag full of objects—you cannot pull the *same* object out of the bag twice. Consider, for example,

$$\{1.2, \ 3+4, \ "discrete \ math \ is \ fun", \ true \ and \ false, \ \{1,2,3\}\}$$

This **set** has five distinct objects in it, namely, the **string** "discrete math is fun", the **integer** 7, the **floating-point** number 1.2, the Boolean value **false**, and the **set** $\{1,2,3\}$. So, we say that the *cardinality* of the **set** is 5. Compare this **set** with the following **set**:

$$\{8-1, \ "discrete \ ma"+"th \ is \ fun", \ 7, \ 1.2, \ false,$$
$$\{3,2,1\}, \ 14 \ div \ 2\}$$

Even though there appear to be more items in this list, the cardinality of this **set** is also 5. What are the 5 distinct objects in this **set**? Hopefully, you noticed that **8-1**, **14 div 2**, and **7** all represent the same object, namely, the **integer** 7. Thus, these two **set** expressions are equal, since not only do they have the same cardinality, they contain the same collection of objects.

Frequently, the elements in a set are all of the same type, such as the **sets**

$$\{"a", "e", "i", "o", "u", "y"\}$$
$$\{1, -9078, 45, 0, 123456789\}$$
$$\{\{\}, \{1\}, \{2\}, \{1, 2\}\}$$

or the set of all people in the world who were alive in the twentieth century, or the set of all objects in the sky that are stars, or the set of all **floating-point** numbers. Some sets of integers are used so often that ISETL has special notation to denote them. For instance, in ISETL

$$\{1..100\}$$

denotes the **set** of all **integers** between 1 and 100, while

$$\{3,\ 6..28\}\ \text{and}\ \{27,\ 24..2\}$$

both denote the **set**

$$\{3,\ 6,\ 9,\ 12,\ 15,\ 18,\ 21,\ 24,\ 27\}$$

What general rules are being used to form these **sets**? Given two **integers,** m and n, the notation

$$\{\texttt{m..n}\}$$

represents the **set** that is given by

$$\{\texttt{m, m+1, m+2,} \ldots, \texttt{n}\}, \quad \text{if}\ m \leq n$$

and

$$\{\}, \quad \text{if}\ m > n$$

So,

$$\{\texttt{1..1}\}$$

is the **set** consisting of a single object, namely, the **integer** 1, while

$$\{\texttt{-12..12}\}$$

is the **set** of all **integers** between -12 and 12 inclusive, but

$$\{\texttt{12..-12}\}$$

is the empty **set**. Another way of thinking about the notation $\{\texttt{m..n}\}$ is to think of n as being an *upper bound* of the **set**—that is, every **integer** in the **set** must be less than or equal to n. Then, to determine which **integers** are in the **set**, first consider the **integer** m. If m is less than or equal to n, then add it to the **set**. Now look at m + 1. Add it to the **set** if it is less than or equal to n. Continue the process as long as you are considering **integers** less than or equal to the upper bound n. Notice that if initially m were greater than n, then you wouldn't add m to the **set**. In fact you wouldn't add any **integer** to the **set**, and the **set** would be empty.

But, what if you want to define a **set** of **integers** that goes down instead of up? What if you wanted to define a **set** that jumps (up or down) by something other than 1? In these cases, you use the general form

$$\{\texttt{m,r..n}\}$$

If m is less than r, then this is a set that can be associated with an increasing sequence of integers. The difference between r and m, namely, r - m, tells the size of the (positive) jump between successive elements. As in the notation {m..n}, n is the upper bound for the set. On the other hand, if m is greater than r, then this is a set corresponding to a decreasing sequence of integers, with downward jumps of m - r. In this case, n is a *lower bound*. Thus, to determine what integers are in the set, you consider each successive member of the decreasing sequence of integers defined by m and r, and as long as the integer currently being examined is greater than or equal to the lower bound n, you add it to the set.

For example, which integers are in the set {-20, -18..21} ? The first two items, -20 and -18, tell you to consider the integers in the increasing sequence -20, -18, -16, -14, and so on, as candidates for items in the set. As long as the integer under consideration is less than the upper bound, which in this case is 21, the integer is added to the set. Therefore, this is the set of all even integers between -20 and 20 inclusive. However, the notation

$$\{100, 90..-1\}$$

tells you to look at the decreasing sequence 100, 90, 80, and so on, comparing each integer to the lower bound, -1. As long as an integer in the sequence is greater than or equal to -1, it belongs in the list representing the set.

Notice that 0 is also a lower bound of this last set, and that there do not exist any other lower bounds larger than 0. Since this is the case, 0 is said to be the *greatest lower bound* of this set. The integer 100, on the other hand, is an upper bound of this set. In fact, 105, 103, and 110 are upper bounds, too. However, 100, is the *least upper bound* since it is the smallest of all the upper bounds.

At this point, you should be able to explain why

$$\{50, 40..45\}$$

is the set containing a single object, namely, the integer 50, while

$$\{45, 50..40\}$$

is the empty set. What sequences of integers do you consider in each case? What is the size of the jump? Do the sequences increase or decrease? What are the least upper and greatest lower bounds?

Of course, just like any other mathematical object a set can have a name, or an identifier, to represent it. In Chapter 1, when we were defining the integer, floating-point, and Boolean operations, we used **N** to denote the set of all nonnegative integers, **F** for the set of all floating-point numbers, and **B** for the set of truth values, {*true, false*}. Similarly, in ISETL, we might define sets S and T as follows.

```
S := {2,7..46};
T := {"abc", 5 - 90, true or false, {1..10}};
```

Then, if you entered

$$S; \qquad T;$$

ISETL would return the values of S and T represented by a list. Try it.

What is the cardinality of S? Of T? You can check your responses by typing in the above lines and then using the # operator in ISETL, which returns the cardinality of a **set**. That is, you can enter:

$$\#(S); \qquad \#(T);$$

to find the cardinality of S and T. In ISETL syntax, the parentheses can be omitted when # is used, so #S and #T are also legal.

What do you think the following lines return?

```
#{1..100};
#({});
#{"A", "B", "A", "B"};
```

Now you can define a **set**, assign it a name, and find its cardinality. Furthermore, you can test whether or not an item is in a **set** by using the **Boolean** operators in and notin. For example, if S and T are defined as given previously, then the following ISETL expressions will all evaluate to *true*:

```
12 in S;     3 notin S;      -85 in T;      "ab" + "c" in T;
```

while these will evaluate to *false*:

```
false in T;     46 in S;      10 in T;
```

In Mathematics, we write $x \in S$ to indicate that x is an element of the set S, and we write $x \notin S$ whenever x is not an element of S.

A *subset* of a given set, say S, is a set where every item in the new set is also in S. A subset may contain all of the objects in S, some of them, or maybe even none. For instance, if

```
S := {"happy", false impl true, {1, 2}, 3.4};
```

then

```
{{2, 1}}
{"happy", true}
{}
{3.4, "happy", true, {1, 2}}
```

are subsets of S. Pause for a moment and check that each set in the list really is a subset of S—make sure you are convinced that every element in each of these **sets** is also an element in S. The mathematical notation for subset is $\subseteq$, so you write

$$\{\text{"happy"}, true\} \subseteq S$$

which you read as "the set containing the string **"happy"** and the proposition *true* is a subset of S." In ISETL, you indicate this fact by using the **Boolean** operator **subset**. So, if you assigned S to be the **set** above, and then entered

$$\{\{2,\ 1\}\} \text{ subset S};$$

ISETL would return *true*. On the other hand, the **sets**

$$\{1,\ 2\}$$
$$\{false\}$$
$$\{\{\text{"happy"}\}\}$$

are not subsets of S. (Why not?) Therefore, you write

$$\{1..3\} \nsubseteq S$$

using the mathematical notation for "not a subset of," or in ISETL if you typed

$$\text{not } (\{1..3\} \text{ subset S});$$

ISETL would return *true*.

The collection of all subsets of a given set S is called the *power set* of S and is denoted in ISETL by **pow(S)**. Notice that the power set is a set where each of its members is also a set. In the above example, the cardinality of **pow(S)** is 16. Try to list the 16 items in **pow(S)**, then check your answer by entering ISETL and typing

```
S := {"happy", false impl true, {1, 2}, 3.4};
pow(S);
```

Up to this point in each of our examples, all the elements of our sets have been expressions involving constants. Sets can also contain variables, as long as each variable in the set has a value. Consequently, you can enter

```
x := false;
y := true;
B := {x, y, {x or y}, (y impl x) = (not y or x)};
```

What does ISETL return if you type

```
#(B);        B;
```

A few final comments about sets. Sets are formed from *previously existing* values; therefore, a set can never contain itself. Furthermore, OM is not a valid element of a set. If one of the values in a set expression has the value OM, then the value of the set expression is OM. For example, if you entered the statements,

$$C := \{3, z\}; \qquad C;$$

without first assigning z a value, ISETL will return OM, since z is itself undefined. Finally, how do mathematical sets and ISETL sets differ? In Mathematics, it's legal to have a set, such as the set of all integers or the set of all real numbers, even if it is impossible to write all the members of a set down in a list. A computer, however, has only a finite amount of space to store the items in a set. So, in ISETL, we can deal only with finite sets.

Summary of Section 3.2

A set is a collection of objects. A set itself is an object, so it can be an element of some other set. In this section, we represented a set by a list—a list that is unordered and has no repetitions. We introduced the notation for sets of integers in ISETL that are defined by arithmetical progressions. We discussed the concepts of set membership, the empty set, subset, power set, equal sets, and the cardinality of a set. The purposes of the following exercises are for you to become more familiar with these ideas and to help you to start thinking about forming new sets from old sets by iterating through the elements in a given set.

Exercises

3.2.1 In ISETL, sets can be heterogeneous (elements can be of different data types), and also, one set can be an element of another set. Suppose the following statement is executed by ISETL:

```
S := {2.0, 3, {"P impl Q", "Q impl P", {-1, 2..5}},
                                          "ISETL", {}};
```

    a. What is the cardinality of the set S?

    b. Enter this set in ISETL and then use the ISETL function # to print its cardinality. Was your answer to part a correct?

3.2.2 Sets defined using arithmetic progressions. List the elements in the following sets. Use ISETL to check your responses. If the set is empty, explain why. Give an upper bound, the least upper bound, a lower bound, and the greatest lower bound for each nonempty set.

    a. {2..12};

    b. {4..4};

    c. {10..1};

    d. {-2, 4..38};

    e. {0, 3..41};

    f. {0, 3..-1};

    g. {100, 90..-5};

    h. {100, 90..100};

    i. {100, 90..101};

    j. {10, 9..0};

    k. {4,4..8};

3.2.3 Enter the following lines in ISETL. Predict what ISETL will return before you hit the return key.

```
a. T := {"XYZ", 3+4, 8.9};     7 in T;     "X" in T;
   (1+7.9) in T;

b. ("X"+"Y"+"Z") notin T;     7.0 notin T;

c. T = {7*1, "XYZ", 14 div 2, "XYZ", 17.8/2.0, "XYZ"};

d. T /= {};     T /= T;     {} subset T;     T subset T;
   T subset {};     not({"xyz"} subset T);

e. {7, 1+6, 2+5, 3+4, 4+3, 5+2, 6+1, 7+0} subset T;

f. #(T);     pow (T);

g. W := {{"tom", "dick"}, "harry"};     W;

h. "tom" in W;     "harry" in W;     {"dick"} in W;
   "dick" notin W;

i. {} in W;     {"dick", "tom"} in W;     {} subset W;

j. "harry" subset W;
   { "dick", "tom", "dick"} subset W;

k. pow(W);     #(pow(W));

l. SixSet := {6, {6, {6}}};     #(SixSet);

m. 6 in SixSet;     {6} in SixSet;
   {{6}} notin SixSet;

n. 6 subset SixSet;   {6} subset SixSet;
   {{6}} subset SixSet;

o. SixSet = {6, 6, {6, 6, {6, 6}}, 6};

p. pow({100..103});
```

q. S := {1..6};   #(pow(S));

r. {arb(S), arb(S), arb(S), arb(S), arb(S)};

s. arb(S) in S;   6 = arb(S);

t. x := 10;   y := 7;
   R := {x, y, sqrt(x*y), x=y, {even(x) impl odd(y)}};

u. x in R;   odd(x) in R;   {y, x} subset R;
   arb(R);   #(R);   R;

3.2.4 Consider the sets Y and Z, where

Y := {"a", 6.9, {2..10}, {{true impl false}, false},
        (10 div -4)+16, {5 in {3, 6..9}}, {{}}};
Z := {{false or true}, 28 mod 2, {10,2,9,3,8,4,7,5,6},
        "A", {}, true impl false, {false}, abs(-6.9)};

  a. For each item in the following list, determine whether the item
is a member of **set** Y, **set** Z, both **sets**, or neither **set**.

    i) **true**
    ii) **false**
    iii) {**true**}
    iv) {**false**}
    v) **13 + 1**
    vi) {**2, 3..10**}
    vii) **"a"**
    viii) {}
    ix) {{}}
    x) {{{}}}
    xi) {**10, 9..2**}

  b. List every object that is in both Y and Z.

  c. List every object that is in either Y or Z or both.

  d. List every object that is in Y but is not in Z.

  e. For each **set** in the following list, determine whether the **set**
is a subset of Y, Z, both Y and Z, or neither Y nor Z:

    i) {**true**}
    ii) {**false**}
    iii) {{**false**}, {**true**}}
    iv) {**false, true**}
    v) {{**2, 3..10**}, {**2..10**}, {**10, 9..2**}}
    vi) {**14**,{**false and true**}, **7.0 - 0.1**}

      vii) {}

     viii) {{}}

      ix) {{{}}}

       x) {{10, 9..1}}

      xi) Y

     xii) Z

3.2.5 Find the power set of each of the following sets:

    a. $\{110, 100..90\}$

    b. $\emptyset$

    c. $\{\{false\}\}$

    d. $\{1, \{1\}\}$

3.2.6 Explain why $\#(pow(S)) = 2^k$ whenever $\#(S) = k$.

3.2.7 Given a set S and a nonnegative integer n, $npow(S, n)$ is the set containing all the subsets of S that have cardinality n. For example, if $S = \{1..4\}$, then

$$npow(S, 2) = \{\{1,2\}, \{1,3\}, \{1,4\}, \{2,3\}, \{2,4\}, \{3,4\}\}$$

while

$$npow(S, 5) = \{\}, \text{ and } npow(S, 4) = \{\{1,2,3,4\}\}$$

Suppose
$$S = \{\{0, 2..6\}, 2, 4, 6, 0\}$$

    a. Find $npow(S, n)$ for $n = 0, 1, 2, 3, 4, 5, 6$. (Note: In the exercises for Section 3.4, you will be asked to write an ISETL **func** defining $npow$.)

    b. Find a formula for the cardinality of $npow(S, n)$ in terms of n and $\#(S)$.

3.2.8 Suppose you have the following **set** definitions:

```
S := {2, 5..100};
T := {"eric", "john", "andy", "sally", "sarah"};
W := {1.2, 3.45, -7.896, 0.0, -89.987};
R := {true impl false, "123", 3.414, "false", 2+3};
X := {-10..10};
```

Recall that you can iterate through the members of a **set** using an ISETL **for** loop. In particular, if S is a given **set**, then the statement

```
for x in S do
    Statement list
end;
```

will execute the statements inside the body of the **for** loop one time for each element in S. Write an ISETL **for** loop that prints:

a. all the items in S that are even.

b. all the items in T that begin with a vowel.

c. all the items in R that are of type **integer**.

d. all the items in X that are odd and positive.

e. all the items in X that are also in S.

f. the square of each member of X.

g. the square root of each member of S times -6.45.

h. $x**5 + 7.0 * x**3 - 4.5 * x + 1.0$ for each x in W.

i. *name* + "*smith*" for each *name* in T.

j. a random integer between 0 and $|n|$ for each n in X.

k. the cubes of all the odd **integers** in S.

l. the square of each **floating-point** number in W that is greater than 3.0 or less than -51.0.

m. the negation of each **Boolean** object in R.

n. the length of each **string** in T that begins with "s".

o. the absolute value of each **integer** in X that is divisible by 4.

p. the product of each member of S with each member of X.

q. the concatenation of each **string** in T with every other **string** in T, including itself.

NOTE: What if you wanted to express the answers to (a)-(q) as a **set** instead of printing an ordered list, which may have repetitions? This process is called **set** formation. We will explain how to do it in Section 3.4.

3.3   Set Operations

Whenever you think about a mathematical object, you not only think about the object itself but also about the operations that are associated with it. For example, when you think about the concept "integer" you think about the actual integers themselves, and also the various operations that you can perform on integers, such as addition, multiplication, exponentiation, equality, less than, max, or min. What operations do we associate with the object "set"? In Section 3.2, you were introduced to the set operations equality, cardinality, membership, and power. In this section, we talk about the basic set operations that can be used to combine sets and form new ones. We also discuss the relationships between the set operations and the Boolean operators.

What are the simplest ways of combining two sets? How can you form a new set from old ones? Given two sets, several possibilities come to mind: you can find the collection of all objects that are in one set or the other, the collection that are in both sets, or the collection of items that are in one set but not in the other. These possibilities represent the notions of *union*, *intersection*, and *difference*, respectively. We considered these operations briefly in Section 1.7, when we discussed the different ways of combining intervals of **floating-point** numbers. For example, we used the notation $A \cup B$ in Mathematics and **A + B** in ISETL to represent the union of two intervals, that is, the set of all **floating-point** numbers in the interval A, in the interval B, or in both. We were thinking of an interval of **floating-point** numbers as a set—the set of all numbers that lie between two given numbers. Thus, if A is the interval $[-1.5, 1.25]$ and $B = (0.0, 2.7]$, then A is the set of all **floating-point** numbers that are greater than or equal to -1.5 but less than or equal to 1.25, while B is the set of numbers strictly greater than 0.0 but less than or equal to 2.7. So, their union (using standard mathematical notation) is

$$A \cup B = [-1.5, 2.7]$$

The three set operations, union, intersection, and difference, are closely related to the Boolean operations, **or**, **and**, and **not**. Let's look at their precise definitions and note the relationships. Suppose S and T are two sets, then

- The *union* of the two sets is the set of all elements that are in S or T (or both). We represent this new set by **S + T** or **A union B** in ISETL, and $S \cup T$ in Mathematics.

- The *intersection* of the two sets is the set of all elements that are in S and T. We use **S * T** or **S inter T** in ISETL to represent the new set, and $S \cap T$ in Mathematics.

- The *difference* of the two sets, $S - T$, is the set of all elements in S that are not in T. The notation for set difference is the same in ISETL and Mathematics.

We can represent these definitions using pictures called *Venn diagrams* as shown in Figure 3.1.

The diagrams are interpreted by looking at the shaded areas. For instance, when representing **S + T** by a Venn diagram, you shade in all of **S**

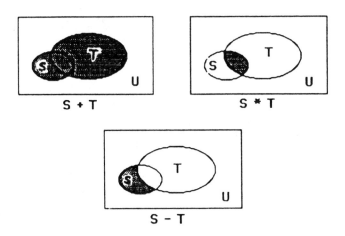

Figure 3.1: Union, intersection, and difference

to indicate that every element in S is in S + T, and you shade in all of T since S + T also contains every element in T.

In the Venn diagrams, the "big" set U, which contains S and T, is usually referred to as the *universal set*. Sometimes instead of talking about the difference of two specified sets, you might want to talk about the set of all objects that are not in a given set, say S, that is, the set of all objects that are in the universal set but are not in S. This set is called the *complement* of S and is denoted in Mathematics by $\overline{S}$. Of course, the complement of S is just $\mathcal{U} - S$, but frequently the exact definition of $\mathcal{U}$ is only implied and not precisely stated, so mathematicians use $\overline{S}$ instead. How might you express $\overline{S}$ using a Venn diagram? What is the complement of the complement of S?

Now let's think about how ISETL might evaluate a **set** expression that uses the binary operators *, +, or -. For example, suppose you entered

    {"abc", true, 1.234, 7} - {false impl true, "g", 1+3};

Then, ISETL would return

    {1.234, "abc", 7};

or some rearrangement.

How might you describe a process, or algorithm, that gives this response? One way might be to begin with the first **set**. That is, your initial "guess" for S - T is S. Now consider an object in the second **set**, say the proposition *true*. If the object being examined appears in your current guess for S - T, then remove it. Repeat the process for each object in the second set. The result is the set S - T. How would you describe the process of finding the union of two sets? The intersection?

Let's look again at the relationships between the Boolean operators and the set operators. Suppose P and Q are Boolean expressions, S and T are set expressions, and $\mathcal{U}$ is a universal set. We list the correspondences between the operations in Table 3.1.

| Boolean operations | | Set operations | |
ISETL	Math	ISETL	Math
P and Q	$P \wedge Q$	S * T	$S \cap T$
P or Q	$P \vee Q$	S + T	$S \cup T$
not P	$\neg P$	U - S	$\mathcal{U} - S$
P and not Q	$P \wedge \neg Q$	S - T	$S - T$
true	*true*	U	$\mathcal{U}$
false	*false*	{}	$\emptyset$

Table 3.1: Comparing Boolean and set operations

In Chapter 2, page 79, we listed the laws of equivalences for Boolean expressions. It probably doesn't surprise you that for all of the laws that involve and, or, and not there are corresponding rules for sets that involve *, +, and -. For example, one of DeMorgan's laws for Boolean expressions is

$$\text{not (P or Q) = (not P) and (not Q)}$$

This corresponds to DeMorgan's rule for sets, namely,

$$\text{U - (S + T) = (U - S) * (U - T)}$$

It can be expressed using standard mathematical notation by

$$\mathcal{U} - (S \cup T) = (\mathcal{U} - S) \cap (\mathcal{U} - T)$$

or

$$\overline{S \cup T} = \overline{S} \cap \overline{T}$$

which is read as

> The complement of the union is the intersection of the complements.

To prove each of the equivalence laws for Boolean values, you had to show that the two sides of the equality have the same truth value for all combinations of truth values for P and Q. One way to do this was to give the truth table. Another was to use reasoning. You can illustrate a rule

for sets by looking at the Venn diagram for each side and observing that
they are the same, or you can prove a rule by using reasoning.

For example, to illustrate the fact that the complement of the union is
the same as the intersection of the complements, you might first construct
the Venn diagram corresponding to the left-hand side of the equation, as
in Figure 3.2,

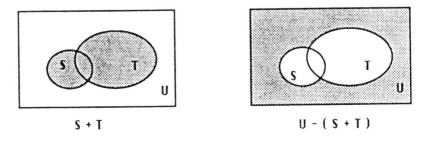

$$S + T \qquad\qquad U - (\ S + T \ )$$

Figure 3.2: Union, complement of union

and then construct the diagram for the right-hand side of the equation, as
in Figure 3.3.

You can then see that whenever the sets S and T, their intersection,
and the complements are all nonempty the two set expressions are equal
since the diagrams for both sides of the equation are the same. Notice that
to use Venn diagrams to give a rigorous proof of this fact you would have
to consider the Venn diagrams for every other possible case, such as (1)
$S \neq \emptyset$, $T \neq \emptyset$, $S \cap T = \emptyset$; (2) $S \neq \emptyset$, $S \subset T$; (3) $S = \emptyset$ or $S = U$; and so on.

You can also use ISETL to illustrate that a rule holds for two given
sets. When using ISETL, however, you are only showing that the rule
holds for specific **sets** (namely, the **sets** you input to test the rule) and
not for every possible **set**.

To actually prove (versus "illustrate") that a rule holds, you can use
reasoning or the "element" approach. That is, you can show that the sets
to the left of the equal sign and to the right are equal by showing that
every element in the set on the left is a member of the set on the right and
vice versa. For instance, to show that DeMorgan's rule for sets,

$$\overline{S \cup T} = \overline{S} \cap \overline{T}$$

holds, you can argue as follows:

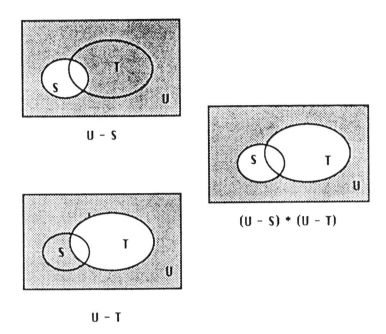

Figure 3.3: Complements, intersection of complements

First, you can show that every item in $\overline{S \cup T}$ is in $\overline{S} \cap \overline{T}$. Assume x is in $\overline{S \cup T}$. Then, x is not in $S \cup T$. Thus, x is not in S and x is not in T. So, x must be in $\overline{S}$ and x must be in $\overline{T}$. Therefore, x is in $\overline{S} \cap \overline{T}$ as desired.

Similarly, you can show that every item in $\overline{S} \cap \overline{T}$ is also in $\overline{S \cup T}$. Assume x is in $\overline{S} \cap \overline{T}$. Then, x is in $\overline{S}$ and x is in $\overline{T}$. Thus, x is not in S and x is not in T. So, x cannot be in $S \cup T$. Therefore, x must be in $\overline{S \cup T}$ as desired.

Therefore, the two sets are equal.

We conclude this section with a list of all the set operations, which is given in Table 3.2. We use **P**, **N**, and **B** to denote, respectively, the set of all **sets** that can be represented in an implementation in ISETL, the set of all **integers** that can be represented, and the set of **Boolean** values ($\{true, false\}$). We use **U** to denote the set of all (ISETL) objects that can be a member of a **set**.

infix (operators)			
+, union	:	$\mathbf{P} \times \mathbf{P} \longrightarrow \mathbf{P}$	union
*, inter	:	$\mathbf{P} \times \mathbf{P} \longrightarrow \mathbf{P}$	intersection
-	:	$\mathbf{P} \times \mathbf{P} \longrightarrow \mathbf{P}$	difference
with	:	$\mathbf{P} \times \mathbf{U} \longrightarrow \mathbf{P}$	returns a set obtained from the given set by inserting the given object
less	:	$\mathbf{P} \times \mathbf{U} \longrightarrow \mathbf{P}$	returns a set obtained from the given set by removing the given object
in	:	$\mathbf{U} \times \mathbf{P} \longrightarrow \mathbf{B}$	test for membership
notin	:	$\mathbf{U} \times \mathbf{P} \longrightarrow \mathbf{B}$	test for membership
subset	:	$\mathbf{P} \times \mathbf{P} \longrightarrow \mathbf{B}$	test for inclusion as a subset
=	:	$\mathbf{P} \times \mathbf{P} \longrightarrow \mathbf{B}$	test for equality
/=	:	$\mathbf{P} \times \mathbf{P} \longrightarrow \mathbf{B}$	test for inequality

prefix (functions)			
#	:	$\mathbf{P} \longrightarrow \mathbf{N^+} + \{0\}$	cardinality
arb	:	$\mathbf{P} \longrightarrow \mathbf{U}$	random member of the given set
pow	:	$\mathbf{P} \longrightarrow \mathbf{P}$	power set
is_set	:	$\mathbf{P} \longrightarrow \mathbf{B}$	test for type set

Table 3.2: Set operations

Summary of Section 3.3

Associated with sets are set operations, such as union, intersection, difference, and complement. We gave the definition of each of these operations and pointed out the close relationships that exist between the operations and the Boolean operators. Because of these relationships, it's not surprising that corresponding to many of the laws of equivalences for propositions is a rule for sets. We list the rules in the exercises and ask you to prove some of them (using element arguments or other rules) and illustrate others (using Venn diagrams or ISETL). Then, just as the laws of equivalences can be used to show that two Boolean expressions are equivalent or to simplify a complicated Boolean expression, the set rules can be used to show the equivalence of two set expressions and also to simplify a messy expression. In the exercises, you'll have a chance to do some of these, to think about some other set operators, to prove some important properties of sets, and to use sets to answer questions in a data base problem.

Exercises

3.3.1 Give an algorithm (describe a process) for determining:

   a. S + T
   b. S * T

3.3.2 Enter the following lines in ISETL. Before you hit the carriage
   return key, predict what ISETL will return by thinking about how
   ISETL will evaluate the statement. If your prediction differs from
   the output of ISETL, explain why.

   a. S := {2, 6..50};   S;
   b. T := {50, 45..-10};   T;
   c. S + T;
   d. T union {};
   e. S + S;
   f. S union {1..10};
   g. S * T;
   h. S inter S;
   i. T * {};
   j. S - T;
   k. T - S;
   l. {} - T;
   m. S - T = T - S;   S + T = T + S;   S * T = T * S
   n. #pow(S + T);   {10, 20..40} subset S * T;
   o. R := {"a", "b", {{7.0, 3+4, 43 mod 9}, 7}};
      #R;   R;
   p. #(R + {7});   R + {7};
   q. #(R * {"a", "b", "c", 7});   R * {"a", "b", "c", 7};
   r. W := {1..6};   arb(W);   arb(W);

3.3.3 ISETL operators **with** and **less**. Let S be any ISETL **set** and x be
   an expression. Then,

$$S \text{ with } x$$

   is a **set** that contains every element that is in S and also contains
   the value of **x**, while

$$S \text{ less } x$$

is a **set** that contains every item that is in S except the value of **x** (if **x** was a member of S). As an example, look carefully at the ISETL session

```
>  S := {1, 2, 3};
>  S with 5;      S with 1;        S;
{3, 2, 5, 1};
{2, 3, 1};
{1, 3, 2};      $ Notice that "with" does NOT change S
>  S less 2;      S less "1";      S;
{3, 1};
{1, 2, 3};
{3, 2, 1};      $ Notice that "less" does NOT change S
>  T := S with 10;      T;      S;
{1, 2, 3, 10 };
{2, 1, 3};
>  S := S less 3;      S;
{2, 1};
>  S with 4 with 9;  S with 4 less 2;
{1,2,4,9};
{1,4};
>  !quit
```

a. Express **with** and **less** in terms of the union and difference operators.

b. The operation **with** lets you construct a new **set** by adjoining one item at a time.

Look again at Exercise 3.2.8 in the last section. Part a., for example, asked you to write a **for** loop that would print the list of all items in S = {2, 5..100} that are even. If you wanted to construct a **set**, say T, containing these items, you could use the ISETL operator **with**:

```
T : = {};
for x in S do
     if even(x) then
            T := T with x;
     end;
end;
```

Redo Exercise 3.2.8, parts b. - q., to return a **set** in each case instead of printing a list.

3.3.4 **Laws of Set Expressions.** The following is a list of equivalent set expressions. You are already familiar with some of these based on the discussion in the text and on your own intuition. For the sake of completeness, we list them all here. Notice that S, T, and R are arbitrary sets while $\mathcal{U}$ denotes the universal set.

Commutative laws

$$S \cup T = T \cup S$$
$$S \cap T = T \cap S$$

Associative laws

$$(S \cup T) \cup R = S \cup (T \cup R)$$
$$(S \cap T) \cap R = S \cap (T \cap R)$$

Distributive laws

$$S \cup (T \cap R) = (S \cup T) \cap (S \cup R)$$
$$S \cap (T \cup R) = (S \cap T) \cup (S \cap R)$$

DeMorgan's laws

$$\mathcal{U} - (S \cup T) = (\mathcal{U} - S) \cap (\mathcal{U} - T)$$
$$\mathcal{U} - (S \cap T) = (\mathcal{U} - S) \cup (\mathcal{U} - T)$$

Law of negation

$$\mathcal{U} - (\mathcal{U} - S) = S$$

Law of the excluded middle

$$S \cup (\mathcal{U} - S) = \mathcal{U}$$

Law of contradiction

$$S \cap (\mathcal{U} - S) = \emptyset$$

∪-simplification laws

$$S \cup S = S$$
$$S \cup \mathcal{U} = \mathcal{U}$$
$$S \cup \emptyset = S$$

∩-simplification laws

$$S \cap S = S$$
$$S \cap \mathcal{U} = S$$
$$S \cap \emptyset = \emptyset$$

Look back at the laws of equivalences on page 79 in Chapter 2. Notice the similarities between the laws of equivalences and the set rules that result from the relationships between the Boolean operators and the set operators.

a. Rewrite each of the above rules using ISETL notation for union, intersection, and difference.

b. Use the Venn diagram approach to illustrate the following rules whenever S is not empty:

   i) U-simplification laws

   ii) ∩-simplification laws

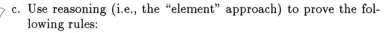 c. Use reasoning (i.e., the "element" approach) to prove the following rules:

   i) Distributive laws

   ii) Law of negation

d. To illustrate the commutative law for union, you might run the following ISETL **func** on a variety of **sets**:

```
comm_union := func (S, T);
              if is_set (S) and is_set (T) then
                  return S + T = T + S;
              end;
          end;
```

   i) What would **comm_union** return if

      1. S and T do commute? don't commute?

      2. S is not a **set**? T is not a **set**?

      3. S is the empty **set**?

      4. both S and T are empty?

      5. S = T? S * T = {}? S * T /= {}?

   ii) Write ISETL **funcs** to illustrate that the following rules hold:

      1. Commutative law for intersection

      2. Associative laws

3.3.5 Use the rules for sets to simplify the given set expressions to one of the following: U, {}, S, T, S + T, or S * T, where U denotes the set of all ISETL objects.

a. (S * T) + (S * (U - T))

b. (S + (T + S)) + (U - T)

c. (S + T) + (U - S)

   d. $((U - S) + T) * (T + S)$

   e. $U - (U - (U - ((U - S) + (U - T))))$

3.3.6 Use reasoning to show that the two given set expressions are equivalent, where $\mathcal{U}$ is the universal set:

   a. $S - T$ and $S \cap (\mathcal{U} - T)$

   b. $(S - T) \cup (T - S) \cup (S \cap T)$ and $S \cup T$

   c. $(S \cup T) - ((S - T) \cup (T - S))$ and $S \cap T$

3.3.7 Symmetric difference of two sets: given two sets S, and T, the *symmetric difference* of S and T is the collection of all elements that are in S or T but not in both.

   a. Draw a Venn diagram that depicts the symmetric difference of S and T.

   b. Show that the following are equivalent set expressions for the symmetric difference:

      i) $(S \cup T) - (S \cap T)$

      ii) $(S - T) \cup (T - S)$

      iii) $(S \cap (\mathcal{U} - T)) \cup (T \cap (\mathcal{U} - S))$

      iv. $(S \cup T) \cap (\mathcal{U} - (S \cap T))$

   c. The **set** operators $\cup$, $\cap$, and $-$ are closely related to the **Boolean** operators **or**, **and**, and **not**, respectively. Explain why the Boolean operator *xor* can be associated with symmetric difference.

   d. Write an ISETL **func** called **sym_diff** that accepts two **sets** and returns their symmetric difference. Test your **func** on a variety of **sets** using both functional and . notation.

3.3.8 Cardinality exercises.

   a. Express #(S + T) in terms of #(S), #(T), and #(S * T).

   b. Express #(S + T) in terms of #(S) and #(T - S).

   c. Express #(S * T) in terms of #(S + T), #(S - T), and #(T - S).

   d. Express #(S * T) in terms of #(S), #(T), and #(S + T).

   e. Express #(S - T) in terms of #(S) and #(S * T).

   f. Express #((S - T) + (T - S)) in terms of #(S + T) and #(S * T).

   g. Express #(S - T) in terms of #(S) and #(T) where T is a subset of S.

3.3.9 Data base problem. Suppose you have the following population data based on a census sample of 1000 people:

* 544 live in a city with a population greater than 100,000.

* 312 are members of a family with more than 2 children.

* 87 live in a city with a population greater than 100,000 and bought a new car in the last year.

* 100 live in a city with a population greater than 100,000 and are members of a family with more than 2 children.

Assume U is the set of people in the sample, P is the set of all people in the sample who live in a city with a population greater than 100,000, F is the set of all people in the sample who are members of a family with more than two children, and C is the set of all people in the sample who bought a new car in the last year.

Express the answer to each of the following questions as the cardinality of a set expression depending on U, P, F, and C. Simplify your expression. Find the cardinality of the set whenever possible.

a. How many have not bought a new car in the past year or do not live in a city with a population greater than 100,000?

b. How many live in a city with a population greater than 100,000 or are members of a family with more than 2 children?

c. How many live in a city with a population less than or equal to 100,000 and are members of a family with more than 2 children but did not buy a new car during the last year?

d. How many live in a city with a population greater than 100,000 and are members of a family with less than 3 children or live in a city with population greater than 100,000 and bought a new car within the past year?

e. How many have either bought a new car in the past year or have not bought a new car in the past year?

f. How many are members of a family with more than 2 children but less than 3 children?

g. How many did not buy a new car in the past year and are not members of a family with more than 2 children?

3.3.10 Prove the following:

a. A is a subset of B if and only if $A \cap B = A$.

b. B is a subset of A if and only if $A \cup B = A$.

c. $A \cap (B - A) = \emptyset$

d. $(\overline{A} \cap B) \cup A = A$

3.3.11 What will appear on the screen after the following lines are entered in ISETL?

```
G := func (N, M);
        return {1, N..M} + {- M, -N..1};
     end;
G(2, 4);      G(2, 3);      G(3, 2);
```

3.4 Set Formation

As you may have noticed in some of the earlier exercises in this chapter, sometimes, instead of simply listing the elements in a set, you needed to be able to build a new set from an old one by giving a precise definition of what elements belong in the new set. This construction process is called *set formation*. For instance, suppose I is a given set of integers, such as $\{1..100\}$ or $\{2, 6, 45, -40, -221, 0, 100\}$, and you want to define

(1a) the set consisting of the square of each member of I.

(2a) the set consisting of all the items in I that are divisible by 5.

(3a) the set consisting of the square of each positive integer in I that is divisible by 5.

In forming each of these sets you need to undertake a *loop-test-evaluate* procedure. That is, you *loop* or iterate through each of the items in the old set (in this case I), *test* whether or not the item being considered meets an optional condition (e.g., Is it divisible by 5? Is it divisible by 5 and positive?). Then, if the condition is satisfied, *evaluate* some expression that depends on the item (e.g., its square) and add the result to the new set. This loop-test-evaluate process is exactly what is implied by ISETL **set** former notation. Let's look at the **set** formers for the sets described in (1a)-(3a) above:

(1b) `{x*x :  x in I}`

(2b) `{x :  x in I | x mod 5 = 0}`

(3b) `{x*x :  x in I | (x > 0) and (x mod 5 = 0)}`

Spend a moment looking at each of these **set** expressions, noting their similarities and their differences. Can you see how each one corresponds to its verbal description?

In general, an ISETL **set** former has the form

$$\{\texttt{E(x)} : \quad \texttt{x in S} \mid \texttt{P(x)}\}$$

where **E** is an expression depending on **x** and **P** is an optional proposition depending on **x**. We read this as

$$\begin{array}{ll} \{ & \text{``The set of all} \\ \quad \texttt{E(x)} & \text{expressions E evaluated at } \mathbf{x} \\ \qquad : & \text{where} \\ \qquad\texttt{x in S} & \mathbf{x} \text{ is a member of S} \\ \qquad\quad | & \text{such that} \\ \qquad\qquad\texttt{P(x) }\} & \mathbf{x} \text{ satisfies the condition P.''} \end{array}$$

We form the new **set** by having the variable x iterate through the members of the given set S, which is called the *domain* of x. This process of iteration is just a **for** loop in disguise. For example, suppose Guys is the **set** of **strings**:

```
Guys := {"tom", "eric", "adam", "andy", "peter", "sam"};
```

Now define T to be the **set** of all first letters of the items in Guys that begin with a vowel. You can do this with a **for** loop:

```
T := {};
for name in Guys do
      if name(1) in "aeiou" then
            T := T with name(1);
      end;
end;
```

or you can use a **set** former:

```
T := {name(1) :  name(1) in Guys | name(1) in "aeiou"};
```

The **set** former notation tells you exactly how the **set** T is constructed—loop through the character **strings** in Guys; test whether the **string** begins with a vowel; if it does, find the first character in the **string** and adjoin it to T. Given the current value of Guys, after the assignment has been executed, T has the value {"a", "e"}. Agree?

One important thing to keep in mind is how ISETL iterates or loops through the members of a **set**. As you know, there is no ordering to the members of a set. Consequently, there is no specific order for the looping process either. ISETL chooses a member from the **set** and then does the checking and evaluating for this member. It then chooses another item and repeats the process until it has examined all the members in the **set**. In the previous example, you know that ISETL will execute the loop 6 times—once for each item in Guys. However, since there are 6 possibilities for the value of name during the first pass through the loop, and 5 during the second pass, and so on, there are actually $6 \cdot 5 \cdot 4 \cdot 3 \cdot 2 \cdot 1$ or 720 different ways ISETL could evaluate the loop. Each one assigns T exactly the same value, namely, {"a", "e"} .

Let's look at another example. Remember back in Section 1.5.2, when we wanted to find the greatest common divisor of two positive integers, say a, b? Well, one way we did this was to construct the set of all common divisors of a and b, and then find the maximum item in this set. How did we construct this set? First, we needed to think about the domain—that is, where should we look for possible common divisors? We chose the set

$$\{1..\min(a,b)\}$$

since any common divisor must be greater than or equal to 1 and less than or equal to both a and b. (Notice that there are other possibilities for the domain. The one we chose just happens to be the one that is most efficient.) Second, we needed to decide what condition must be satisfied in order for an integer n to be a common divisor of a and b. We chose the Boolean expression

$$(a \bmod n = 0) \text{ and } (b \bmod n = 0)$$

since this is true exactly when n divides both a and b evenly. Then, we used set former notation to define the set of all common divisors of a and b. That is,

$$\{n : n \text{ in } \{1..\min(a,b)\} \mid (a \bmod n = 0) \text{ and } (b \bmod n = 0)\}$$

As we described, to determine the items in this set, ISETL iterates through the elements in the domain set, checks the Boolean expression for each element, and if it is satisfied, adds the element to the set.

A related question to finding the greatest common divisor of two positive integers is to find their least common multiple. One way to do this is to look at a set of common multiples of a and b and then find the smallest item in this set. What could you use as a domain? What Boolean expression must be satisfied by an integer in your domain in order to guarantee that the integer is a multiple of both a and b? Write out the set former for this set. *Think* about how ISETL might evaluate the expression you have written for various inputs. What items are in your set with $a = 3$, $b = 2$? $a = 2$, $b = 2$? $a = 2$, $b = 4$? As you think about the process of forming your set, check that the smallest element in your set actually is the least common multiple of a and b.

What if you want to construct a new set using the elements from several sets? For instance, suppose A and B are given sets of strings and you want to define Concat to be the set containing the concatenation of each item in A with each item in B. So, if

$$A := \{"A", "B", "C"\}; B := \{"+", "", "-"\};$$

(note that the set A and the string "A" are completely different), then Concat would be

$$\{"A+", "A", "A-", "B+", "B", "B-", "C+", "C", "C-"\}$$

We can use a set former to define Concat by writing

```
Concat := {letter + sign :  letter in A, sign in B};
```

Suppose

$$S := \{-2..2\}; \qquad T := \{-5, -3..6\};$$

and you wanted to find the **set** containing the product of each positive **integer** in S with each odd **integer** in T. You might write

```
SomeProd := {x*y :  x in S, y in T | x>0 and odd(y)};
```

How do you "think ISETL" as you evaluate this **set** former? You choose an item in S, and then with this item, you loop (in your mind) through all the elements of T. You choose another item in S and repeat the process until you have examined every possible combination of elements from S and T. (By the way, how many possible combinations are there?) The **set** former is implementing the action of a nested **for** loop, and thus, you could have defined **SomeProd** by

```
SomeProd:= {};
for x in S do
    for y in T do
        if x>0 and odd(y) then
            SomeProd := SomeProd with x*y;
        end;
    end;
end;
```

Of course, you can have any number of variables in your domain. Moreover, the definition of the variables to the right can depend on those to the left but not vice versa. For example, suppose you wanted to define a set where each element in the new set is itself a set that contains two integers between 1 and 10. One way to do this is

```
Pairs1 := {{i,j} :  i in {1..10}, j in {1..10} | i /= j};
```

It takes 100 iterations to evaluate this **set** expression. As the **set** Pairs1 is being evaluated, the values of i and j are *bound to* successive elements in their respective domains. Thus, i and j are called *bound variables*. Furthermore, the evaluation of **Pairs1** is actually a double **for** loop. That is, for each fixed value of i, you iterate through all possible values of j. The domain of j, however, is allowed to depend on the value of the bound variable i. Therefore, a more efficient way of defining **Pairs1** would be to say

```
Pairs2 := {{i,j} :  i in {1..10}, j in {(i+1)..10}};
```

How many iterations does it take to evaluate **Pairs2**? Of course, the domain of the first variable, in this case i, cannot depend on the value of the bound variable j. Why not?

One final comment about the domains of **set** former variables: if two or more successive variables have the same domain, you can list the variables (separated by commas) and then give their common domain. Thus, since the variables i and j have the same domain when evaluating the **set** former for Pairs1, namely, {1..10}, we could have written

$$\text{Pairs1} := \{\{i,j\} :\quad i, j \text{ in } \{1..10\} \mid i \mathrel{/=} j\};$$

What happens when we apply the **set** operations to two **sets** expressed in terms of **set** formers, which have the same domain? How can we find the result? For instance, assume I is a given **set** of **integers** and define S to be the **set** of all elements in I which are divisible by 2 or divisible by 5. That is,

$$\text{S} := \{x :\quad x \text{ in I} \mid (x \bmod 5 = 0) \text{ or } \mathtt{even}(x)\};$$

Furthermore, let T be the **set** that contains all the members of I except the positive elements in I that are odd. So,

$$\text{T} := \{x :\quad x \text{ in I} \mid x{>}0 \text{ impl } \mathtt{even}(x)\};$$

Find S + T. Recall that this means we want the **set** of all elements in S or T. In this example, this is the **set** of all elements in I such that the condition for S holds or the condition for T holds. Thus, S + T is the **set** of all elements in I such that

$$((x \bmod 5 = 0) \text{ or } (\mathtt{even}(x))) \text{ or } ((x{>}0) \text{ impl } (\mathtt{even}(x)))$$

This can be simplified using the laws of equivalences (see Exercise 3.3.4) as follows:

$$((x \bmod 5 = 0) \text{ or } (\mathtt{even}(x))) \text{ or } ((x{<=}0) \text{ or } (\mathtt{even}(x)))$$

That is,

$$(x \bmod 5 = 0) \text{ or } (\mathtt{even}(x)) \text{ or } (x{<=}0)$$

Therefore, the union of S and T is the **set** of all elements in I that are divisible by 5, divisible by 2, or negative, which is the **set**

$$\{x:\quad x \text{ in I} \mid (x \bmod 5 = 0) \text{ or } (\mathtt{even}(x)) \text{ or } (x{<=}0)\}$$

In general, if

$$\text{S1} = \{x :\quad x \text{ in W} \mid P(x)\};$$

and

$$\text{S2} = \{x :\quad x \text{ in W} \mid Q(x)\};$$

then

$$\text{S1} + \text{S2} = \{x :\quad x \text{ in W} \mid P(x) \text{ or } Q(x)\}$$

How would you express S1 * S2 using a set former that involves the conditions P and Q? How about S1 - S2?

Summary of Section 3.4

This section introduces you to **set** former notation. We tried to encourage you to think like ISETL—to think *loop-test-evaluate*—when you construct and evaluate a **set** former expression. Because a **set** former says in one line something that takes a number of lines of code to say with a loop, we call an ISETL **set** former expression a *one-liner*.

We also talked about performing **set** operations on **sets** expressed in **set** former notation.

Exercises

3.4.1 For each of the following **sets**:

1. Give a verbal explanation concerning what objects are in the **set**. For example, you might describe the **set**

   $$\{x**2 \; : \quad x \text{ in } \{2..10\} \mid x \text{ mod } 2 = 0\}$$

   as "the set of all squares of the even integers in the set $\{2..10\}$."

2. Think carefully about how ISETL evaluates the **set** expression. That is, think *loop-test-evaluate*.

3. List the elements in the **set**.

4. Use ISETL to check that your answer to part 3. is correct.

a. $\{x \; : \quad x \text{ in } \{ \; 2, \; 5..10 \; \}\}$;

b. $\{r \; : \quad r \text{ in } \{2, \; 5..100\} \mid r \text{ mod } 5 = 0\}$;

c. $\{t**4 + t**2 \; : \quad t \text{ in } \{-6..6\} \mid \text{even}(t \text{ div } 3)\}$;

d. $\{y>10 \; : \quad y \text{ in } \{25, \; -12, \; -67, \; 34, \; 100\} \mid \text{odd}(y)\}$;

e. $\{\text{even}(n) \; : \quad n \text{ in } \{-3, \; -1..11\}\}$;

f. $\{m \; : \quad m \text{ in } \{\text{"hi"}, \; 1.234, \; 9.0, \; \text{false}\} \mid \text{is\_set}(m)\}$;

g. $\{\{2*z\} \; : \quad z \text{ in } \{\text{"lb"}, \; \text{"aa"}, \; \text{"ca"}, \; \text{"da"}, \; \text{"ba"}\}$
   $\mid z>\text{"c"}\}$;

h. $\{\text{abs } (x*y) \; : \quad x, \; y \text{ in } \{-4, \; -3, \; 0, \; 3, \; 4\} \mid x<y\}$;

i. $\{\{s, \; t\} \; : \quad s \text{ in } \{10,8..4\}, \; t \text{ in } \{5..s\}$
   $\mid (s+t)\text{mod } 2 = 0\}$;

j. $\{\text{max}(k,10) = k \; : \quad k \text{ in}$
   $\{20, \; \text{"20"}, \; (20<0) \text{ impl } (20>0), \; 20.0\}$
   $\mid \text{is\_integer}(k)\}$;

k. $\{(p \text{ and } q) = (q \text{ and } p) \; : \quad p, \; q \text{ in } \{\text{true}, \; \text{false}\}\}$;

3.4.2 If a **set** is constructed using a **for** loop instead of using a **set** former, how many times would ISETL execute the body of the loop? Consider the **set** formers in Exercise 3.4.1, parts a.-k. Determine how much work must be done by ISETL to construct each **set**. For example, the first **set**,

$$\{x : \quad x \text{ in } \{2, 5..10\}\}$$

could be constructed as follows:

```
A := {};
for x in {2, 5..10} do
      A := A with x;
end;
```

Construction of this **set** requires that the body of the loop be executed three times (when x = 2, x = 5, and x = 8).

3.4.3 Since the objects in a set are not ordered, when ISETL iterates through a **set** while executing a **for** loop or a **set** former, the only thing you are certain of is that the loop will be executed once for each element in the domain **set**—you cannot predict what order ISETL will choose elements from the domain **set**. Thus, for instance, there are 6 (or $3 \cdot 2 \cdot 1$ or 3!) ways to construct the set $\{x : \quad x \text{ in } \{2, 5..10\}\}$ namely, x = 2, then 5,then 8; x = 2,8,5; x = 5,2,8; x = 5,8,2; x = 8,2,5; and x = 8,5,2. Of course, evaluation of the **set** expression gives you the same **set** each time, { 8, 2, 5 }. Look again at the **set** formers in Exercise 3.4.1. For each **set** in parts b.-k. determine how many different ways ISETL can construct the given **set**.

3.4.4 **Set** former expressions for the **set** operations and **sets** of **integers** defined by arithmetic progressions.

   a. Suppose $\mathcal{U}$ is the universal set and S and T are subsets of $\mathcal{U}$. Then, you can use set formers to define the set operations. For example, you can define set union by

$$S \cup T = \{x : x \in \mathcal{U} \mid (x \in S) \; \vee \; (x \in T)\}$$

   Use set formers to define
      i) the intersection of S and T.
      ii) the difference of S and T.
      iii) the complement of S.
      iv) the symmetric difference of S and T (see Exercise 3.3.7).

b. You can also use set formers to define sets of integers based on arithmetic progressions. For instance, if you notice that each member of the set $\{2, 5..67\}$ has a remainder of 2 when it is divided by 3, then

$$\{2, 5..67\} = \{x : x \in \{2..67\} \mid x \bmod 3 = 2\}$$

Use set formers to define

   i) $\{10, 15..101\}$

  ii) $\{98, 95..0\}$

 iii) $\{-21, -19..22\}$

 iv) $\{10, 6.. -51\}$

3.4.5 Write ISETL one-liners to return the following:

a. The set of all cubes of the members of a given set of integers, S, that are positive and odd.

b. The set containing all possible sums of elements from a given set of integers S with elements from a given set of integers T.

c. The set of all composite numbers between 2 and 100. Note: An integer is said to be a *composite number* if it can be written as the product of two integers greater than 1. Therefore, 6 is a composite number since $6 = 2 \cdot 3$.

d. Use the set you constructed in part c. to define the set of all prime numbers between 2 and 100. Note: An integer is *prime* if its only divisors are 1 and itself. Therefore, a prime number is *not* composite.

e. Use the set you constructed in part d. to define the set of all primes between 2 and 100 that are 1 more than a multiple of 4.

f. The set of all integers between 2 and 100 that are the product of two different primes.

g. The set of all values of the Boolean expression $(p \implies q) = (\neg p \lor q)$ for $p, q \in \{true, false\}$.

h. The set of all values of the proposition `h>5 impl odd(h)` for h in the set $\{2, 9, 4, 3, 0, -1, 15, 99\}$.

i. The set of all two-element sets $\{V, n\}$, where V is a vowel and n is a nonnegative, even integer less than 20.

3.4.6 For each of the following **set** expressions:

1. List the members of the **set**.

2. Rewrite the **set** expression so that the domain of the second bound variable depends on the domain of the initial bound variable.

3. Compare the number of iterations needed to evaluate the given expression with the number needed to evaluate your expression.

a. S := {{x}+{y} : x, y in {-5..-2}};

b. T := {r+s : r in {0..3}, s in {-3..3} | s<r};

c. Z := {p*sqrt(q) : q in {1..4}, p in {-5..5} |
                      abs(p)<q and p/=0};
   Hint: If y is a positive integer, then $abs(x) < y$ if and only if $-y < x$ and $x < y$.

3.4.7 ISETL **funcs** that return a **set**.

a. Define a **func** called common_divisors that accepts two positive **integers** n and m and returns the **set** containing all their common divisors.

b. Define a **func** called odd_int that accepts an arbitrary **set** S and returns the **set** containing all the members of S that are nonnegative, odd **integers**.

c. Define a **func** called npow that accepts a **set** S and a nonnegative **integer** n and returns the **set** containing all the subsets of S with cardinality n. (See Exercise 3.2.6.) Notice that this **func** may be predefined in your version of ISETL, but there is no harm in redefining a predefined function.

3.4.8 **Sets** whose elements are ISETL **funcs**. Suppose the following has been given to ISETL as input:

```
F := func(P, Q);
          return P impl Q;
      end;
G := func(P, Q);
          return (P and Q) or P;
      end;
```

a. What is the meaning of the following two assignments?

   i) S := {F};

   ii) T := {F, G};

b. What will appear on the screen as output if the following input is given to ISETL? (Note: T is defined in part a.)

```
for x in T do
        print x(true, true);
        print x(false, true);
end;
```

   c. After all of the above ISETL statements have been entered, what will appear on the screen if the following input is given to ISETL?

     i) `arb(S)(true, false);`

     ii) `arb(T)(false, true);`

3.4.9 For each of the following pairs of sets, find $S \cup T$, $S \cap T$, and $S - T$. Express your answers in terms of formal set expressions. That is, describe formally the elements in each of these sets using the conditions that are satisfied by the the elements in S and the elements in T. Simplify the condition associated with the new set if possible.

   a. $S = \{x : x \in \{-100..100\} \mid x > 0 \Rightarrow (x \bmod 2 = 0)\}$
     $T = \{x : x \in \{-100..100\} \mid x \le 0 \lor odd(x)\}$

   b. $S = \{r : r \in \{\text{``}a\text{''}, \text{``}b\text{''}, \text{``}c\text{''}, \text{``}d\text{''}, \text{``}e\text{''}\} \mid (r \in \text{``}bye\text{''}) \lor (r \in \text{``}aeiou\text{''})\}$
     $T = \{v : v \in \{\text{``}a\text{''}, \text{``}e\text{''}, \text{``}c\text{''}, \text{``}b\text{''}, \text{``}d\text{''}\} \mid$
                            $(v \in \text{``}aeiou\text{''}) \lor (ord(v) = 10)\}$

3.4.10 Assume I is a fixed set of integers.

   a. Give a verbal description of the set P, which is defined by the following set expression:

$$P = \{x : x \in I \mid \neg((x < -3) \land (x \bmod 2 \ne 0))\}$$

   b. Let R be the set of all elements of I that are divisible by 2 if they are greater than 10. Write a formal set expression for R.

   c. Assume in Figure 3.4 that the rectangle represents the set P from part a. and the circle represents the set R from part b. Write a *formal set expression* for the set represented by the shaded portion of the diagram.

3.5 Tuples and Tuple Operations—Strings (Revisited)

First, let's recall a few important facts about sets:

- Sets are unordered. The set $\{\text{``}a\text{''}, \text{``}b\text{''}\}$ is equal to the set $\{\text{``}b\text{''}, \text{``}a\text{''}\}$.

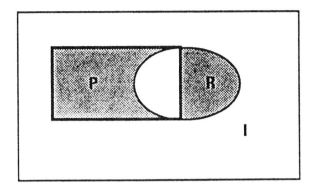

Figure 3.4: Sets P and R

- A set can contain a particular object at most one time. The set $\{1, 2, 3\}$ is equal to the set $\{1, 2, 2, 3, 3, 3\}$.

- Every item in a set must be previously defined. The expression $\{1.23,\ \text{OM}\}$ has the value OM.

Tuples, on the other hand, are ordered, can have repetitions, and can contain OM. We use square brackets ([]) to denote a tuple as opposed to curly brackets ({}) for a set. So, the following are valid tuple assignments in ISETL:

```
T1 := [1.23, "hi", OM, 9, OM, {1..3}, 9, true];
T2 := [5 mod 2, max(-4,1), 4 div 4, 1, abs(-1), -10+11];
T3 := [];
```

If you entered these tuple assignments in ISETL, and then entered

T1;   T2;   T3;

ISETL would return

```
[1.23, "hi", OM, 9, OM, {3, 2, 1}, 9, true];
[1, 1, 1, 1, 1, 1];
[];
```

One way to think about a tuple is as an infinite sequence that has a finite number of non-OM elements. Thus, there exists a certain point in the sequence after that every element is OM, and the trailing OMs are simply not listed. Actually, this is exactly how ISETL handles OMs that appear at the end of a tuple. Furthermore, the operator # applied to a tuple returns the position of the last non-OM item in the tuple, ignoring any OMs that may be tacked onto the end of the `tuple`. With these ideas in mind, consider the following ISETL terminal session:

```
> Tup := ["a", OM, "b", OM, "c", OM, OM, OM, OM];
> #(Tup);
5;
> Tup;
["a", OM, "b", OM, "c"];
```

You can use the `Boolean` operations `in` and `notin` to test for membership. For instance, with `Tup` defined as before,

$$\text{"a" in Tup;} \qquad \text{2.3 notin Tup;}$$

both return *true*. You can use the `Boolean` operators `=` and `/=` to test whether or not two `tuples` are the same. So,

```
[true impl false, 1+2, "a"+"bc"] = [false, 3, "abc"];
```

returns *true*, as do

```
[1, OM, OM, 3] = [1, OM, OM, 3, OM, OM, OM];
```

and

```
["guys", "gals"] /= ["gals", "guys"];
```

Since `tuples` act like vectors, we usually refer to an item in a `tuple` as a *component*. Then, since `tuples` are ordered, it makes sense to talk about the first component in a `tuple`, or the second, or the third, and so on. To access the component in the ith position of a `tuple` T you write `T(i)`, where i is referred to as the *index* of the component. So, continuing the above, we have

```
> Tup(1);   Tup(2);   Tup(#(Tup));
"a";
OM;
"c";
```

Of course, things can get complicated and you can have `tuples` of `tuples`, where `T(5)(4)` is the fourth component of `T(5)`—that is, the fourth component of the fifth component of the `tuple` T. For example,

```
> Family := [["Bob", "Mary", "Peter"],
>> ["John", "Sue", "Todd", "Heather"]];
> Family(1);
["Bob", "Mary", "Peter"];
> Family(2);
["John", "Sue", "Todd", "Heather"];
> Family(1)(2);
```

<u>"Mary"</u>;
> Family(2)(5);
<u>OM;</u>

Thinking back to the discussion about **strings** in Chapter 1, Section 1.9, it may seem that **tuples** are a lot like **strings**, and that's right. The major difference is that every component of a **string** must be defined and in particular must be a character, while a component in a **tuple** can be of any valid ISETL type, including **OM**. Recall that the operations on **strings** include length (#), concatenation (+), replication (*), tests for equality and inequality (=, /=), tests for lexicographic order (<, <=, >, >=), and a test for substring (in). In addition, you can access an item in a **string** just as you access an item in a **tuple** by indicating the position, or index, of the desired item. Thus, in the **tuple** Family,

$$\text{Family(1)(2)(4)};$$

will return "y".

What does ISETL return if you enter Family(2)? Family(2)(2)? Family(2)(2)(2)? Family(2)(1) < Family(1)(2)? "Ed" in Family(1)? "a" in Family(2)(4)?

Just as you can concatenate and replicate **strings**, you can concatenate and replicate **tuples**, using the **tuple** operations + and *, respectively. There is a list of all the **tuple** operations at the end of the section, and you will have the opportunity to get an idea as to how they behave when you do the exercises.

You can define **tuples** of **integers** that increase or decrease by a fixed jump using notation very similar to the notation for **sets**. The general forms for useful **tuples** of integers are

$$[\text{m..n}]$$

and

$$[\text{m, r..n}]$$

Try to guess what the following **tuple** expressions in ISETL return:

$$[-3..6];$$
$$[6..3];$$
$$[-3..-3];$$
$$[-2, 1..12];$$
$$[100, 98..89];$$
$$[100, 98..102];$$

If you guessed the **tuples**

```
[-3, -2, -1, 0, 1, 2, 3, 4, 5, 6];
[];
[-3];
[-2, 1, 4, 7, 10];
[100, 98, 96, 94, 92, 90];
[];
```

then you're right. However, if you're having some trouble understanding any of these examples, look back at the discussion concerning {m..n} and {m, r..n} in Section 3.2, noting the difference between sets and tuples that are defined using an arithmetic progression is that with tuples the result is an *ordered* sequence.

As you probably suspect, we have tuple formers just like the set formers in Section 3.4. Let's look at some specific examples. Suppose

```
F := [x*x :   x in [1..10] | x mod 2 = 0];
```

Then, the value of F is

```
[4, 16, 36, 64, 100]
```

As when evaluating a set former expression, ISETL does a loop-check-evaluate routine—in this case, ISETL iterates through the components of the tuple [1..10] (*in order*), checks if the component is even, and if it is, evaluates its square and puts the result in the tuple. In a tuple former the domain of the variable x can be a set instead of a tuple, but then you might have a number of different possible values for the tuple. Why? Think about how ISETL evaluates the tuple expression

```
FS := [x*x :   x in {1..10} | x mod 2 = 0];
```

Again it is a loop-test-evaluate procedure, but since {1..10} has no order, you have no idea which element in {1..10} ISETL will consider first. All you know is that it will eventually consider them all. As a matter of fact, depending on how ISETL iterates through the elements in {1..10}, there are 5 possible values for the first component of FS, 4 for the second, and so on. Thus, FS has $5 \cdot 4 \cdot 3 \cdot 2 \cdot 1$ or 120 different possible values.

What if you use a tuple as the domain for a set former variable? Does this modify the value of the set? For example, what does ISETL return if you enter the following lines?

```
SetDomain := {x*x :   x in {1..10} | x mod 2 = 0};
TupDomain := {x*x :   x in [1..10] | x mod 2 = 0};
SetDomain = TupDomain;
```

If you are not sure of the answer, try it!

The general form for a tuple former is

```
[exp(x) :   x in T | P(x)]
```

where **exp** is an expression and the domain of **x**, namely, **T**, is a **set**, a **tuple**, or maybe even a **string**. More generally,

```
[exp(x1,x2,...,xn) :  x1 in T1, x2 in T2,...,xn in Tn |
                           P(x1,x2,...,xn)]
```

where **Ti** is a **set**, a **tuple**, or a **string** and may depend on **x1,...,xi-1**.

We conclude this section with a list of the **tuple** and **string** operations (see Tables 3.3 and 3.4). We use **N**, **B**, **S**, and **T** to denote, respectively, the set of all **integers** that can be represented in an implementation of ISETL, the set of **Boolean** values, {*true, false*}, the set of all character **strings**, and the set of all **tuples**. We also use **U** to denote the set of objects in ISETL.

infix (operators)			
+	:	$\mathbf{T} \times \mathbf{T} \longrightarrow \mathbf{T}$	concatenation
*	:	$\mathbf{N^+}+\{0\} \times \mathbf{T} \longrightarrow \mathbf{T}$	replication
=	:	$\mathbf{T} \times \mathbf{T} \longrightarrow \mathbf{B}$	equality test
/=	:	$\mathbf{T} \times \mathbf{T} \longrightarrow \mathbf{B}$	inequality test
in	:	$\mathbf{U} \times \mathbf{T} \longrightarrow \mathbf{B}$	membership test
notin	:	$\mathbf{U} \times \mathbf{T} \longrightarrow \mathbf{B}$	membership test
with	:	$\mathbf{T} \times \mathbf{U} \longrightarrow \mathbf{T}$	concatenate the given object to the given **tuple**
prefix (functions)			
#	:	$\mathbf{T} \longrightarrow \mathbf{N^+}+\{0\}$	index of last non-**OM** component
arb	:	$\mathbf{T} \longrightarrow \mathbf{U}$	random member of **tuple**
is_tuple	:	$\mathbf{T} \longrightarrow \mathbf{B}$	test for type **tuple**

Table 3.3: Tuple operations

Summary of Section 3.5

In this section, you were introduced to the **tuple** and its operations. You thought about the similarities and differences between **sets**, **tuples**, and **strings**. You learned about **tuple** formers and were encouraged to think about how ISETL evaluates a given expression. The following exercises are designed to give you lots of practice using these ideas and to introduce you to slicing definitions, assignment operators, and the statements **take...frome** and **take...fromb**. The exercises also include some very important pointers to things to come, namely, pointers to the concepts of induction, functions and relations, and formal languages.

infix (operators)		
+ :	$S \times S \longrightarrow S$	concatenation
* :	$N^+ + \{0\} \times S \longrightarrow S$	replication
= :	$S \times S \longrightarrow B$	equality test
/= :	$S \times S \longrightarrow B$	inequality test
in :	$S \times S \longrightarrow B$	substring test
notin :	$S \times S \longrightarrow B$	nonsubstring test
< :	$S \times S \longrightarrow B$	less than (lexical order test)
<= :	$S \times S \longrightarrow B$	less than or equal to
> :	$S \times S \longrightarrow B$	greater than
>= :	$S \times S \longrightarrow B$	greater than or equal to

prefix (functions)		
# :	$S \longrightarrow N^+ + \{0\}$	length
arb :	$S \longrightarrow S$	random character in string
ord :	$S \longrightarrow N^+ + \{0\}$	index of character (machine dependent)
char :	$N \longrightarrow S$	character with given index
is_string :	$S \longrightarrow B$	test for type string

Table 3.4: String operations

Exercises

3.5.1 **Tuple** operations, slicing and indexing definitions, and assignment operators.

In both expressions and assignment statements, you can access the ith component of a **tuple** T by writing T(i). You can access the ith through jth components with T(i..j), the ith component through the last non-OM component by entering T (i..), and the first through the jth component with T(..j). Look carefully at what happens in the following ISETL session:

```
>  T := [2,5,7,8];
>  T(1);                    $ returns the first
2;                          $ component of T

>  T(2) := -10;    T;       $ assigns the second
[2,-10,7,8];                $ component to be -10

>  T(1..3);                 $ returns the tuple
[2,-10,7];                  $ [T(1),T(2),T(3)]
```

```
>  T(4..4);                    $ returns the tuple [T(4)]
[8];
>  T(3..);                     $ returns the tuple
[7, 8];                        $ [T(3),T(4)]
>  T(..2);                     $ returns the tuple
[2, -10];                      $ [T(1),T(2)]

>  T(2..3):=["a","b","c"];     $ replaces T(2) through T(3)
>  T;                          $ with "a","b","c"
[2,"a","b","c",8];
>  T(3..5):=[3.1414];    T;    $ replaces T(3) through T(5)
[2,"a",3.1414];                $ with 3.1414
>  T(2..):=[om,1,om,6];    T;  $ replaces T(2) through T(#T)
[2,OM,1,OM,6];                 $ with om,1,om,6
>  T(..2):=[1,8,9];    T;      $ replaces T(1) through T(2)
[1,8,9,1,OM,6];                $ with 1, 8, 9
>  !quit
```

With these ideas in mind, enter the following lines in ISETL. Before you hit the carriage return key, predict what ISETL will return by thinking about how ISETL will evaluate the statement. If your prediction differs from the output of ISETL, explain why.

a. T := [10, om, 30, om, 50, om, 70, om];   T;

b. T(1);   T(4);   T(4..4);   T(2..6);   T(2..);
   T(3..#(T))=T(3..);

c. T(..4);   T(..5);   #(T(..2));   T(1..5)=T(..5);

d. T(2) := "CS131";   T(7) := 123.45;   T;

e. T(2..4) := [1, 2, 3, 4];   T;

f. T(1..3) := ["a", "b"];   T;

g. T(3..5) := [];   T;

h. T(3..)   := [1.0, 11.0, 111.0, 1111.0];   T;

i. T(..3) := [1..3];   T;

j. S := T + [2..4];   S;

k. S := 2*[3, 5..12];   S;

l. 2 in S;   2 notin S;
   S=[11, 9, 7, 5, 3]+[3, 5, 7, 9,11];

m. W := [100, 95..77];   W;

n. W with 95;   W;

o. W := W with 95;   W;

p. arb(W);   arb(W);   arb(W);

q. [45, 48..40];

3.5.2 Tuple formers.

   a. Think about how ISETL evaluates the following tuple expressions. Give the value of the tuple.

     i) [x**2 :   x in [1, 3..10]];

     ii) [{i, j} :   i, j in ["L", "M", "N"] | i /= j];

     iii) [[1..r] :   r in [0, 2..6]];

     iv) [N+2 < 2**N : N in [0..10]];

     v) [u*v :   u in [-5..0], v in [-5..(u+1)] |
                                   (u+v) mod 2 = 0];

   b. In the following tuple expressions, the domain specification in the tuple former is a set. How many possible values does the tuple have?

     i) [6*t + 5 :   t in {0, 2..10}];

     ii) [3*s :   s in {"m", "a", "t", "h"} | s > "cs"];

   c. Construct the following tuples.

     i) Assume T is a tuple that has been assigned a value. Construct a tuple whose components are the components of T in reverse order.

     ii) Assume T is a tuple each of whose components is also a tuple. Construct a tuple such that each component is the length of the corresponding component of T.

3.5.3 Tuples of propositions.

   a. Assume T is a tuple of Boolean values. Construct a tuple whose ith component is the value of the implication, T(i) impl T(i+1), where i is in [1..#T-1].

   b. Write an ISETL tuple former to construct a tuple B having 10 components each corresponding to the value of the statement $x > 7$ where x is a random integer between 0 and 9. Evaluate B several times. What can you say about the value of B?

   c. Consider the following proposition:

$$2^N > 2N^2 - N - 2$$

     i) Use an ISETL for loop to print the truth or falsity of the given proposition for $N = 1, \ldots, 20$.

     ii) One may consider the statement of the task in part a. as setting up a tuple of 20 propositions. Use an ISETL tuple former to construct a tuple whose Nth component has the value of the given proposition for $N = 1, \ldots, 20$.

iii) Define a **func** called P that accepts an **integer** N and returns the value of the given proposition. Construct a second **tuple** of length 19 whose Nth component is the value of the implication P(N) impl P(N+1). (Look again at part a.)

3.5.4 ISETL **funcs** having **tuples** as parameters.

Run the following **funcs** on a variety of ISETL inputs:

a. Define an ISETL **func** that accepts a **tuple** and returns the number of non-OM components in the **tuple**.

b. Define a **func** that accepts a **tuple** of **integers** or **floating-point** numbers and returns the maximum item in the **tuple**.

c. Define a **func** called tup_sum that accepts two **tuples**, V and W, that have the same length and whose components are numbers, and returns the **tuple** corresponding to the sum of V and W "component-wise," that is, the ith component of the output equals the sum of the ith components of V and W. Call your **func** using both functional and . notation.

d. Define a **func** called scalar_prod that accepts a **tuple** V whose entries are numbers and a number r and returns the **tuple** associated with the scalar product of r and V, that is, the **tuple** whose ith component equals the ith component of V multiplied by r.

3.5.5 Sets of ordered pairs.

a. Given two sets A and B, the *Cartesian product* of A and B is the set of all ordered pairs (or two-**tuples**) of elements of A with elements of B. For example, if $A = \{1, 2, 3\}$ and $B = \{$"a", "c"$\}$, then the Cartesian product of A and B is

$$\{[1, \text{"a"}], [1, \text{"c"}], [2, \text{"a"}], [2, \text{"c"}], [3, \text{"a"}], [3, \text{"c"}]\}$$

In Mathematics, we denote the Cartesian product by $A \times B$.

Define a **func** called c_prod that accepts two **sets** A and B and returns their Cartesian product. Run your **func** on a variety of **sets** using both the functional and the . notation.

b. Define an ISETL **func** called D that accepts a **set** of ordered pairs (2-**tuples**), F, and returns the **set** containing all the first components of the members of F. Run your **func** on a variety of **sets**. Note: The **set** of all the first components of a given **set** of ordered pairs, F, is called the *domain* of F.

c. Define an ISETL **func** called **Im** that accepts a **set** of ordered pairs and returns the **set** containing all the second components of the elements in the given **set** of ordered pairs. Again, run your **func** on a variety of **sets**. Note: The **set** of all second components of a given set of ordered pairs, **F**, is called the *image* of **F**.

d. Define an ISETL **func** called **Val** that accepts a **set** of ordered pairs, F, and any ISETL object, x, and returns the **set** containing all the second components of the members of F that have x as their first component. As usual, evaluate **Val(F, x)** with a variety of values for F and x. What happens if no member of F has x as its first component?

3.5.6 **Strings.** Indexing. Iterating through the members of a **string**.

a. Review the elementary **string** operations by repeating Exercise 1.9.1 in Chapter 1.

b. The slicing and index definitions and the assignment operators for **strings** are defined exactly as they are for **tuples** (see Exercise 3.5.1.). Enter the following lines in ISETL. Think about how ISETL will evaluate each line and predict what ISETL will return. Hit the carriage return key. Does your prediction correspond to the output of ISETL?

```
  i) h := "hi_there"; #(h); h;
 ii) "h" in h; "the" notin h;
iii) h(6); h(8); h(9);
 iv) h(1..2); h(3..3); h(1..#(h)); h(4..); h(..2);
  v) h(1..2) := "hello"; h;
 vi) h(6..)  := " my friend!"; h;
vii) h(6..)  := ""; h;
viii) h(..5) := "cheers"; h;
 ix) bf := [["brian", 44], ["mark", 22],
            ["karin", 19], ["freckles", 9]];
  x) bf(1);
 xi) bf(2)(1); #(bf(4)(1));
xii) bf(3)(1)(4);
xiii) bf(2..);
xiv) bf(2..)(3);
 xv) bf(2..)(1)(1);
xvi) bf(3..)(2)(1)(7);
xvii) bf(2..3)(1)(2);
xviii) bf(..3)(2)(1) > "ed";
```

c. In ISETL, not only can you iterate through the members of a **set** and a **tuple**, but you can also iterate through the members of a **string**. The loop is executed in the order in which the characters occur in the **string**. Assume **str** is a character **string** that has been assigned a value. Using a **for** loop, iterate through the characters in **str** and print each member of **str** that is a vowel.

d. Since you can iterate through the members of a **string**, a **string** can serve as the domain of a **set** or a **tuple** former variable. Assume **str** is a character **string** that has been assigned a value.

   i) Construct the **tuple** that contains each of the vowels in **str** in the order in which they occur in **str**.

   ii) Construct the **tuple** each of whose components is itself a **tuple** of length 2, where the first component is the character from **str** and the second component is the (machine dependent) index of the character. **Hint: Use the ISETL function ord.**

3.5.7 Definition of the **string** primitives.

The **string** primitives generally have the form

operation_name (scanned_string, pattern_string)

Each operation attempts to match a portion of its scanned_string with its pattern_string. If a portion of scanned_string is successfully matched, it is removed from (broken off) scanned_string and the matched portion and the new value of scanned_string are returned in a **tuple**. Otherwise, the operation returns **OM**.

Write ISETL **funcs** to implement each of the following **string** primitives, where we have used **ss** to denote the scanned_string and **ps** for the pattern_string:

a. **span(ss, ps)** : Finds the longest initial segment of **ss** that consists entirely of characters from **ps**.

b. **any(ss, ps)** : Breaks off the first character of **ss** if it belongs to **ps**.

c. **break(ss, ps)** : Scans **ss** from the left up to but not including the first character that does not belong to **ps**. This part of **ss** is broken off.

d. **len(ss, n)** : Has an **integer** second parameter. If #ss >= n, then **len** breaks off ss(1..n). Otherwise, returns **OM**.

e. **match(ss,ps)** : If #ps <= #ss and ps = ss(1..#ps), then this part of **ss** is broken off.

f. lpad(ss, n) : Has an **integer** second parameter. Returns the **string** obtained by padding **ss** out to length n by adding as many blanks to the left of **ss** as necessary. If #ss >= n, then lpad returns **ss**. Note: lpad does not change the value of **ss**.

g. rpad(ss, n) : Behaves like lpad but adds blanks on the right.

3.5.8 The ISETL statements take...fromb..., take...frome..., and **with**. Let T be a **tuple** and x an ISETL variable name. Then, execution of the statement

take x fromb T;

assigns x the value of the first component of T, and all the components of T are shifted left one position. Execution of the statement

take x frome T;

assigns x the value of the last defined (non-OM) component in T and replaces the component with OM. On the other hand,

T with x;

is an expression that evaluates to a **tuple** whose initial components are the components of T and whose last component is the value of x. Note that **with** does not change the value of T itself. For example, consider the ISETL terminal session

```
>  T := [2..5]; T;
[2, 3, 4, 5 ];
>  take y fromb T;
>  y; T;
 2;
[3, 4, 5];
>  take y frome T; y; T;
5;
[3, 4];
>  S := [om, 5.6, om, 7**2, om, om]; S;
[OM, 5.6, OM, 49];
>  take x fromb S; take z frome S; x; z; S;
OM;
49;
[5.6];
>  W := S with "a"; W; S;
[5.6, "a"];
[5.6];
>  quit!
```

a. A *stack* is an ordered list from which you can remove (*pop*) items and into which you can insert (*push*) items at one end called the *top*. A stack is said to be *empty* if it does not contain any elements. A `tuple` can serve as a natural home for a stack.

   i) Initialize a stack to be empty.

   ii) Implement the stack operations push and pop using the `tuple` operations `take...frome...` and `with`.

   iii) Write a `Boolean` valued ISETL `func` called `empty` that accepts a stack (represented by a `tuple`) and returns *true* if the stack is empty and *false* otherwise.

   iv) Write an ISETL `func` called `stack_top` that accepts a stack and returns the value of the top item in the stack without removing it from the stack.

b. You can use a stack to determine whether or not a given algebraic expression is *well formed*. For example, the expressions $((A+B)-C)$ and $(X+Y)*Z$ are well formed, while $((X+Y)*Z$ and $(V/W)*U)$ are not. Define a `Boolean` valued ISETL `func` that accepts a `string` containing an expression and returns *true* if the `string` is well formed and *false* if it is not. The general idea is to initialize a stack to be empty and then iterate through the members of the `string`. If you see a (, then push it onto the stack. If you see a), then pop the matching parenthesis off the stack. If at any point you try to pop an empty stack, or if when you are done iterating through the `string` the stack is not empty, then the expression is not well formed.

3.6 One-Liners

3.6.1 Proofs

When talking about propositional calculus in Chapter 2, we discussed ways of

- determining whether or not a Boolean expression is a tautology.

- verifying that an argument is valid.

- doing proofs: enumeration, direct, and indirect.

Each of these ideas involved iterating through sets of Boolean values, {*true, false*}, testing, and evaluating Boolean expressions. This loop-test-evaluate process is precisely the way ISETL constructs a `set`—in this case a `set` of propositions—using a `set` former. For example, to show that the Boolean expression

```
(not q impl not p) = (p impl q)
```

is a tautology, you had to show that the expression is true for all combinations of values of p and q. One way to do this is to look at the output of ISETL code, such as

```
for p,q in {true, false} do
        print (not q impl not p) = (p impl q);
end;
```

If ISETL returns a list of all *trues*, you know that the expression is a tautology. But, if it returns at least one *false*, then you know that the expression is false for some combination of values of p and q, and, hence, is not a tautology. An alternate proof is to form the **set** of propositions

```
{(not q impl not p) = (p impl q) :  p, q in {true, false}};
```

If this one-liner ISETL **set** expression evaluates to {*true*}, then the **Boolean** expression is a tautology. But, if *false* is in the **set** returned by ISETL, then you know that it is not a tautology.

Now, let's think about how we can use a one-liner to determine whether or not a given argument is valid. In Chapter 2, we proved that the argument

I run to class or it is not late.
If my alarm rang then it is late.
Therefore, it is late or my alarm did not ring or I don't run to class.

is valid by writing ISETL code to do: a proof by enumeration (exhaustion), a direct proof, and an indirect proof. For all three methods, we first expressed the argument in ISETL:

```
((Run or not Late) and (Ring impl Late)) impl
                            (Late or not Ring or not Run)
```

where we used obvious names for each of the variables in the argument. A proof by enumeration (exhaustion) requires that we show that the argument is true for all combinations of values of variables, so we looked at the output of the ISETL code

```
for Run, Ring, Late in {true, false} do
        print ((Run or not Late) and (Ring impl Late)) impl
                (Late or not Ring or not Run);
end;
```

When the output was nothing but *true*, we knew that the argument was valid. But, what about using a one-liner instead? One way would be to look at the value of

```
{((Run or not Late) and (Ring impl Late)) impl
                (Late or not Ring or not Run) :
                        Run, Ring, Late in {true, false}};
```

What can you say about the validity of the argument if the value of the set expression is {*true*}? What if it is {*true, false*} or {*false*}?

A direct proof involves showing that the conclusion of the argument is always true whenever the hypothesis is true. Again, we used a loop-test-evaluate process to do this in Chapter 2. That is, we looped through all the possible combinations of truth values, tested whether or not the hypothesis was true, and if it was true, evaluated the conclusion. You can do this with a one-liner, too.

```
{Late or not Ring or not Run :
        Run, Ring, Late in {true, false} |
        (Run or not Late) and (Ring impl Late)};
```

Again, the argument is valid if the output is {*true*}. Why?

What about using a one-liner to do an indirect proof? In this case you want to show that the hypothesis of the argument is always *false* whenever the conclusion is *false*. What one-liner would you consider? The argument is valid if the output of ISETL is {*false*}. Write a one-liner that does an indirect proof of the argument we've been discussing. Think about how ISETL evaluates your **set** former. Compare your **set** former to the loop method we used in Section 2.3. Make sure you are convinced that the two methods are the same.

3.6.2 Prime Numbers

The concept of a prime number is important in Mathematics. In the exercises in Section 3.4, you constructed the set of all prime numbers between 2 and 100 by thinking about a prime number as something that is not a composite number. But, how could you get this set of primes without first building the set of composite numbers? To do this, you need to find conditions that are satisfied if and only if a given integer is prime. Now, 3 is prime, 29 is prime, and 53 is prime. What do each of these numbers have in common? One thing you might notice is that the only divisors of each of these numbers is 1 and the number itself. Therefore, you can construct the set of all prime numbers in {2..100} by iterating through the elements of {2..100}, and for each element p, test if its set of divisors equals {1, p}, and if it does, add the integer to the new set. But, what's the set of divisors of some integer p? It's all the integers between 1 and p that divide into p evenly or

$$\{x : x \text{ in } \{1..p\} \mid p \text{ mod } x = 0\}$$

So, p is prime if and only if the ISETL expression

$$\{x : x \text{ in } \{1..p\} \mid p \text{ mod } x = 0 \} = \{1, p\}$$

has the value *true*. Using this **Boolean** expression, you can now define the **set** of primes between 2 and 100 as

```
primes := {p :  p in {2..100} |
                ({x :  x in {1..p} | p mod x = 0} = {1,p})};
```

Here, we are iterating through the integers between 2 and 100, and testing whether or not the set of divisors of the integer being examined contains just 1 and itself. If it does, the integer is added to the set primes.

Suppose p is a prime number. Requiring that the only divisors of p are 1 and itself, however, is equivalent to requiring that each integer greater than 1 and less than p does not divide p evenly, or for all x in {2..(p-1)}, it must be the case that p mod x /= 0. That is, the following Boolean expression must be *true*:

$$(p \text{ mod } 2 \text{ /= } 0) \text{ and } (p \text{ mod } 3 \text{ /= } 0) \text{ and}$$
$$\cdots \text{ and } (p \text{ mod } (p-1) \text{ /= } 0)$$

But, how can we express this condition in ISETL? One way is to use the quantifier, forall:

$$\text{forall x in } \{2..(p-1)\} \mid p \text{ mod } x \text{ /= } 0$$

Can you describe how ISETL finds the truth value of this Boolean expression? As you can probably guess, it iterates through the elements in the domain {2..(p-1)} and tests whether or not the condition p mod x /= 0 is *true*. If the condition is *true* for *every* member of the domain, then ISETL returns *true*, otherwise, it returns *false*. Thus, another way of representing the primes is

```
primes := {p :  p in {2..100} |
                (forall x in {2..(p-1)} | p mod x /= 0)};
```

Let's think about this a little more. An integer p is prime if no element in {2..(p-1)} divides p evenly, that is,

$$(\text{not}(p \text{ mod } 2 = 0)) \text{ and } (\text{not}(p \text{ mod } 3 = 0))$$
$$\text{and}\ldots\text{and } (\text{not}(p \text{ mod } (p-1) = 0))$$

or

```
not((p mod 2 = 0) or (p mod 3 = 0) or ...or (p mod (p-1) =0))
```

must be *true*. Thus, for p to be prime, it must be the case that there does not exist an x in {2..(p-1)} such that p mod x = 0. You can use the existential quantifier exists to express this condition in ISETL by

$$\text{not (exists x in } \{2..(p-1)\} \mid p \text{ mod } x = 0)$$

So, a third way of constructing the set of primes between 2 and 100 is

```
primes := {p :p in {2..100} |
              (not exists x in {2..(p-1)} | p mod x = 0)};
```

You can use your **set primes** to construct other **sets**. For example, suppose you wanted to find the **set** of all primes that are 1 more than a multiple of 4. To form this **set**, you want to examine each prime number p and test whether or not p-1 is a multiple of 4. That is, does 4 divide p-1 evenly? If it does, add p to the **set**, otherwise, look at another member in primes. So, using **set** former notation

```
Mult4Plus1 := {p :  p in primes | (p-1) mod 4 = 0};
```

Summary of Section 3.6

This section discusses using ISETL one-liners to do proofs and to find **sets** of prime numbers. We also talked about the quantifiers **exists** and **forall**—you'll be reading a lot more about these in Chapter 5.

One-liners let us write compact expressions for complicated statements. As you do the following problems, think like a mathematician and not like a programmer! That is, think *loop-test-evaluate* as you develop the **set** constructs but don't use any **for** loops.

Exercises

3.6.1 Write one-line ISETL expressions to construct the following **sets**:

   a. Assume S is a given **set** of **integers**. Construct the **set** consisting of the largest and the smallest elements in S without using the operations **max** and **min**.

   b. A *diphthong* is a double vowel. Write a program that will read a **string** and then construct the **set** of all diphthongs in the given **string**. For example, if the input to ISETL were "conscientious", then the output would be {"ie", "io", "ou"}.

3.6.2 A **tuple** of positive **integers**, [a, b, c], is a *Pythagorean triple* if and only if $a^2 + b^2 = c^2$. That is, a and b are sides and c is the hypotenuse of a right triangle. For example, [3,4,5] and [5,12,13] are Pythagorean triples.

   a. Write an ISETL statement to print the **set** of all Pythagorean triples where a and b are less than or equal to 30.

   b. Print the **set** of all Pythagorean triples where a and b are less than or equal to 30, eliminating all triples for which a > b. That is, include the triple [3,4,5] in your **set** but not the triple [4,3,5], since these two triples represent the same right triangle.

c. Notice that if $[a, b, c]$ is a Pythagorean triple then so is $[n \cdot a, n \cdot b, n \cdot c]$ for $n = 2, 3, \ldots$. Print the **set** of all Pythagorean triples where $a \le b \le 30$. Eliminate all triples for which a and b have a common divisor, since these are simple multiples of smaller **integer** right triangles. For example, include the triple $[3,4,5]$ in your **set** but not $[6,8,10]$, $[9,12,15]$, and so on.

3.6.3 Counting problems.

a. A student takes four courses for which he or she can receive the grades A, B, C, D, or F. Print the **set** representing all the possible grade distributions (i.e., the number of As, Bs, Cs, Ds, and Fs) the student might receive.

b. An urn has 3 red balls, 2 white balls, and 1 green ball. Print the **set** describing all possible outcomes if 3 balls are drawn in succession without returning the ball drawn. For example, drawing 2 red balls and 1 green ball is one possible outcome.

c. Ignoring suits, how many possible 5-card poker hands are there in a deck of 52 cards? Assume a deck of cards has 4 each of 1(ace), 2, 3,..., 10, 11(jack), 12(queen), and 13(king). Write an ISETL expression that constructs the **set** containing all possible 5-card poker hands. Use # to find its cardinality.

d. Ignoring suits and assuming ace is low, how many possible straights are there in a deck of 52 cards, where a straight is a sequence of 5 cards in a row? For example, a hand consisting of 3,4,5,6,7 is a straight. Write an ISETL expression that constructs the **set** of 5-card poker hands that are straights. Print its cardinality.

e. How many possible 5-card poker hands have exactly one pair? For example the hand 3,3,2,1,13 has exactly one pair. Write an ISETL expression that constructs the **set** of all 5-card poker hands that contain exactly one pair. Find its cardinality.

3.6.4 More on primes. Assume the **set primes100** of all primes between 2 and 100 has been constructed.

a. The pair $[p, p + 2]$ is said to be a *prime pair* if and only if both p and $p + 2$ are prime. Write an ISETL expression that constructs the **set** of all prime pairs up to 100.

b. The *Goldbach conjecture* asserts that every even integer greater than 2 can be written as the sum of two primes. Write a one-line ISETL statement that verifies the Goldbach conjecture for the first 50 positive even integers.

c. Assume x has been assigned a positive integer value less than 200. Write an ISETL expression that evaluates the following proposition:

The number of distinct primes that divide x is 2.

d. Use ISETL to construct the **set** of all **integers** between 2 and 100 that are the product of exactly 2 (not necessarily distinct) primes.

e. Use ISETL to construct the **set** of all positive **integers** less than 100 that are divisible by at least 2 distinct primes.

f. For a given **set** S of positive **integers**, construct the **set** that contains for each nonprime **integer** in S a **tuple** consisting of the **integer** and the **set** of all its prime factors. For instance, if

$$S = \{3, 2, 9, 16, 44, 30\}$$

then the output would be

$$\{[16, \{2\}], [44, \{2, 11\}], [30, \{5, 2, 3\}],$$
$$[9, \{3\}]\}.$$

3.6.5 Let N and M be fixed, positive integers.

a. Construct a **set** called **Zeroes** that contains all **strings** of 0s with length less than or equal to N, including the empty **string**. *Hint: Use the* **string** *operation* *.

b. Define the **set** **Ones** to be the **set** of all **strings** of 1s with length less than or equal to M, including the empty **string**.

c. Use the **sets** constructed in parts a. and b. to construct a **set** called **Words** that contains all the **strings** formed by concatenating a **string** with 0 to N 0s with a **string** containing 0 to M 1s. For example, if $N = 2$ and $M = 3$, then **Words** equals

$$\{"", "0", "00", "01", "001", "011", "0011", "0111",$$
$$"00111", "1","11", "111"\}$$

3.6.6 Using **sets** of propositions to do proofs.

a. Use a one-line ISETL **set** expression to determine whether or not each of the following **Boolean** expressions is a tautology:

i) (not p impl not q) = not (p impl q)

ii) (not (p impl (q and r)))
     = (p and (not q or not r))

iii) (p and (q or r)) = ((p and q) or r)

iv) (p impl q) = (not (q impl p))

    v) `((p impl q) and (q impl r)) = (p impl r)`

b. Construct an ISETL **set** expression that uses proof by enumeration to determine whether or not the following arguments are valid. Recall that, in general, if `hyp(p,q)` is the hypothesis of the argument and `conc(p,q)` is the conclusion, then the the argument is valid if and only if `hyp(p,q) impl conc(p,q)` is *true* for all values of **p** and **q**. Thus, the argument is valid if and only if the expression

```
{hyp(p,q) impl conc(p,q) :
                p,q in {true,false}} = {true};
```

evaluates to *true*.

    i) If it rains, then the picnic will be canceled and James will be grumpy. The picnic is not canceled. Therefore, James is not grumpy.

    ii) If the labor market is perfect, then the wages of persons with the same job will be equal. But, it is always the case that wages for such persons are not equal. Therefore, the labor market is not perfect.

c. Construct an ISETL **set** expression that determines whether or not the following arguments are valid, using a direct proof. Since the general idea underlying a direct proof is to show that the conclusion is true whenever the hypothesis of the argument is true, if the expression

```
{conc(p,q) : p,q in {true,false} | hyp(p,q)}
  = {true};
```

returns *true*, then the argument is valid. Otherwise, it is not.

    i) If it is before two o'clock, Joe can go visit Julie and take the bus home afterward. If it isn't earlier than two, then Joe doesn't have time to visit Julie and he must take the train. Joe visits Julie. Therefore, it is before two o'clock.

    ii) If GM located its new plant in Smalltown, PA, then property values in Smalltown would be rising and Smalltown's population would be increasing. Smalltown's property values are rising, but its population is decreasing. Therefore, GM did not locate its new plant in Smalltown.

d. Construct an ISETL **set** expression that simulates an indirect proof to determine whether or not the following arguments are valid. When using an indirect proof, you must show that the hypothesis of the argument is false whenever the conclusion is false. Therefore, you can consider the value of the expression

```
{hyp(p,q) :   p,q in {true,false} |
                        not conc(p,q)} = {false};
```

If the expression evaluates to {*true*}, then the argument is valid; otherwise, it is not.

i) If investments fail to increase each period, then income will decline several periods later. But, investments are higher this period than last. Therefore, no new decline is indicated.

ii) If a country is poor, then it cannot devote much of its time to technological development. Moreover, if a country cannot devote much of its time to technological development, then its income will not change. Therefore, if a country is poor, then its income will not change.

iii) If Dad praises me, then I can be proud of myself. Either I do well in sports or I cannot be proud of myself. If I study hard, then I cannot do well in sports. Therefore, if Dad praises me, then I do not study hard.

3.6.7 In proving relationships between set expressions such as the equivalences in Exercise 3.3.4, we have insisted that the equality be established for every possible set that could appear in the expression. Thus, in showing, for example, that

$$(S \cup T) \cup R = S \cup (T \cup R)$$

it is necessary to prove that this equality holds for every possible set S, T, and R. What would happen if we only checked the relation for two special values $\mathcal{U}$ (the universal set), and $\emptyset$ (the empty set)? Consider this for every equality in Exercise 3.3.4.

3.6.8 Suppose n and m are positive integers with assigned values. Write a one-line ISETL statement that evaluates to *true* if n and m are relatively prime (that is, they have no common divisors) and *false* otherwise.

3.7   The Compound Operator %

Up to this point, we have talked about binary operators, such as the union of two sets, the conjunction of two propositions, or the sum of two integers. But, what if you want to apply a binary operator again and again? For instance, what if you wanted to add together the members of the **set**

```
{i**2 :   i in {1..100}}
```

or find the product of the components in the **tuple**

$$[j**3 - j : j \text{ in } \{-3..4\}]$$

or find the union of the **sets** that are elements of the **set**

$$\{\{1..k\} : k \text{ in } [1..n]\}$$

where **n** has an assigned value? That is, what if you wanted to find

$$1^2 + 2^2 + 3^2 + \cdots + 100^2$$

$$((-3)^3 - (-3)) \cdot ((-2)^3 - (-2)) \cdots (4^3 - 4)$$

or

$$\{1\} \cup \{1..2\} \cup \cdots \cup \{1..n\}$$

without actually writing out every term?

You can do this in ISETL by using the compound operator **%**, which indicates that a specified binary operation is to be repeatedly applied to the members of a **set** or a **tuple**. The general form for applying the compound operator is

% binary_op S

where **binary_op** is an ISETL binary operator and **S** is a **set** or a **tuple** containing the items to which the specified operator is to be applied. As usual, the difference between running an operation over a **set** and an operation over a **tuple** is one of order—an operation is applied to the elements of a **set** without regard for order, while it is applied to the components of a **tuple** in the order in which they occur. If the operator is not commutative, that is, if **a** .**binary_op** **b** is not equal to **b** .**binary_op** **a** for all values of **a** and **b**, then, of course, the order that the operator is applied can make a difference. All our examples above, however, concern operators that are commutative, so we can write them in ISETL (where we have used a mixture of **sets** and **tuples**) as follows:

```
%+ {i**2 :  i in {1..100}};
%* [j**3 - j :  j in {-3..4}];
%+ {{1..k} :  k in [1..n]};
```

where **n** has an assigned value.

Let's think about how ISETL might evaluate expressions such as these. In the first expression, for example, the first step in the process might be to construct the **set** {i**2 : i in {1..100}} by looping through **integers** in {1..100}, evaluating the expression **i**2**, and adjoining it to the **set**. Then, the second step would be to apply the operation of **integer** addition to the members in this **set** (in some order), that is, add **1** to **4**, add **9** to the result, and so on.

We can do all this using mathematical notation, too. In the case of the above examples, you can denote the sum of the members in the first set by

$$\sum_{i=1}^{100} i^2$$

the product of the components of the **tuple** by

$$\prod_{j=-3}^{4} (j^3 - j)$$

or, letting $S_k = \{1..k\}$, the union of the given set of sets by

$$\bigcup_{k=1}^{n} S_k$$

As you examine each of these mathematical expressions, think about the corresponding ISETL expressions and imagine how ISETL evaluates each one.

Let's look at an application of the compound operator. Suppose G is a **tuple** that contains all your grades for this course. So, assuming things are going well for you, G might be something like [90, 97, 88, 92]. What if you wanted to find your average grade? The idea, of course, is to add up the items in G and divide by the total number of items. You can convert this algorithm directly to ISETL by writing

$$(\%+ \ G) \ / \ \#G;$$

How could you modify this expression to find the average with the lowest grade dropped? How about the average of your first three grades? We'll give you a chance to think about these questions in the exercises.

As a final example, let's apply the **Boolean** operators **and** and **or** to a given **tuple** of propositions. Suppose you entered the following lines:

```
EvenSet := {2, 4..10};
PropTup := [t < 6 :  t in EvenSet];
%and PropTup;
```

How does ISETL evaluate the expression %and PropTup? What does ISETL return? Since this is a **Boolean** expression, you know that the value of %and PropTup must be *true* or *false*. Furthermore, based on the definition of conjunction, you know that %and PropTup is *true* if and only if every member of PropTup is *true*, that is, if and only if for all t in EvenSet, t is less than 6; otherwise, %and PropTup is *false*. So, %and PropTup is *false*. But what about the value of %or PropTup?

Summary of Section 3.7

Sometimes you want to apply a binary operator repeatedly. You can do this in ISETL by using the compound operator % followed by the binary operator itself. In Mathematics, there is usually a special symbol corresponding to a particular operation, such as the summation sign for repeated addition.

Exercises

3.7.1 Rewrite the following expressions:

1. As a list with the binary operation repeatedly applied.

2. Using mathematical notation, if special notation exists.

For example, given %+ [i : i in {1..100}] you might write

$$1 + 2 + 3 + \cdots + 100$$

as the answer to part 1. and

$$\sum_{i=1}^{100} i$$

as the answer to 2.

    a. %+ [i**2 : i in [-2..4]];

    b. %+ [5..25];

    c. %+ {2, 4..100};

    d. %+ {{n-1..n+1} : n in {-5..5}};

    e. %* {k : k in {1..n}};
       (where n is assumed to have a value)

    f. %* {1, 3.. 59};

    g. %* [{1..k} : k in {1..n}];
       (where n is assumed to have a value)

    h. %+ T;
       (where T is a **tuple** of **integers** with an assigned value)

    i. %* T(5..10);
       (where T is a given tuple of **integers** and #T >= 10)

    j. %and [x in S : x in [1..100]];
       (where the **set** S is assumed to have a value)

    k. %or {abs(r)<5 : r in {-11, -8..11}};

    l. `%max Temps;`
       (where `Temps = {-23.4, 0.0, 56.9, 23.5, 88.2}`)

   m. `%min {100, 90..1};`

3.7.2 Suppose $S_k = \{k \ldots 2 \cdot k\}$, for $k = 1, 2, \ldots$. We say that the set of all S_k is an indexed collection of sets. Find the values of the following expressions:

    a. $\bigcup_{k=1}^{5} S_k$

    b. $\bigcap_{j=5}^{10} S_j$

    c. `%- [{k..2*k} : k in [10, 9..1]];`

    d. `%- [{k..2*k} : k in [1..10]];`

    e. Compare the answers you got for parts c. and d. Why aren't they the same?

    f. Explain why the following expression is not necessarily *true*:

```
%- [{k..2*k} :   k in {1..5}]
= %- [{k..2*k} :   k in [1..5]];
```

3.7.3 Suppose `G` is a **tuple** of **integers** that contains a list of grades. Give an ISETL expression whose value is

    a. the highest grade.

    b. the lowest grade.

    c. the average with the lowest grade dropped.

    d. the average with the highest grade included twice.

    e. the average of the highest and lowest grades.

    f. the sum of all the grades that are greater than 75.

3.7.4 Solve each of the following problems by defining an appropriate ISETL **func**. In each case, the body of the **func** should consist simply of a **set** or **tuple** former expression.

    a. Define a **func** that accepts a **tuple** of **integers** I and returns a **tuple** containing all the *partial sums* of I. That is, the *j*th component of the **tuple** returned by the **func** is the sum of the first through the *j*th components of I.

    b. Noting that an item can occur in a **tuple** more than one time, define a **func** that accepts a **tuple** of **integers** t and returns the **set** containing the indices of the largest item in t.

c. Given a vector $v = (v_1, v_2, \ldots, v_n)$, the *magnitude* of v, which is denoted by $|v|$, is the square root of the sum of the squares of the components of v. That is,

$$|v| = \sqrt{v_1^2 + v_2^2 + \cdots + v_n^2}$$

Define a func that accepts a floating-point vector (stored as a tuple) and returns its magnitude.

3.7.5 Consider the tuple of Boolean values

```
Boolean_Tup := [x < 10 :   x in S];
```

What condition(s) must the members of the set S satisfy to guarantee that

a. %and Boolean_Tup is *true*?

b. %and Boolean_Tup is *false*?

c. %or Boolean_Tup is *true*?

d. %or Boolean_Tup is *false*?

3.7.6 More about the compound operator %.

a. Initialization: to give an expression containing % an initial value, we can use the general form

```
x %binary_op S
```

where S is a set or a tuple. If #S = 0, then the value of this expression is x; otherwise, assuming that the elements in S are x1,...,xn, the value is:

```
x binary_op x1 binary_op ... binary_op xn
```

associating to the left. Find the value of:

i) 5 %+ [1..10];

ii) 5 %max {};

iii) %* [];

iv) {"a","b","c"} %+ {{n..n**2} : n in {0..5}};

b. User defined funcs of two variables: the compound operator can be used not only with a predefined binary operator but also with a user-defined func of two variables as long as the value the func returns is suitable as a left argument. (The common case is where the two parameters and the result are the same type.)

For example, consider the ISETL terminal session

```
>  plus := func (x,y); return x + y; end;
>  plus (4, 5);
9;
>  %plus [1.2, 3.4, -7.89, 100.0];
96.71;
>  0 %plus {};
0;
>  !quit
```

Define a **func** that accepts two values T1 and T2 and returns a 2-tuple whose first component is T1 and second component is T2. Use the compound operator to apply your **func** repeatedly. Describe the result in English.

Chapter 4

Functions

To the Instructor

The purpose of this chapter is to help the student develop her or his concept of function. Early in a young person's study of Mathematics, he or she may think of a function as nothing more than an expression or formula that sits, statically, on a piece of paper or on the blackboard. This is not a very powerful notion, and it is essential that the student develop a dynamic view of a function as a process that takes an object of some sort and transforms it into another object.

Psychologically, this means that the student must acquire the ability to respond to the description of a function by *re-presenting* (or, for short, *representing*) this description as an internal mental process. The process acts on objects that have been constructed in the mind of the student. They can be representations of physical objects, such as triangles or words in an English text, or mathematical objects, such as real or **floating-point** numbers. The objects can even be functions that, although originally understood as processes, have been encapsulated by the student so as to become mental objects.

Beyond thinking about the process of a particular function, the student should begin to think about this process in general, independently of specific examples. The idea of the process should be sufficiently robust that the student can imagine composing two processes or running one in reverse—to obtain the inverse function. Again, these should be understood initially with examples and eventually as abstract concepts. Finally, all this should exist both in terms of the function process idea, which is dynamic, and the ordered pairs point of view which is static. The student should be comfortable in going back and forth from one to the other.

Our intention in Section 4.2 of this chapter is that by studying various ISETL representations of the function process, the student will first develop the ability to construct the process for a function on the com-

163

puter and, as a result, begin to make corresponding constructions in her or his mind. Learning the syntax for the different ISETL data types that represent functions; going back and forth between verbal descriptions and computer representations of functions; thinking carefully about domains, ranges, and images; evaluating functions at single points or sets of points—all of these are designed to get the student to construct mental processes corresponding to individual functions.

In Section 4.3, we expect the student to use this new foundation for the function concept to construct the standard operations with functions that will be extremely useful in further study of Mathematics. At the end of that section, we also do a few things that may start the student on developing the ability to encapsulate the process of a function to make it a mental object.

Finally, in Section 4.4, we offer a sampling of applications of functions. We hope that the student will derive pleasure from a new ability to understand some nontrivial ideas and will have experiences of which future work may provide echoes of familiarity.

4.1 Preview

A function is more than an expression. An expression just sits there, waiting for something to happen. A function is a "doer." It makes things happen. It takes numbers or other kinds of objects and changes them. It is a process.

In this chapter, you will work on your understanding of functions. For you to understand a function, you have to construct some mental representation of its process. For a computer to "understand" a function, someone has to construct something like a program that corresponds to the function. The most essential thing for our purposes is that the activity of figuring out representations of functions on a computer (which is done consciously) helps you represent them mentally (which is usually done with less awareness), and this will make it easier for you to work with functions. Also, the specific ways in which you use functions in writing programs will affect the nature of the functions that live inside your mind. Therefore, in order to learn to use your head to work with functions, you begin by using the computer to represent them and to manipulate them.

Once you develop the ability to "construct functions in your head," a lot of the ideas surrounding the function concept will become much easier for you. Evaluating functions, images, and pre-images; adding and multiplying functions; and composing functions, the inverse function, and functions of several variables; all of these are very important in Mathematics. Because ISETL has several powerful constructs for representing functions, these and other operations with functions can be implemented conveniently on the computer. You will write ISETL code to make these implementations and as a result you will see how they work and have little difficulty in understanding them. The main thing to remember as you work with these ideas is to always think about the process of a function or functions you are working with. Try to imagine the processes and even run through them in your mind.

Finally, we hope you will enjoy working with some interesting applications at the end of the chapter. These will give you an indication of how useful functions can be and how they can often make very complicated situations more clear and more easy to work with.

4.2 Representations of Functions

You have already been working quite a bit with functions in ISETL. ISETL programs, funcs, tuples, and strings are all ways of representing functions in the computer. We will talk about these ISETL representations in some detail and also discuss a new way of doing it—with smaps.

4.2.1 Programs

Here is an example, similar to one you worked with in Chapter 1, of an ISETL program.

```
program rel_prim;
      rp := om;
      read x;
      read y;
      if is_integer(x) and is_integer(y)
                       and x>0 and y>0 then
          rp := true;
          for i in [2..(x .min y)] do
              if (x mod i = 0) and (y mod i = 0) then
                      rp := false;
              end;
          end;
      end;
      file_id := openw("rp.file");
      print x, y, rp to file_id;
      close(file_id);
  end;
```

An ISETL program has three parts: header, body, and end. The header is a single phrase beginning with the keyword program, followed by an identifier, which is the name of the program, followed by a semicolon. The body is a list of statements. The end is a single phrase consisting of the keyword end followed by a semicolon.

A **program** will generally have **read** statements to obtain input data and **print** statements to present output data. A **program** may be thought of as a *process* that transforms input data into output data. Thus, in this example, there are two pieces of input data, **x** and **y**. The **program** transforms them into three pieces of output data, **x**, **y**, and **rp**. The first two repeat the input data and the third is a value (*true*, *false*, or **OM**), determined by the **program**.

Strictly speaking, a **program** will often have additional "output" that is not explicitly contained in a **print** statement. Anything in the entire environment whose value can be affected by the **program** is really output. Thus, in this example, the values of **x**, **y**, **file_id**, and **rp** are affected, not to mention the external file, **rp.file**, and various internal registers. All these are referred to as *side effects*. In thinking of a **program** as a function, we normally consider only the explicit input data that the **program** transforms explicitly to output data.

A **program** will not necessarily accept every value for its input data. Indeed, in the above example of **rel_prim**, if **x** or **y** has other than positive **integer** values, then **rp** will be **OM**. Moreover, if the **is_integer** tests were not present and other than integer values were given as input data, then a run-time error would occur. The set of all possible input values for which there will be no error message and the **program** will compute output values different from **OM** is referred to as the *domain* of the function that the **program** represents. In the case of **rel_prim**, the domain is the set of all pairs of positive integers.

One can also discuss the *image* of a function. This is the set of all possible output values. If a function is represented by a **program**, then its image is determined by its domain and the specified computations. So, in principle (although it can sometimes be difficult in practice), you can figure it out. In the case of **rel_prim**, the image is the set of all triples whose first two components are positive **integers** and whose third component is, based on the first two, *true* or *false*.

Sometimes you will see the term *range* used where we have used image. There is a slight difference. The range of a function is any set that includes all values that the function will produce. It may include values that the function will not produce.

The range should be given as part of the specification of the function. This is for functions that are in your mind. (If the range is not specified explicitly, then it is usually taken to be equal to the image.) When you represent a function in ISETL as a **program**, there is no way to use the statements to represent its range, so it *must* be taken to be equal to the image. It is possible to attach a comment that will tell another person what you intend the range to be, but ISETL will not pay attention to it.

In Chapter 1, we introduced a special notation for a function with its domain and range. If F is a function (for example, the one represented by rel_prim), D is its domain set, and R its range set, then we write

$$F : D \longrightarrow R$$

With this notation, we represent the image of the function as $F(D)$, read "F of D." For example, suppose we set up (in Mathematics, we say "define") a function by taking the domain to be the set of all positive floating-point numbers, the range to be the set of all floating-point numbers, and the process to be the act of taking the square root. In Mathematics, we would write this as follows.

$$
\begin{aligned}
D &= \{\text{all positive floating-point numbers}\} \\
R &= \{\text{all floating-point numbers}\} \\
F &: D \longrightarrow R \quad \text{by } F(x) = \sqrt{x}
\end{aligned}
$$

Then, the image, $F(D)$, of this function is the set of all *positive* floating-point numbers, which is different from its range, the larger set of all floating-point numbers, positive or not.

On the other hand, if we represent this function as a program, it would look like the following:

```
program sqrt_function;
        read x;
        if is_floating(x) and x > 0 then
            print sqrt(x);
        end;
end;
```

The domain is represented in this function by the test for positive floating-point numbers (it could have been nonnegative, which would have given a different function), and the image of the function is represented in the sense that with some thought one can figure out what it is; but there is no way that a program can represent the range of the function.

To summarize, the act of taking a description of a function, together with its domain and representing it as a program is what you do when you "write a program to implement a function." On the other hand, reading a program to figure out what values it will accept and what it will do with them, is to convert the *representation* of the function as a program to a mental construction corresponding to the function. The domain and image are carried along in both cases, but the range is not involved in these conversions. Sometimes, in discussing such things informally, we speak of the program and the function as if they were the same. It is all right to do this as a matter of convenience, but be sure you are capable of separating the two ideas if necessary.

4.2.2   Funcs

The syntax for an ISETL `func` is very similar to the syntax for a `program`.
It can be described as follows.

```
func (list of parameters);
     local list-of-local-ids;
     value list-of-global-ids;
     statements;
end
```

The list of parameters can be empty, but the pair of parentheses must be
present. The `local` declaration, which is optional, declares the identifiers
in the list that follows it to be local to the `func`. Thus, a name that appears
outside the scope of a `func` can be declared `local` and used within this
scope and the two names will be treated as different variables. The `value`
declaration, which is also optional, declares that the identifiers in the list
are global and that the value used for these identifiers will be their value
at the time of definition, as long as they are not modified within the body
of the `func`. Any identifier that is neither `local` nor `value` will have its
global value anytime the `func` is executed. (See page 172 for an example
of the use of the `value` declaration.)

An important statement used in a `func` is `return` *expression*. Its effect,
when a `func` is being executed, is to evaluate the expression, terminate
execution of the `func`, and return this value. A `func` should always have
at least one `return` statement. If execution comes to the `end` of the `func`
without having encountered a `return` statement or if there is no expression
after the `return` statement, then the value returned by the `func` will be OM.
You should not allow a `func` to have any side effects. Thus, in principle,
execution of a `func` should return a value or OM and not change any global
variables.

It is important to make a distinction between the *value* of a `func` and
the *result of execution* of a `func`. A `func` is a representation in ISETL
of a function. As we saw in the previous section, a function is a process
that transforms input data to output data. To execute a `func`, that is, to
perform this process, the `func` must be called and actual values given to
each of its parameters. The *value* of a `func`, on the other hand, is this
process itself. In order to execute a `func`, its *value* can first be assigned
to some variable, say F. Then, an expression that consists of F followed by
a list, enclosed in parentheses, of actual values of the parameters for the
`func` will have as *its* value the result of execution of the `func` using the
given values for its parameters.

Thus, the value of a `func` is a process that is performed when the `func`
is called. The result of calling a `func` is the quantity evaluated in the
`return` statement that terminates the `func` or OM if it terminates without
encountering a `return` statement. You can think of the act of assigning

(the value of) a **func** to an identifier as "teaching that identifier how to execute the **func**," or you can think of it as storing the representation of a function, which has been given as a list of statements in a **func**, in a single identifier.

This is one difference between a **program** and a **func** in ISETL. A **func** is both a *process* that can be executed and an *object* on which operations can be performed. A **program** is only a process and cannot be manipulated.

Let's consider an example. Suppose you were thinking of a function G whose process applies to real numbers and transforms the number x into the value of the following expression, if x is positive,

$$\frac{x^2 + x - 2}{x - 1}$$

and into the fixed number 2, if x is less than or equal to 0. Clearly, this process cannot be applied to the number 1 so we would have to take the domain A of this function to be, for example, the set of real numbers other than 1. We could take its range to be the set of all real numbers, which we might call B. (Notice that this is different from the image. What is the image of this function?) Then, we would write, in Mathematics,

$$G : A \longrightarrow B \ \text{ by } \ G(x) = \begin{cases} \frac{x^2 + x - 2}{x - 1} & \text{if } x > 0 \text{ and } x \neq 1 \\ 2 & \text{if } x \leq 0 \end{cases}$$

We can convert this representation of our function into ISETL, but the domain will have to be cut down from the set of reals to the set of **floating-point** numbers.

```
func(x);
    if is_floating(x) and x /= 1.0 then
        if x>0 then
                return (x**2+x-2.0)/(x-1.0);
        else
                return 2.0;
        end;
    end;
end
```

This code is an *expression* whose value is a *process*. If you entered this expression in ISETL as is and followed it with a semicolon, then ISETL would evaluate it and return some symbol that corresponded to its internal representation of this process, such as !FUNC#365!;. This would not be of much value to you. It is the same as what would happen if you entered an expression corresponding to the sum of the first 12 odd integers,

1+3+5+7+9+11+13+15+17+19+21+23;

or

```
%+[1,3..23];
```

Either of these would return **144**; however, as in the case of the above
func, *you* could look at the value, but ISETL could not do anything with
it. It is true that the two situations differ in that **144** is meaningful to you
while !FUNC#365! is not, but this has to do with you, not ISETL. ISETL
cannot multiply the **integer** by 2 or evaluate the function or do anything
with either of them.

In order to make an ISETL value such as an **integer** or a **func** available
to ISETL for combination with other values, it must be part of an expression or stored somehow—in a **set** or as the value of a variable. Thus, if we
enter

```
y := %+[1,3..23];
```

then the value has been assigned to **y**, and if we enter

```
3*y+2;
```

ISETL will print the value

```
434;
```

Similarly, if we enter

```
G := func(x);
        if is_floating(x) and x /= 1.0 then
        if x>0 then
                return (x**2+x-2.0)/(x-1.0);
        else
                return 2.0;
        end;
    end;
end;
```

then the value of the **func** has been assigned to **G**, and the result of entering

```
G(2.5); 0.5*G(3.0)-G(-4.0); G(true);
```

will be

```
4.5;
0.5;
OM;
```

There are other things you can do with **funcs**. For example, you can put
one or more of them into a **tuple** as in the following code.

```
fn_list := [
        func(n);
            if is_integer(n) and n>=0 then
                return 2**n > n**2+2*n-2;
            end;
        end, $ End of first func
        func(n);
            if is_integer(n) and n>=0 then
                return (2**n > n**2+2*n-2) impl
                       (2**(n+1) > n**2+4*n+1);
            end;
        end
    ];
```

This one is pretty complicated. Before we explain it, try to understand for yourself what is going on. For example, after the above lines have been entered, can you figure out what would be the value of the following expressions?

```
fn_list(1)(6); fn_list(2)(1);
not(fn_list(2)(5) or fn_list(1)(3));
```

Think about it. Try it on the computer. Any effort you spend on your own trying to understand will be a big help when you are trying to follow the explanation.

Here is the explanation. fn_list is a tuple with two components, each of which is a func that represents a function. (N.B.: Semicolons follow statements, and the funcs in the tuple are expressions; therefore there are no semicolons after the end of the funcs.) In both components, the domain of the function is the set of nonnegative integers and the range is the two-element set {*true, false*}. The first component represents a function that transforms the integer n into the value (*true* or *false*) of the assertion

$$2^n > n^2 + 2n - 2$$

The second component represents a function that transforms the integer n into the value of a more complicated assertion. This one is an implication between two inequalities, so the value will be *false* if the hypothesis

$$2^n > n^2 + 2n - 2$$

is *true*, while the conclusion

$$2^{n+1} > n^2 + 4n + 1$$

is *false*; otherwise the value will be *true*.

Thus, an expression like fn_list(1) has for its value the first of these funcs, and the value of fn_list(1)(6) is the result of evaluating that func with the actual parameter n=6, which gives the result *true*. Similarly, the value of fn_list(2)(1) is the result of the second func evaluated with n=1, which gives *false*. Finally, the third expression,

not(fn_list(2)(5) or fn_list(1)(3));

is evaluated by first evaluating the second func for n=5 to obtain *true* and then the first func at n=3, which gives *false*, and then calculating

not(true or false)

which gives the result *false*.

Any difficulty of thinking about these things is worth struggling with because once you have achieved some understanding you will have a powerful tool for thinking and calculating. Suppose, for example, that you wanted to study the properties of all (or at least many) of the functions $F_1, F_2, F_3, \ldots$ each of which is defined on the unit interval $[0, 1]$ of real numbers and whose process transforms the real number x into

$$F_n(x) = x^n - 1/n$$

You could use ISETL to work with 100 of these functions, for example, by constructing the tuple

```
F := [func(x); value n; return x**n-1/n; end
    :  n in [1..100] ];
```

Then, F(n) represents F_n, and for each **floating-point** number in $[0, 1]$, F(n)(x) has the same value as $F_n(x)$. Notice that we have been a little sloppy in omitting the test for x being in the domain of this function. This means we will have to be a little careful in using the tuple of funcs.

Now, suppose we wanted to find the **set** of all values that these functions produce when $x = 1/2$. We can simply write

{f(0.5) : f in F};

Or suppose you wanted to find all pairs $[F_j, F_k]$ of these functions such that j and k differ by at least 10, but the difference of the values of the two functions at 3/7 was less than 0.0001. This can be translated almost directly into ISETL as follows:

```
{[F(j), F(k)]:  j, k in [1..100] | abs(j - k) >= 10.0
                and abs(F(j)(3/7) - F(k)(3/7)) < 0.0001};
```

As a simple exercise, notice that this expression will produce both [F(j), F(k)] and [F(k), F(j)]. How could you change the expression so that only one of them will appear?

4.2.3   Tuples and Strings

The **tuple**, which you studied in Chapter 3, is another data structure that can be used to represent functions in ISETL. The domain of any function represented by a **tuple** is most naturally taken to be a finite set of positive integers—the integers corresponding to the indices of the non-**OM** components of the **tuple**. Thus,

$$[5,2..-15]$$

represents a function that we can describe mathematically as follows.

$$D = \{1,2,3,4,5,6,7\}, \quad R = \text{the set of integers}$$

$$F : D \longrightarrow R \text{ by } F(x) = 8 - 3x$$

The image of any function represented by a **tuple** is the set of values that occur in the non-**OM** components. As with **programs** and **funcs**, there is no way that a **tuple** can represent the range of a function.

You can also use **tuples** to help you think about a function whose domain is the set of *all* positive integers. Such a function is called a *sequence* in Mathematics. Expressions representing the values returned by execution of the function are called *terms*. Of course, you can never represent an entire sequence in ISETL with a **tuple**, but, given the fact that the length of a **tuple** can always be increased just by assigning values to more components, you might think of a **tuple** as representing "part" of a sequence; or, you might think of a sequence for which you only know finitely many of its values, but you have the ability to calculate any additional terms that you want to know. Then, the **tuple** represents the terms of the sequence that you know.

For example, suppose you were thinking about a particular sequence whose nth term is the set of all positive integers less than or equal to n. What is the intersection of all terms of this sequence from the nth term on?

You can begin to answer this question by thinking about a **tuple** in which the first component is the set $\{1\}$, the second component is the set $\{1,2\}$, the third component is the set $\{1,2,3\}$, and so on. You might write

```
T := [{1},{1,2},{1,2,3},{1,2,3,4},{1,2,3,4,5},{1,2,3,4,5,6}];
```

In ISETL you have to stop, but in your mind you can go on forever. Thus, you can easily check that the following expressions evaluate to *true* in ISETL (that is, the equalities hold):

```
%*T = T(1);
(%*[T(i):  i in [2..#T]] = %*T(2..))  and (%*T(2..)  = T(2));
(%*[T(i):  i in [3..#T]] = %*T(3..))  and (%*T(3..)  = T(3));
(%*[T(i):  i in [4..#T]] = %*T(4..))  and (%*T(4..)  = T(4));
```

In other words, this is a **tuple** of **sets** with the property that, if you take the intersection of all the components from the nth one on, the result is the **set** which is the same as the nth component—because all of the other **sets** involved in the intersection contain that one.

After thinking through this situation, which, although not exactly the same as the given problem, is concrete and understandable, it is easy to see that the same thing is the answer to the original question.

Sequences of numbers (integers, real, complex) are very important in Mathematics, and there are a few simple ideas that are worth mentioning because it is easy to think about them in terms of **tuples**. In the next few paragraphs, we will speak loosely and talk about **tuples** as if they were the same as sequences. Conceptually, a sequence is a **tuple** with infinite length, and so, unless the length is mentioned, there really is no difference. The value of the ith component of a **tuple, t,** is written in ISETL as **t(i)**. In Mathematics, we speak of the ith term and write t_i. The entire sequence is often written as $(t_i)_i$.

Let $x = (x_i)_i$ and $y = (y_i)_i$ be sequences that we will also think of as **tuples**, X and Y, respectively. That is,

$$x_i = \texttt{X(i)}, \quad y_i = \texttt{Y(i)}$$

Then, we define the *coordinate-wise sum* and *product* of the two sequences as

$x + y = (x_i + y_i)_i$ or   X .cplus Y
                              = [X(i)+Y(i):i in [1..min(#X,#Y)]]

$x \cdot y = (x_i y_i)_i$ or   X .ctime Y
                              = [X(i)*Y(i):i in [1..min(#X,#Y)]]

where **cplus** and **ctime** are the names of two **funcs** that have been defined in ISETL by a user.

If c is a number, then we define the *scalar product* of c and x by

$c \cdot x = (cx_i)_i$ or   c .sc_prod X = [c*X(i) : i in [1..#X]]

We say that the sequence x is *increasing* if

$x_i \leq x_{i+1}\ i = 1, 2, \ldots$ or   forall i in [1..#X-1] | X(i) <= X(i+1)

In Chapter 7, where we will study mathematical induction, we will work a lot with *sequences of propositions*, that is, sequences that, as functions, have for their range the two-element set $\{true, false\}$. A statement such as

$$2^n > n^2 + 2n - 2, \quad n = 1, 2, \cdots$$

or

Using only \$3 bills and \$5 bills, it is possible
to make change for any amount of dollars.

can be thought of as a *Boolean* or *proposition valued function of the positive integers* and (a finite part of it) can be represented in ISETL as a `tuple` of Boolean values. The value of such a function, for a given positive integer n, is the truth or falsity of the statement when the value of n is plugged in.

Strings in ISETL are very similar to `tuples`, and they can be used to represent a function whose domain is a finite set of consecutive positive integers, or, in the spirit of the previous few paragraphs, a sequence. The value of a component of a `string` must be a `string` of length 1—that is, a character—and so a function represented by a `string` must have for its range a set of characters. This is the only place in ISETL where it is possible to represent the range of a function as something other than its image.

4.2.4 Smaps

The last construct for representing functions in ISETL is what we refer to as an `smap` (single-valued map).[1] Suppose you had a `set` called `cells` in ISETL, and you wanted to linearly order its elements, string them out, one after the other in some arbitrary order. Conceptually, you can imagine doing this with a function whose domain and range are both the `set cells` and whose process acts on a cell (a natural name for an element of `cells`) and transforms it to its immediate successor. If we called this function *next*, then the value of *next*(*c*) would be the successor of *c*. Moreover, the last cell in the string would be the one for which *next* is not defined (so we have to adjust the domain to exclude this one), and the first cell would be the one that was in the domain of *next* but not in its image.

To fully appreciate the power of the ISETL construct for representing such a function, it is important to understand that the process of this kind of function is not very dynamic nor is there necessarily any computational way of determining the value of *next*(*c*)—it could be completely arbitrary. The only thing that is going on is that to each element in the domain there has been assigned (in a perhaps arbitrary way) a single element in the range. This information can be recorded by simply making a list or table. Each entry in the list would have two items—a domain element and a range element. Moreover, each domain element must appear once and only once. Finally, we note that the order of the entries in the list carries no information about the function, and so it can be ignored.

Well, by now you must begin to see that what we have been describing can easily be implemented in ISETL. For the entries in the list, we take `tuples`, whose first two components and no others are defined—these are called *pairs*—and for the list, we take the `set` of all the pairs. This object is called a `map` in ISETL, and it will be important in Chapter 8, where we discuss relations. In this chapter, however, we add an additional

---

[1] Pronounced es-map.

requirement—that each element of the domain must appear exactly once
as the first component of one of the **tuples**. This guarantees that there is
a definite answer to the question: Given something that appears as a first
component in some **tuple**, what is *the* value that appears in the second
component? A **map** that satisfies this criterion is called an **smap**, and it has
the property that if **id** is an identifier whose value is this **set** of pairs then

$$id(x);$$

will return the value **y**, provided the **set** contains a pair of the form [x,y],
that is, the value is the second component of a **tuple** whose first component
is **x**. It has the value **OM** if **x** does not appear as the first component of any
tuple in **id**.

   Both the domain and image of a function represented by an **smap** are
very easy to determine. For the domain, you just take the set of first
components of the **tuples** in the **smap** (which, remember, is a **set**) and
for the image you take the second components. In fact, there are ISETL
functions **domain** and **image** that do just that. If the value of **id** is
an **smap**, then **domain(id)** returns its domain and **image(id)** returns its
image.

   To illustrate these ideas, let's consider a simple example. Suppose S is
the following set of ordered pairs.

$$\{["a",9], [6,"hi"], [2.3,9]\};$$

Then S is an **smap**, since the first component in each ordered pair in S,
for example "a", does not appear as the first component of any other pair
in S. Thus, S represents a function with domain $\{"a", 6, 2.3\}$ and image
$\{9, "hi"\}$. Notice how static the process of this particular function is—it
assigns to each element in its domain an apparently arbitrary element in
the range, but the **smap** representation does not give us any indication as
to how this process occurs. Now look carefully at the following ISETL
terminal session and try to describe what is happening.

```
> S := {["a",9], [6,"hi"], [2.3,9]};    S;
{[6,"hi"], [2.3,9], ["a",9]};
> S(6);
"hi";
> S("b");
OM;
> domain(S);
{"a", 2.3, 6};
> image(S);
{"hi", 9};
> S("b") := 2.3;
> S;    domain(S);    image(S);
{["a",9], ["b",2.3], [6,"hi"], [2.3,9]};
```

{"a", "b", 6, 2.3};
{2.3, "hi", 9};

Notice that "b" was not contained in the original domain of S. By making the assignment

$$S("b") := 2.3;$$

we *extended* the domain of S to include "b". If, on the other hand, you wanted to *restrict* the domain of S by removing for instance "a" from the domain, you could do this with the assignment

$$S("a") := om;$$

Now let's return to our initial example of linearly ordering a **set**. Suppose we had a **set** that consisted of the first ten words of the third paragraph in this subsection. Thus, we might execute the following ISETL statement.

```
words := {"well", "by", "now", "you", "must", "begin", "to",
          "see", "that", "what"};
```

One way to store these words in alphabetical order would be to put them in a **tuple** rearranging them so that the component order was the alphabetical order. Another way would be to construct the following **smap**.

```
next := {["well","what"], ["by","must"], ["now","see"],
         ["must","now"], ["begin","by"], ["to","well"],
         ["see","that"], ["that","to"], ["what","you"]};
```

Then, for example, the value of next("to") is "well" and next("you") has the value OM, which tells us that "you" is the last word in this ordering.

You might think that using a **tuple** to represent the ordering is simpler than using a function represented by an **smap**. The **tuple** certainly takes up less space and is simpler to read and write. But, suppose a new word came in, for example, "we", and you wanted to insert it in the list between "to" and "well". With a **tuple**, you would have to move every word from "well" on down one component before placing "we" in the component vacated by "well". This can be done easily in ISETL but could take a long time to run if the list were very long.

Since with an **smap**, there is the possibility (as with **tuples** and **strings**) to make an assignment of a single value, we can make the insertion with just two lines of ISETL code, no matter how large the **set** is, and we only need to know that "we" is to follow "to"—we don't use the fact that "we" is to come before "well":

```
next("we") := next("to");
next("to") := "we";
```

This very simple solution to the insertion problem is one example of the power of the function concept. There are many others.

Summary of Section 4.2

In this section, we discussed five different ISETL constructs for representing a function: `program`, `func`, `tuple`, `string`, and `smap`. There is also the *mathematical* representation, which consists of specifications of the domain and range sets along with a description (either in terms of a mathematical formula or in English, or a combination of both) of the process of the function. We also discussed the domain, range, and image of a function. One important case occurs when the domain is the set of positive integers, in which case the function may be called a sequence.

Exercises

4.2.1 Each of the following is an ISETL representation of a function. Give a mathematical description of the function—that is, define its domain and range and describe its process.

```
a.  program example;
            print "Please enter a positive integer";
            read x;
            if is_integer(x) and x > 0 then
                base_2 := "";
                while x /= 0 do
                    if x mod 2 = 0 then
                          base_2 := "0" + base_2;
                    else
                          base_2 := "1" + base_2;
                    end;
                    x := x div 2;
                end;
                print base_2;
            else
                print "That is not a positive integer";
            end;
    end;
```

b.
```
program sample;
          read S;
          if is_set(S) then
              ans := true;
              for x in S do
                  if not is_func(x) then
                      ans := false;
                  end;
              end;
          end;
          print ans;
end;
```

c.
```
func(p,q);
    local rp;
    if is_integer(p) and is_integer(q)
                 and p>0 and q>0 then
        rp := %max{i:  i in [1..min(p,q)] |
                       (p mod i = 0)
                       and (q mod i = 0)};
        return [p div rp, q div rp];
    end;
end;
```

d.
```
func(x);
    local y, n;
    if is_number(x) then
        y := abs(x);
        n := fix(y);
        return sgn((y-n)**2 + n/(n+1));
    end;
end;
```

e.
```
func(x);
    if is_number(x) then
        return (x**3-2*x**2-5*x+6)/(x**2-4);
    end;
end;
```

f. ```
    func(x);
        if is_number(x) then
            if x > 0 then
                return x**2+1;
            elseif x = 0 then
                return 1;
            else
                return x+1;
            end;
        end;
    end;
    ```

g.  ```
    func(f,x0); $ f is an ISETL representation of a
               $ function whose domain is a set of
               $ floating-point numbers.
               $ x0 is a number in the domain of f.
        return func(h);
                   if is_floating(h) and h/=0 then
                       return (f(x0+h)-f(x0))/h;
                   end;
               end;
    end;
    ```

h. Assume that **def_int** has been defined as a **func** that represents a
    function that will calculate (an approximation to) the definite
    integral of a function (whose domain and range are sets of
    real numbers) from a to b (where a and b are real numbers).
    Thus, **def_int** accepts three parameters: a representation of a
    function whose domain and range are **sets** of **floating-point**
    numbers, a **floating-point** number, and a second **floating-point** number. The value returned by **def_int** is a **floating-point** number.

    ```
    func(f);
        return func(x);
                   return def_int(f,0,x);
               end;
    end;
    ```

i. ```
    func(t);
        if is_tuple(t) and #(t)=3
            and (forall x in t | is_integer(x)) then
        return [t(3), t(1)+t(2), t(1)];
        end;
    end;
    ```
    ```

j.
```
f := func(x);
        if is_number(x) then
            if -1<=x and x<1 then
                return -x;
            elseif x>=1 then
                return 2.0*f(x-2);
            else
                return 0.5*f(x+2);
            end;
        end;
    end;
```

k.
```
g := func(x);
        if is_floating(x) then
            if 1.0<=abs(x) and abs(x)<=3.0 then
                return abs(x);
            elseif x=0.0 then
                return 3.0;
            elseif abs(x) < 1.0 then
                return g(3.0*x);
            else
                return g(x/3.0);
            end;
        end;
    end;
```

l.
```
{["Smith",{["Sue",11],["Jim",13]}],
 ["Jones",{["Albert",1],["Anna",3],["Ron",9]}],
 ["Skallagrim",{["Thorolf",7],["Egil",5],
                ["Asgerd",4]}]}
```

m. `{[float(x),1/(1+x**2)] : x in [1..1000]}`

n. `{[p/q,((p/q)**2+1)/(p/q)] : p,q in [1..30]}`

o.
```
{["011","111"],["101","100"],["001","011"],
 ["100","110"],["110","010"],["000","001"],
 ["010","000"]}
```

p. `"18734537465849302938476384950489544594859485843330"`

q. `[(n**3+(n+1)**3+(n+2)**3) mod 9 = 0 : n in [1..25]]`

r. `[(exists x,y in [0..n]| 5*x+9*y=n) : n in [1..30]]`

s. `[(exists x,y in [0..n]| 6*x+10*y=n) : n in [1..30]]`

t. `[2**n > n**2+2*n-2 : n in [1..10]]`

u.
```
[%+{3*i**2-3*i+1 : i in [1..k]} = k**3 :
                        k in [1..45]]
```

v. `["000","001","011","111","101","100","110","010"]`

w. "fantasamagoriosiepticomis"

x. {[x**2+x-10, abs(x)/3+1-2*x/3] : x in [-1,-2..-5]}

y. {[w+4, abs(3*w)-w] : w in [-2..2]}

4.2.2 Each of the following is a mathematical description of a function. Use an ISETL func to represent this function (with restricted domain if necessary). If the domain and range are not specified, then you should define them explicitly.

a. The domain and range are both the set of real numbers. The process takes a real number y and changes it to 2^y.

b. The domain is the set of real numbers other than $1/2$, -1 and the range is the set of real numbers. The process takes a number u in the domain and changes it to the value of the following expression.

$$\frac{u^3 + u^2 - 3u - 3}{2u^2 + u - 1}$$

c. The domain is the set of real numbers in the interval $[-2,1]$ and the range is the set of all real numbers. The value of the function for a number x in the domain is $x^3 + x$ if x is negative and x^2 otherwise.

d. The domain is the set of positive real numbers and the range is the set of real numbers. The function transforms a number x in its domain into $x - 1$ if $0 < x \leq 2$, and the other values are determined by the condition that if x is increased by 2 then the value of the function is doubled. For example, the value of the function at 2.75 is twice the value at 0.75 or $2(-0.25)$ since $2.75 = 0.75 + 2$.

e. The domain and range of the function g is the set of real numbers. We have

$$g(0) = 3, \qquad g(x) = |x| \text{ for } 1 \leq |x| \leq 3$$

and the rest of the values are determined by the relation

$$g(3x) = g(x)$$

which is to hold for all x.

f. The domain and range are the set of of all points in the square with vertices labeled A, B, C, and D, sides of length 1, and situated in a coordinate system with the base parallel to the horizontal axis and the center at the origin. The action of the function is to flip the square along its diagonal from the upper left to the lower right vertex.

g. The domain and range are the set of all points in the square with vertices labeled A, B, C, and D, sides of length 1, and situated in a coordinate system with the base parallel to the horizontal axis and the center at the origin. The action of the function is to rotate the square counterclockwise through an angle of 180°.

h. The domain of the function is the **set** of all pairs **[f,g]** of ISETL **funcs**, where each **func** represents a function whose domain and range are the set of real numbers. The range of the function is the **set** of all ISETL **funcs** that, as in the domain, represent functions whose domain and range are the set of real numbers. The value of the function applied to a pair **[f,g]** is the **func** whose value at a point is the sum of the values of **f** and **g** at that point.

i. The function takes a **4-tuple** of positive integers and adds up their cubes.

j. The function determines the number of digits in an integer.

k. The function determines the diameter of a circle, given its area.

4.2.3 Write an ISETL **func** called **is_smap** that accepts a **set** and returns true if the **set** is a collection of 2-**tuples** representing a function and false otherwise.

4.2.4 Each of the following is a mathematical description of a function. Use an ISETL **smap** to represent this function (with restricted domain if necessary). If the domain and range are not specified, then you should define them explicitly.

a. The domain and range of the function is the set of integers greater than 1 and less than 100, and the value of the function at an integer is the largest prime less than or equal to that integer.

b. The domain and range of the function is the set of integers from 1 to 1000, and the value of the function is given by the following expression:

$$3^n + n^4 - \frac{n(n+1)(n+2)}{3}$$

c. The domain is the set of words in the last three lines of the "To the Instructor" section at the beginning of this chapter. The range is the set of all words in that section. The value of the function applied to a word is the word that precedes it in the text.

d. The domain is the set of letters in the alphabet and the range is the set of nonnegative integers. The function records the number of times each letter occurs in the text of the "To the Instructor" section at the beginning of this chapter.

e. There are 2^n binary integers of length n.

f. The number of factors in each of the first 20 positive integers.

g. The number of characters in each of the words in this sentence.

4.2.5 Each of the following is a mathematical description of a function that is a sequence. Use an ISETL **tuple** to represent this function with its domain restricted appropriately.

a. The sum of the cubes of the first n odd integers is $n^2(2n^2 - 1)$.

b. The sequence of positive integers (in ascending order) that are not primes.

c. The sequence whose ith term is the set of factors of i.

d. $\frac{1}{n^3+1}$, $n = 1, 2, \ldots$

4.2.6 Converting **funcs** to **smaps**.

a. Write an ISETL **func** that will take a **func** and a finite subset of the domain of the function it represents as input and return an **smap** that represents the same function with its domain restricted to this finite subset.

b. Apply your **func** to the **funcs** given in the following exercises.

  i) Exercise 4.2.1, part c.
 ii) Exercise 4.2.1, part d.
iii) Exercise 4.2.1, part e.
 iv) Exercise 4.2.1, part f.
  v) Exercise 4.2.1, part i.
 vi) Exercise 4.2.1, part j.
vii) Exercise 4.2.1, part k.

4.2.7 Converting **smaps** to **funcs**.

a. Write an ISETL **func** that will take an **smap** as input and return a **func** that represents the same function.

b. Apply your **func** to the **smaps** given in the following exercises.

  i) Exercise 4.2.1, part l.
 ii) Exercise 4.2.1, part m.
iii) Exercise 4.2.1, part n.
 iv) Exercise 4.2.1, part o.

      v) Exercise 4.2.1, part x.

     vi) Exercise 4.2.1, part y.

4.2.8 Calculate the images of the functions specified in the following exercises. In each of your answers, state clearly the set that you are taking as the domain of the function.

    a. Exercise 4.2.1.

    b. Exercise 4.2.2.

    c. Exercise 4.2.3.

    d. Exercise 4.2.4.

4.2.9 Sketch the graph of the function represented in the following exercises.

    a. Exercise 4.2.1, part d.

    b. Exercise 4.2.1, part e.

    c. Exercise 4.2.1, part f.

    d. Exercise 4.2.1, part j.

    e. Exercise 4.2.1, part k.

    f. Exercise 4.2.1, part m.

    g. Exercise 4.2.1, part n.

4.2.10 Sketch the graph of the function given in the following exercises.

    a. Exercise 4.2.2, part a.

    b. Exercise 4.2.2, part b.

    c. Exercise 4.2.2, part c.

    d. Exercise 4.2.2, part d.

    e. Exercise 4.2.2, part e.

4.2.11 Calculate the number of functions with domain D and range R where D and R have the following values.

    a. $D = \{\text{``}A\text{''}, \text{``}B\text{''}\}, \quad R = \{0, 1, 2\}$

    b. $D = \{2, 0, 1\}, \quad R = \{0\}$

    c. $D = \{\}, \quad R = \{\text{``}A\text{''}, \text{``}B\text{''}, \text{``}C\text{''}, \text{``}D\text{''}, \text{``}E\text{''}\}$

    d. $D = \{\text{``}A\text{''}\}, \quad R = \{2\}$

    e. $D = \{\text{``}A\text{''}, \text{``}B\text{''}, \text{``}E\text{''}\}, \quad R = \{0, 2\}$

    f. $D = \{\text{``}B\text{''}\}, \quad R = \{0, 1, 2, 3\}$

    g. $D = \{\text{``}A\text{''}, \text{``}B\text{''}\}, \quad R = \{0, 1\}$

    h. $D = \{\}, \quad R = \{\}$
    i. $D = \{\text{``}A\text{''}, \text{``}B\text{''}\}, \quad R = \{\}$

4.2.12 Let $x = (x_n)_n$ be a sequence of numbers. Represent this function (with appropriate domain restriction) as an ISETL **tuple** and write an ISETL expression that represents the sequence whose nth term is the maximum of the first n terms of the function x.

4.2.13 Let $y = (y_n)_n$ be the sequence that is the solution to the previous problem. Representing this function (with appropriate domain restriction) as an ISETL **tuple**, write an ISETL expression that represents the sequence whose nth term is the minimum of the first n terms of this function. Can you say anything clever about this function?

4.2.14 Let $z = (z_n)_n$ be a sequence. Given positive integers k and r, write a formal expression that asserts that from the kth term of the sequence on the values of the sequence repeat a cycle of length r.

4.2.15 A certain university has six times as many students as professors. Define a function in mathematical notation that will give the number of professors if the number of students is known.

4.2.16 The vendor at the ball game sold 2 hamburgers for every 3 hot dogs. Define a function in mathematical notation that will give the number of hot dogs sold if the number of hamburgers sold is known.

4.3 Function Operations

4.3.1 Evaluation and Equality of Functions

We use the notation

$$F : D \longrightarrow R$$

to denote a function. Here, D is the domain of the function, R is the range of the function, $\longrightarrow$ stands for the process by which elements of the domain are transformed into elements of the range, and F is the *name* of the function. It is very helpful, when working with functions, to think about one of the ISETL representations: **program**, **func**, **smap**, **tuple**, or **string**. Sometimes it is even useful to think about two or more representations of the same function simultaneously.

    A function can be *evaluated* at a point in its domain. If x is an element of the domain of the function $F : D \longrightarrow R$, then $F(x)$ is called the *value of F at x*. It is the result of applying the function's process to x.

    If the function is represented as a **func**, then this value is the value returned when the **func** is called with x passed as its actual parameter. If the function is represented as an **smap**, then this value is obtained by finding the unique pair whose first component has the value x and returning the second component as the result of the evaluation. If no first component has the value x, then OM is returned. If the function is represented as a tuple or string, then the value is the value of the xth component or character. An important feature of ISETL syntax is that an expression such as F(x) means the same thing whether F is a func, smap, tuple, or string.

If we compare two functions for equality, then before we are willing to say that they are equal, it is necessary not only that the processes of the two functions are the same, but also that their domains as well as their ranges are identical. Consider, for example, the following function from the previous section:

$$
\begin{aligned}
D &= \{\text{all positive floating-point numbers}\} \\
R &= \{\text{all floating-point numbers}\} \\
F &: \quad D \longrightarrow R \qquad \text{by } F(x) = \sqrt{x}
\end{aligned}
$$

and suppose we wanted to compare it with

$$
\begin{aligned}
A &= \{\text{all nonnegative floating-point numbers}\} \\
R &= \{\text{all floating-point numbers}\} \\
G &: \quad A \longrightarrow R \qquad \text{by } G(x) = \sqrt{x}
\end{aligned}
$$

These are different functions because the domain D of F is different from the domain A of G. Operationally, this means that, if both functions were represented in ISETL, F(0) would have the value OM whereas G(0) would have the value 0. If you think of these functions as represented as smaps, then the pair [0,0] is in G but not F. (Actually, since D and A are infinite, representation as smaps is not possible, but you can do it in your mind and our point still makes sense.)

Although these two functions are not equal, they are related. We say that F is the *restriction* of G to the domain D and that G is an *extension* of F. This is written

$$
F = G_{|D}
$$

The same thing can be done with the range. The function

$$
\begin{aligned}
A &= \{\text{all nonnegative floating-point numbers}\} \\
B &= \{\text{all nonnegative floating-point numbers}\} \\
H &: \quad A \longrightarrow B \qquad \text{by } H(x) = \sqrt{x}
\end{aligned}
$$

is different from both F and G. It differs from G because its range is different, and it differs from F because both its range and domain are different. On the other hand, we say that H is the *restriction* of G to the range B, and we say that H is the *extension* of F to the domain A.

4.3.2   Images and Pre-images

If $F : D \longrightarrow R$ is a function, then we have already mentioned the image $F(D)$ of this function, which is the set of all values that would be returned if the function were applied to every element of the domain. This set can be expressed in ISETL as follows:

$$\{\texttt{F(x) : x in D}\}$$

More generally, if S is any subset of D, that is $S \subseteq D$, then the *image of S under F*, denoted $F(S)$, is the set of all values that would be returned if the function were applied to every element of S, that is,

$$\{\texttt{F(x) : x in S}\}$$

For example, consider the following function:

$$
\begin{aligned}
A &= \{\text{all } \texttt{integers}\} \\
R &= \{\text{all nonnegative } \texttt{integers}\} \\
F &: \quad A \longrightarrow R \qquad \text{by } F(x) = x^4 + 1
\end{aligned}
$$

If S is the set of $\texttt{integers}$ $\{-2..2\}$, then $F(S)$ is the set $\{17, 2, 1\}$. If the function is represented as a $\texttt{func}$, then to find the image of a set S, you have to execute the $\texttt{func}$ (on the computer or in your mind) for every value in S and the set of values that are returned is $F(S)$.

If the function is represented as an $\texttt{smap}$, for example, the function $\texttt{next}$ of the previous section,

```
next := {["well","what"], ["by","must"], ["now","see"],
         ["must","now"], ["begin","by"], ["to","well"],
         ["see","that"], ["that","to"], ["what","you"]};
```

then to figure out the image of a $\texttt{set}$, for example, $\{\texttt{"well"}, \texttt{"see"}, \texttt{"begin"}, \texttt{"for"}\}$, you just run through all of the pairs in $\texttt{next}$ and every time you see a first component that is in this $\texttt{set}$, you note the value of the second component. When you are finished, the $\texttt{set}$ of all values that you noted is the image. We can implement this calculation with the following ISETL code:

```
{next(s) : s in {"well","see","begin","for"}*domain(next)};
```

(Note that $\texttt{"for"}$ contributed nothing to the result. Since $\texttt{next("for")}$ has the value $\texttt{OM}$, it was necessary to restrict this $\texttt{set}$ former to $\texttt{domain(next)}$.)

It is possible to do all of this in reverse. If $\texttt{M}$ is an $\texttt{smap}$ and $\texttt{S}$ is a $\texttt{set}$, then you could run through all the pairs in $\texttt{M}$ and every time you see a *second* component that is in $\texttt{S}$, you could note the value of the *first* component. When you are finished, the $\texttt{set}$ of all values that you noted is called the *pre-image* or *inverse image of* $\texttt{S}$ *under* $\texttt{M}$. For example, if you took the map $\texttt{next}$ discussed above and the same set, $\{\texttt{"well"}, \texttt{"see"}, \texttt{"begin"}, \texttt{"for"}\}$, then the pre-image of this set under $\texttt{next}$ is the set

$$\{\texttt{"to"}, \texttt{"now"}\}$$

Notice that neither **begin** nor **for** contributed anything to the answer.

If $F : D \longrightarrow R$ is a function and S is a set, then we write the pre-image of S under F as $F^{-1}(S)$. In terms of the process of the function, the pre-image of S is the set of all values in the domain of F that are transformed by F into an element of S. We can express this in ISETL notation as

$$\{\texttt{x : x in domain(F) | F(x) in S}\};$$

In other words, you run the process of F for every element of the domain and include in the pre-image every value in the domain for which the result is in S.

We consider another example of a function, which can be represented in two ways, as a **func** and as an **smap**. The domain of this function will be the **set**

```
words := {"well", "by", "now", "must", "begin", "to",
          "see", "that", "what"};
```

which was also the domain of **next**. The process of the function will be to transform an element of its domain, which will be a **string**, into the positive **integer** that is the length of the **string**. Thus, we can take the range of this function to be the set of nonnegative integers. We can represent the function with the following **func**:

```
func(st);
    if st in words then return #(st); end;
end;
```

We can also represent it with the following **smap**:

```
{["well",4], ["by",2], ["now",3], ["must",4], ["begin",5],
 ["to",2], ["see",3], ["that",4], ["what",4]};
```

If we call this function L, then it is probably a little more convenient to use the **smap** representation to make calculations such as the following.

$$
\begin{aligned}
L(\{\text{``now''},\text{``see''}\}) &= \{3\} \\
L(\{\text{``by''},\text{``must''},\text{``to''}\}) &= \{2,4\} \\
L^{-1}(\{3\}) &= \{\text{``now''},\text{``see''}\} \\
L^{-1}(\{4,5\}) &= \{\text{``well''},\text{``must''},\text{``begin''},\text{``that''},\text{``what''}\}
\end{aligned}
$$

Finally, we would like to mention three important properties that a function could have. We will express them succinctly in precise mathematical language, using the concepts we have discussed in this section. There will be examples in the exercises to help you pin down these ideas.

A function is said to be *one-to-one* (1-1) or an *injection*, if the inverse image of every single-element set in the function's image is again a single-element set. This means that for each element in the image of the function, there is exactly one element of the domain that is transformed into it. The function represented by the `smap next` described previously is 1-1, but the function L is not because, for example, there is more than 1 element in the domain that is transformed to 3.

A function is said to be *onto* or to be a *surjection* if its image is equal to its range. This means that for each element in the range of the function, there is an element of its domain that is mapped into it. The function H defined in the previous section is onto, but G is not because negative `floating-point` numbers are in the range but not the image of G.

A function is called a *one-to-one correspondence* or a *bijection* if it is both 1-1 and onto. The function H of the previous section is a bijection.

4.3.3   Arithmetic with Functions

We will be studying a number of ways in which an operation can be performed on one or more functions to obtain a new function. The simplest ways of doing this, which will be the subject of this section, are to add two functions, multiply them, or multiply a function by a constant. Recall that on page 174 we considered adding two functions and multiplication of a function by a constant for the special case in which the functions were sequences.

If $F, G : D \longrightarrow R$ are two functions having the same domain and range, then their *sum*, $F + G$, is a function with the same domain whose process can be expressed, in terms of the processes of F and G, by

$$(F + G)(x) = F(x) + G(x), \qquad x \in D$$

For example, suppose that F is the square root function defined on page 187 and is represented in ISETL by the following `func`:

```
F := func(x);
        if is_floating(x) and x>0 then
                return sqrt(x);
        end;
    end;
```

Suppose G is a function, with the same domain and range, whose process is expressed by

$$G(x) = \frac{x}{2}, \qquad x \in D$$

so that G is represented by the following `func`:

```
G := func(x);
        if is_floating(x) and x>0 then
                return x/2;
        end;
    end;
```

Then, $F + G$ is a function with the same domain and range whose process is expressed by

$$(F + G)(x) = \sqrt{x} + \frac{x}{2}, \qquad x \in D$$

and it can be represented by the following func:

```
FplusG := func(x);
            if is_floating(x) and x>0 then
                return sqrt(x) + x/2;
            end;
         end;
```

The same thing can be done with multiplication of two functions. This time let's take an example of functions represented by smaps. Suppose that J is an extension of the function discussed on page 189 and is represented by the following smap:

```
J := {["well",4], ["by",2], ["now",3], ["you",3], ["must",4],
      ["begin",5], ["to",2],["see",3], ["that",4], ["what",4],
      ["we",2], ["have",4], ["been",4], ["describing",10] };
```

Suppose further that K is a function whose domain is the same as J and whose process assigns to a word the number of times a word with that number of characters appears in the domain. Then, we could represent K by the following smap:

```
K := {["well",6], ["by",3], ["now",3], ["you",3], ["must",6],
      ["begin",1], ["to",3],["see",3], ["that",6], ["what",6]
      ["we",3], ["have",6], ["been",6], ["describing",1] };
```

Then, the *product* $K \cdot J$ of these two functions is a function whose process gives, for each word in the domain, the total number of characters that are used in writing all the instances of words of that length. It can be represented in ISETL by the following smap:

```
KtimesJ := {["well",24], ["by",6], ["now",9], ["you",9],
            ["must",24], ["begin",5], ["to",6],["see",9],
            ["that",24], ["what",24],["we",6], ["have",24],
            ["been",24], ["describing",10] };
```

It should be clear to you how to do this for the arithmetic operation of multiplying a function by a scalar.

There is an interesting problem that arises in this connection. Could you write a func that accepts two ISETL representations of functions (via funcs or smaps) and returns a representation of the function (as a func or smap) that is the sum of the two original functions? Of course, you would have to do something to make sure that the values in the range of these two functions were objects that could be added. What about the domains? Do they have to be the same? If not, what would be the domain of the new function? You will have a chance to deal with these and other issues in the exercises.

4.3.4 Composition of Functions

Suppose that $F : A \longrightarrow B$ and $G : C \longrightarrow D$ are functions. It is possible to think of "linking these two functions in series" in the sense that you could have a process that consisted of first performing the process of G and then, if possible, performing the process of F using the value that resulted from performing G. This is the idea of *composition*.

For example, suppose we had two functions, F and G, described as follows:

F: The domain of F is the set of all triples (tuples of length 3) of integers. The result of applying F to a triple is the integer obtained by subtracting the first integer in the triple from the second.

G: The domain of G is the set of all triples of integers. The result of applying G to a triple is the triple obtained from the original triple by replacing its second integer by the sum of all three integers.

What is the composition? If we perform G first on a triple, the result is again a triple of integers so we can apply F to it. Thus, we have a process that transforms a triple in two steps. First, it adds all three integers and replaces the second with that sum. Then, in this new triple, it subtracts the first integer from the second. The result is an integer. If you think about it for a minute, you will see that this composition is a new function H, which can be described as follows:

H: The domain of H is the set of all triples of integers. The result of applying H to a triple is the sum of the second and third integers.

Now let's discuss the matter in a little more detail. We have two functions, F and G. We take an element x in the domain of G and evaluate G at that point to obtain $G(x)$. Now, we must be sure that $G(x)$ is a value in the domain of F. The only way to be certain is to impose a restriction and say that we will not even *try* to form the composition of these two functions unless we know that every time we evaluate G the result is in the domain of F. In terms we have defined in this chapter, that is just the requirement that the image of G be contained in the domain of F, that is,

$$G(\text{domain}(G)) \quad \subseteq \quad \text{domain}(F)$$

or

$$\text{image}(G) \quad \subseteq \quad \text{domain}(F)$$

If this is the case, then we can apply F to the value $G(x)$, which resulted from applying G to x, to obtain the final result, $F(G(x))$.

To summarize and put things a little more formally, suppose that we have two functions,

$$F : A \longrightarrow B \qquad \text{and} \qquad G : C \longrightarrow D$$

Suppose further that we know

$$G(C) \subseteq A$$

Then, we define the *composition*, which is denoted by $F \circ G$, as follows.

$$F \circ G : C \longrightarrow B \qquad \text{by} \qquad F \circ G(x) = F(G(x)), \qquad x \in C$$

Let's look at a couple of examples. First, consider the following two functions represented in ISETL by funcs:

```
F := func(x);
          if is_floating(x) and x>0 then
                return sqrt(x);
          end;
     end;


G := func(x);
          if is_floating(x) and x>0 then
                return x + 1/2;
          end;
     end;
```

Notice that the result obtained from G will always be a positive floating-point number, that is, it will be in the domain of F so we can form the composition. Here is a func that represents it:

```
FcompG := func(x);
                if is_floating(x) and x>0 then
                      return sqrt(x + 1/2);
                end;
          end;
```

In other words, the process for $F \circ G$ is expressed by

$$(F \circ G)(x) = \sqrt{x + \frac{1}{2}}$$

To see how it looks when the functions are represented by smaps, let's go back to the function L, defined on page 189,

```
L := {["well",4], ["by",2], ["now",3],
      ["must",4], ["begin",5], ["to",2],
      ["see",3], ["that",4], ["what",4]};
```

Suppose that it takes 4 bytes to represent a character and we want to know how many bytes are used to represent each word. We could define an **smap**, **Bt**, that recorded for words up to 10 characters how many bytes were needed.

$$Bt := \{[i, 4*i] : i \text{ in } [1..10]\};$$

or

```
Bt := {[1,4], [2,8], [3,12], [4,16], [5,20], [6,24],
       [7,28], [8,32], [9,36], [10,40]};
```

Then, the required information would be recorded in the composition $Bto L$. To figure out the **tuples** in the representation of this function as an **smap**, you take an element in the domain of L, for example, "now". You find the second component of the **tuple** in L that has "now" as the first component. It is 3. Then, you look in **Bt** for the **tuple** that has 3 as its first component. It is the **tuple** [3,12]. Then, you form a new **tuple** by taking the first component of the original **tuple** and the second component of the **tuple** that you found in **Bt**. This latter is 12, so the **tuple** that gets placed in the composed function that you are constructing is ["now",12]. Doing this for everything in the domain of L gives, finally,

```
BtcompL := {["well",16], ["by",8], ["now",12],
            ["must",16], ["begin",20], ["to",8],
            ["see",12], ["that",16], ["what",16]};
```

4.3.5   The Inverse of a Function

Another way to make new functions out of old ones is to construct the inverse of a given function. This can only be done if the function is both one-to-one and onto, that is, a bijection. In this case, for each element of the range of a function, there is a unique element of the domain that is transformed into it. If we reverse the direction of our thinking about the transformation, we can say that there is a new function whose domain is the range of the old function and whose range is the domain of the old function. The process of the new function is the same as the process of the old one run in reverse. This new function is called the *inverse* of the old one.

For example, consider the following **smap** representing a function that we discussed in Section 4.2.4:

```
next := {["well","what"], ["by","must"], ["now","see"],
         ["must","now"], ["begin","by"], ["to","well"],
         ["see","that"], ["that","to"], ["what","you"]};
```

whose domain is the **set**

```
words1 := {"well", "by", "now", "must", "begin", "to",
           "see", "that", "what"};
```

If we take the **set**

```
words2 := {"well", "by", "now", "you", "must", "to",
           "see", "that", "what"};
```

to be its range, then the function represented by **next** is both one-to-one and onto, so it has an inverse that we can represent in ISETL with the following **smap**, noting that computing the inverse consists of merely reversing the first and second component of each **tuple**.

```
next_inv := {["what","well"], ["must","by"], ["see","now"],
             ["now","must"], ["by","begin"], ["well","to"],
             ["that","see"], ["to","that"], ["you","what"]};
```

Often we can give an interpretation to the inverse. In this case, if we think of **next** as recording a linear ordering for these words (alphabetical, in fact), then **next_inv** records the reverse order.

As a second example, consider the following function which, as we saw in Section 4.3.2, is a bijection:

$$A = \{\text{all nonnegative \textbf{floating-point} numbers}\}$$
$$B = \{\text{all nonnegative \textbf{floating-point} numbers}\}$$
$$H : A \longrightarrow B \quad \text{by } H(x) = \sqrt{x}$$

The inverse of this function has the set B for its domain, and A is its range. To figure out the process, we simply ask ourselves the question, given an element y in the domain B of this function, what is the element x in its range A such that if the original function, H, were applied to x, the result would be y? That is, given y, we wish to solve the equation

$$H(x) = y \quad \text{or} \quad \sqrt{x} = y$$

for $x \in A$. The answer is y^2, and so the inverse of H, which is denoted by H^{-1}, may be described as follows.

$$A = \{\text{all nonnegative \textbf{floating-point} numbers}\}$$
$$B = \{\text{all nonnegative \textbf{floating-point} numbers}\}$$
$$H^{-1} : B \longrightarrow A \quad \text{by } H^{-1}(y) = y^2$$

4.3.6   The Partial Sum Operator

Let $x = (x_i)_i$ be a sequence of numbers, that is, a function whose domain is the set of positive integers and whose range is some set of numbers. We can define a new sequence, $y = (y_i)_i$, by taking the value of y_i to be the sum of the first i terms of the original sequence x:

$$y_i = \sum_{n=1}^{i} x_n, \qquad i = 1, 2, \ldots$$

For example, if the process of the sequence is expressed by

$$x_n = 2n + 3$$

then we can calculate the ith partial sum by

$$
\begin{aligned}
y_i &= \sum_{n=1}^{i}(2n+3) \\
&= (2 \cdot 1 + 3) + (2 \cdot 2 + 3) + (2 \cdot 3 + 3) + \cdots + (2i + 3) \\
&= 2(1 + 2 + 3 + \cdots + i) + \overbrace{(3 + 3 + \cdots + 3)}^{i \text{ copies}} \\
&= 2\left(\sum_{n=1}^{i} n\right) + 3i \\
&= 2\left(\frac{i(i+1)}{2}\right) + 3i \\
&= i^2 + 4i
\end{aligned}
$$

The sequence $\left(\sum_{1}^{i} x_n\right)_i$ is called the *sequence of partial sums* for the given sequence x. It is very important in questions concerning sums of infinite series. In fact, we sometimes distinguish between the sequence $x = (x_i)_i$ and the *series* $\sum_{1}^{\infty} x_i$ determined by a given function whose domain is the set of positive integers.

You can think of partial sums in terms of tuples, again using only an initial segment of the sequence. Thus, tuple is a *metaphor* for sequence. We can take K to be some very large positive integer—so large that it is beyond any actual indices of the sequence that occur in a particular discussion. (You might want to think of K as being increased every so often.) If t is a tuple of numbers, then the tuple of partial sums can be constructed in ISETL as follows.

```
sum_t := [%+t(1..i) : i in [1..K]];
```

For example, if t is the tuple,

```
[2*i+3 : i in [1..K]]
```

then sum_t is

$$[\%+[2*n+3:n \text{ in } [1..i]] : i \text{ in } [1..K]]$$

which can be simplified, using the above calculations, to

$$[i**2+4*i : i \text{ in } [1..K]]$$

The process of making a new sequence with partial sums makes sense for operations other than addition. You can use multiplication as well or even operations on objects other than numbers, such as conjunction ($\wedge$) and disjunction ($\vee$) on propositions or operations on sets. For example, consider a sequence whose nth term is the set of all integer multiples of n up to $K \cdot n$. We could represent this as the following tuple:

$$[\{j*n:j \text{ in } [1..K]\} : n \text{ in } [1..K]]$$

Then the sequence of *partial intersections* would be the sequence whose ith term was the intersection of the first i sets in the original sequence. We can represent this new sequence as a tuple.

$$[\%*[\{j*n:j \text{ in } [1..K]\} : n \text{ in } [1..i]] : i \text{ in } [1..K]]$$

You can simplify this quite a bit. Think about it and maybe review some ideas from Chapter 1.

4.3.7 Collections of Functions

In earlier parts of this chapter, we have referred to the fact that you can think of a function as either a process that *does* something or an object that *is* something, or both. For example, if you use an smap to represent a function, then you have an ISETL object, a set. It exists, in some sense, inside the computer, and you can do things with it, such as pass it as an argument or take its union with another set. On the other hand, you can evaluate the function that the smap represents at a point x in its domain. ISETL will do this by iterating through the tuples in the smap, looking for the unique tuple that has x as its first component, and returning the second component of that tuple.

You can do the same thing if you represent your function as a func. Execution of the func corresponds to evaluating the function, and, again, the func is an ISETL object—it has a value and can be manipulated.

Since functions are objects, you can conglomerate a bunch of them to form a set or sequence of several (even infinitely many) of them. You can use ISETL to represent (finite) sets and tuples of functions. Once you have, for instance, a set of functions, you can iterate through it, picking out certain functions according to some criteria, thereby forming a subset. You can do the same with tuples. We will illustrate all this in the present section.

Here is an example of an smap that represents a function whose domain is the set {"A","B","C"} and whose range is the set {1,2,3,4}:

$$\{["A",2], ["B",1], ["C",2]\}$$

It is easy to write a **set** former whose value will be the **set** of *all* such functions:

mapset := $\{\{["A",x], ["B",y], ["C",z]\} : x,y,z \text{ in } \{1,2,3,4\}\}$;

and, after executing this statement, the value of mapset will be the **set** of smaps in Figure 4.1.

How many functions does this **set** have? You can count them or you can calculate the number as follows. In defining a function whose domain is the **set** $\{A,B,C\}$, you must determine the value of the function for each of the three characters in the domain. If the range of the function is the **set** $\{1,2,3,4\}$, each of these determinations consists of making a choice of one of these four **integers**. Thus, to define a single function, you must select one of four **integers** three times. Therefore, you can define a function in $4 \cdot 4 \cdot 4 = 64$ ways, so this is the number of elements in the **set** of functions.

Now, suppose you wanted to construct the **set** of those functions (still with the same domain and range) that are one-to-one. First, you would need to think about a condition that determined that an **smap** represented a function that is one-to-one. One reasonable possibility is to require that for each **y** in the **image** of the **smap** the **set** of elements in the **domain** that are mapped to **y** has exactly one element. If m is an **smap**, then the following expression represents this condition:

forall y in image(m) | (#{x : x in domain(m) | m(x) = y} =1)

So, we can use the following expression to construct the subset of **mapset** consisting of all of the 1-1 functions:

{m : m in mapset | (forall y in image(m) |
                    (#{x : x in domain(m) | m(x) = y} =1))};

How many elements does this **set** have?

You can do the same kind of thing if your functions are represented by **funcs**. As an example, let's consider the following **func**, which represents a function whose process is given by evaluating the polynomial $x^4 + x^3 + 1$.

```
func(x);
   if is_number(x) then
        return x**4 + x**3 + 1;
   end;
end;
```

This is a polynomial of degree 4, all of whose coefficients are either 0 or 1. It is easy to construct a **tuple** of **funcs** that represents a sequence of all such functions. Here is the code:

```
{{["A",1], ["B",1], ["C",1]}, {["A",1], ["B",1], ["C",2]},
 {["A",1], ["B",1], ["C",3]}, {["A",1], ["B",1], ["C",4]},
 {["A",1], ["B",2], ["C",1]}, {["A",1], ["B",2], ["C",2]},
 {["A",1], ["B",2], ["C",3]}, {["A",1], ["B",2], ["C",4]},
 {["A",1], ["B",3], ["C",1]}, {["A",1], ["B",3], ["C",2]},
 {["A",1], ["B",3], ["C",3]}, {["A",1], ["B",3], ["C",4]},
 {["A",1], ["B",4], ["C",1]}, {["A",1], ["B",4], ["C",2]},
 {["A",1], ["B",4], ["C",3]}, {["A",1], ["B",4], ["C",4]},
 {["A",2], ["B",1], ["C",1]}, {["A",2], ["B",1], ["C",2]},
 {["A",2], ["B",1], ["C",3]}, {["A",2], ["B",1], ["C",4]},
 {["A",2], ["B",2], ["C",1]}, {["A",2], ["B",2], ["C",2]},
 {["A",2], ["B",2], ["C",3]}, {["A",2], ["B",2], ["C",4]},
 {["A",2], ["B",3], ["C",1]}, {["A",2], ["B",3], ["C",2]},
 {["A",2], ["B",3], ["C",3]}, {["A",2], ["B",3], ["C",4]},
 {["A",2], ["B",4], ["C",1]}, {["A",2], ["B",4], ["C",2]},
 {["A",2], ["B",4], ["C",3]}, {["A",2], ["B",4], ["C",4]},
 {["A",3], ["B",1], ["C",1]}, {["A",3], ["B",1], ["C",2]},
 {["A",3], ["B",1], ["C",3]}, {["A",3], ["B",1], ["C",4]},
 {["A",3], ["B",2], ["C",1]}, {["A",3], ["B",2], ["C",2]},
 {["A",3], ["B",2], ["C",3]}, {["A",3], ["B",2], ["C",4]},
 {["A",3], ["B",3], ["C",1]}, {["A",3], ["B",3], ["C",2]},
 {["A",3], ["B",3], ["C",3]}, {["A",3], ["B",3], ["C",4]},
 {["A",3], ["B",4], ["C",1]}, {["A",3], ["B",4], ["C",2]},
 {["A",3], ["B",4], ["C",3]}, {["A",3], ["B",4], ["C",4]},
 {["A",4], ["B",1], ["C",1]}, {["A",4], ["B",1], ["C",2]},
 {["A",4], ["B",1], ["C",3]}, {["A",4], ["B",1], ["C",4]},
 {["A",4], ["B",2], ["C",1]}, {["A",4], ["B",2], ["C",2]},
 {["A",4], ["B",2], ["C",3]}, {["A",4], ["B",2], ["C",4]},
 {["A",4], ["B",3], ["C",1]}, {["A",4], ["B",3], ["C",2]},
 {["A",4], ["B",3], ["C",3]}, {["A",4], ["B",3], ["C",4]},
 {["A",4], ["B",4], ["C",1]}, {["A",4], ["B",4], ["C",2]},
 {["A",4], ["B",4], ["C",3]}, {["A",4], ["B",4], ["C",4]}};
```

Figure 4.1: A set of smaps that represents the set of all functions F : $\{A, B, C\} \longrightarrow \{1, 2, 3, 4\}$

```
funtup := [
        func(x);
          value a,b,c,d,e;
          ifis_number(x) then
            return a*x**4 + b*x**3 + c*x**2 + d*x + e;
          end;
        end :
        a,b,c,d,e in {0,1}
      ];
```

How many functions are there in this sequence of functions?

Notice that the coefficients a, b, c, d, and e are listed in the **value** declaration. This ensures that when the **tuple** former is evaluated and on each iteration the **func** expression is evaluated to define the function the coefficients will have the values corresponding to the choice made from the **set** {0,1} on that iteration.

Now, you can construct the **set** of those functions in the list that vanish (that is, have the value 0) when x = -1 as follows:

$$\{fn : fn \text{ in } funtup \mid fn(-1) = 0\};$$

How many functions are there in this **set**?

4.3.8   Functions of Several Variables

You can use a **func** or an **smap** to represent a function of more than one variable. For instance, if you want to use a **func** to represent a function of two variables, then you just list two parameters in the **func**'s header. To use an **smap** to represent such a function, the **smap** must consist of pairs, whose first component is a **tuple** of length 2 (for 2 variables) and whose second component is the result of evaluating the function at the values of the two variables in the first component.

Suppose, for example, that $\mathcal{R}$ is the set of reals and $g{:}\mathcal{R} \times \mathcal{R} - \{[0,0]\} \longrightarrow \mathcal{R}$ is a *weighted Newtonian gravity function*. That is, the value of g at a point [x,y] (here, the square brackets indicate a point in the plane, not a **tuple**) is the magnitude of the gravitational force exerted on this point by a point mass at the origin. This value might be given by

$$g(x,y) = \frac{1}{(2x^2 + 3y^2)^{\frac{1}{2}}}$$

Here is a **func** that represents this function (restricted to ISETL **floating-point** numbers):

```
    g := func(x,y);
            if is_floating(x) and is_floating(y)
            and (x/=0 or y/=0) then
                    return 1/sqrt(2*x**2+3*y**2);
            end;
        end;
```

and here is an **smap** that represents the same function with its domain restricted to a few integer points on the plane:

```
g := {[[x,y], 1/sqrt(2*x**2+3*y**2)] : x,y in {-K..K}-{0}}
```

Now, consider the following situation. Suppose that you wanted to be able to construct for each $x \in \mathcal{R}$, $x \neq 0$, a function g_x whose process assigned to each $y \in \mathcal{R}$ the value

$$g_x(y) = g(x,y) = \frac{1}{(2x^2 + 3y^2)^{\frac{1}{2}}}$$

This can easily be implemented in ISETL. The point is that the operation we want to perform is a "higher level" function whose domain is $\mathcal{R}$ and whose range is *the set of all functions from $\mathcal{R}$ to $\mathcal{R}$*. In other words, we want a **func** that takes a **floating-point** input and returns a **func**. The entire operation is sometimes called the *slot function*. We can implement it in ISETL as follows:

```
slot_g := func(x);
            if x=0 then return;
            return func(y);
                    if is_floating(x) and is_floating(y) then
                        return 1/sqrt(2*x**2+3*y**2);
                    end;
                end;
        end;
```

Sometimes, in Mathematics, you will see $g(x,\cdot)$ written for **slot_g(x)**, which is the function obtained for a given value of **x**.

There is a "machine," called *curry* (which is named after a logician who worked on λ-calculus and a logic of functions), that is often used to construct the slot functions. The function *curry* has as its domain a set of functions (of two variables). Its process transforms a function of two variables into a function that takes an element in the domain of the first variable and transforms it into a function whose domain is the domain of the second variable—in other words, the slot function.

That's a mouthful. Let's try to clarify it by implementing *curry* in ISETL and then pointing out that this is nothing more than a formal description of a standard activity, such as computing a double integral by calculating an iterated integral. This example may be a little complicated for you if you have not had a lot of experience with calculus.

Here is the implementation of *curry*, ignoring domain checks:

```
curry := func(g);
            return func(x);
                       return func(y);
                                  return g(x,y);
                       end;
            end;
        end;
```

(Notice that the only requirement for **g** is that it represents a function of two variables. This can be done with a **func** or an **smap**.)

In other words, after executing the above assignment statement,

curry(g)(x)

will be an expression whose value is the function of one variable that we called the "slot function" and wrote $g(x, \cdot)$.

This language can be used to describe the operation of using an iterated integral to calculate a double integral. Suppose you wanted to integrate a function of two variables, for example, the weighted Newtonian gravity function described previously, over the triangle shown in the following figure: An interpretation of the result might be the "average" gravitational force exerted by a point mass at the origin on points in the triangle.

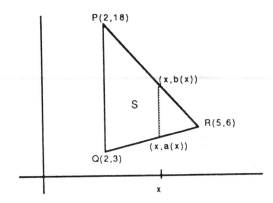

If we call the triangle S and the function g, then we need to evaluate the double integral,

$$\int_S g(x,y)\, dx\, dy = \int_S \frac{1}{(2x^2 + 3y^2)^{\frac{1}{2}}}\, dx\, dy$$

This is usually done with an iterated integral. First, we decide the order of integration. For this set, since one side of the triangle is parallel to the y-axis, it is convenient to first "fix x and integrate with respect to y." Having done this, we then "integrate with respect to x." In both cases, the main task is to figure out the limits of integration. The actual calculation looks like,

$$
\int_S \frac{1}{(2x^2 + 3y^2)^{\frac{1}{2}}} \, dx \, dy
$$

$$
= \int_2^5 \int_{a(x)}^{b(x)} \frac{1}{(2x^2 + 3y^2)^{\frac{1}{2}}} \, dy \, dx
$$

$$
= \int_2^5 \int_{x+1}^{-4x+26} \frac{1}{(2x^2 + 3y^2)^{\frac{1}{2}}} \, dy \, dx
$$

$$
= \frac{\sqrt{3}}{3} \int_2^5 \log\left(y + \sqrt{y^2 + \frac{2}{3}x^2} \right) \Big]_{x+1}^{-4x+26} \, dx
$$

$$
= \frac{\sqrt{3}}{3} \int_2^5 \log\left(\frac{-4x + 2 + \sqrt{\frac{14}{3}x^2 - 16x + 4}}{x + 1 + \sqrt{\frac{4}{3}x^2 + 2x + 1}} \right) \, dx
$$

and tables would be needed to carry the calculation any further.

Let's interpret this calculation in terms of *curry*. As indicated in the picture, the first step is to define a function whose domain is the interval (on the x-axis) from 2 to 5 and whose range is the set of functions with domains consisting of intervals (on the y-axis) and ranges taken as the set of real numbers. This is just $curry(g)$. Now, for a given x in the interval $[2, 5]$, we take the function $curry(g)(x)$ whose domain is the interval $[a(x), b(x)]$ (where a is the function whose graph represents the side QR and b represents PR) or $[x + 1, -4x + 26]$ and we integrate it along this interval. This gives us a number that depends on the original value of x. In other words, the entire operation results in a function from $[2, 5]$ to the reals. We integrate this function to obtain the final result.

In formal notation, it looks like the following.

$$
\int_S g(x, y) \, dx \, dy = \int_2^5 \left(\int_{x+1}^{-4x+26} curry(g)(x)(y) \, dy \right) dx
$$

We can consider even more complicated situations by taking a function of several variables and interpreting it in different ways. As an example, let's consider the following func expression that we used in the previous section:

```
func(x);
  value a,b,c,d,e;
  if is_number(x) then
        return a*x**4+b*x**3+c*x**2+d*x+e;
  end;
end
```

As it stands, this expression cannot be evaluated unless a, b, c, d, and e have **floating-point** or **integer** values. If they all have such values, then the value of this expression will be a polynomial of degree at most 4—that is, a representation of a function whose process is given by such a polynomial. In Mathematics, it would be written

$$ax^4 + bx^3 + cx^2 + dx + e$$

What we are saying is that for each choice of (**floating-point** or **integer**) values for a, b, c, d, and e a (polynomial) function is determined. In other words, we have a function of 5 variables whose values are polynomials. The domain of this function is the set of all 5-tuples, that is, sequences of length 5, of real numbers. We may denote this set by $\mathcal{R}^5$. The range of this function may be taken to be the set of all polynomials, which we may denote by $\mathcal{P}$. We can denote the function by

$$pol : \mathcal{R}^5 \longrightarrow \mathcal{P}$$

The same overall situation can be interpreted in many different ways. For example, you could consider this to be a function of 6 variables: the coefficients a, b, c, d, and e and the "independent variable" x. Then, using a different name (because it is a different function), you could write $q(a,b,c,d,e,x)$ to mean the value at x of the fourth degree polynomial whose coefficients are a, b, c, d, and e. This function can easily be represented as an ISETL **func** as follows (omitting the domain check).

```
q := func(a,b,c,d,e,x);
        return a*x**4+b*x**3+c*x**2+d*x+e;
    end;
```

Now, it is possible to use q to do a number of things. For example, the following **func** will accept a value, a0, for a and return a **func** that will generate all fourth-degree polynomials whose leading coefficient is a0.

```
r := func(a);
        return func(b,c,d,e);
                return func(x);
                        return q(a,b,c,d,e,x);
                    end;
            end;
    end;
```

To test your understanding of these ideas, see if you can tell what the following equations mean and if you agree that they are true.

$$q(a, b, c, d, e, x) = pol(a, b, c, d, e)(x)$$

$$r(a) = q(a, \cdot, \cdot, \cdot, \cdot, \cdot) = pol(a, \cdot, \cdot, \cdot, \cdot)$$

Finally, we point out that a function of two variables may be considered as a *binary operation*. You studied examples of binary operations such as *max*, *min*, *gcd*, and *lcm* in Sections 1.4 and 1.5. You may recall that at that that time you constructed your own binary operation, such as $a6$ and $m6$.

Summary of Section 4.3

This section was the meat of the chapter. The eight subsections of this section were concerned with eight kinds of things you can do with functions, and if you are comfortable with all of these operations, your understanding of the function concept is very strong. Each way of representing a function (mathematical process, **program**, **func**, **smap**, **tuple**, and **string**) carries with it the means of evaluating a function at a point and checking the equality of two functions. You can also find the image of a set under a function by applying it to every element of the set. This can even be reversed to find the inverse image of a set or a single point.

In any future study of Mathematics, it will be impossible to avoid adding, subtracting, multiplying, dividing, and composing functions. You will also find many situations in which it is necessary to calculate the inverse of a given function. A number of mathematical topics will have interpretations as partial sum operators, and you have probably seen functions of several variables in other courses.

Most of the work with functions in this chapter concentrates on the "function process," that is, the transformation of values in the domain into values in the range. Once you understand this process very well, you can think of its totality. That is, focusing on a single function, instead of getting inside it and working with the details of the transformations, you can step back and think of the "whole function." This becomes a single object that you can compare with other such objects, put several of them in a **set** or **tuple**, iterate through them, and so on. The ability to do this is a major step toward understanding some very advanced Mathematics.

Exercises

4.3.1 Let F be the function represented in Exercise 4.2.1, part d. Find two values in the domain of this function for which the values of the function are different.

4.3.2 Let G be the function represented in Exercise 4.2.1, part f. Find

    a. $G(\sqrt{17})$

    b. $G(\sqrt{17} - 5)$

    c. $G(G(\sqrt{17} - 5) + 1)$

    d. $G(G(0) - 1)$

    e. a number x such that $G(x) = x$

    f. $G^{-1}(10)$

    g. $G^{-1}(1)$

    h. the image of the closed interval from -1 to 1

    i. the pre-image of the closed interval from -1 to 1

4.3.3 Let H be the function represented in Exercise 4.2.1, part g. and let K be the function represented in Exercise 4.2.1, part d.

    a. Find $H(K, 0)(1)$.

    b. Using a very rough sketch of the function K, indicate the result of part a. on this graph.

    c. What is $H(K, -2)$?

4.3.4 Let I be the function of Exercise 4.2.1, part h., and let G be the function represented in Exercise 4.2.1, part f. Find

    a. $I(G)(-1)$

    b. $I(2G)(-1)$

    c. $I(G)(0)$

    d. $I(G)(2) - I(G(1))$

4.3.5 In Exercise 4.2.1, part j., find $f(7)$, $f(\frac{1}{2})$, and $f(-3)$.

4.3.6 If L is the function represented in Exercise 4.2.1, part l., what are $L(\texttt{"Smith"})$ and $L(\texttt{"Skallagrim"})(\texttt{"Egil"})^2 - L(\texttt{"Jones"})(\texttt{"Ron"})$?

4.3.7 If M is the function represented in Exercise 4.2.1, part n., find $M(M(\frac{3}{2}))$.

4.3.8 If x is the function represented in Exercise 4.2.1, part o., find $x(x(x(x(\text{``101''}))))$ and $x^{-1}(x^{-1}(\text{``000''}))$.

4.3.9 If y is the function represented in Exercise 4.2.1, part o., find $y(y(y(y(\text{``001''}))))$.

4.3.10 If F is the function represented in Exercise 4.2.1, part p., find $F^{-1}(\text{``8''})$, the image of the set of all even integers and the pre-image of the set of all even digits.

4.3.11 If u and V are the functions represented in Exercises 4.2.1, parts r. and s., respectively, what is the value of $V(7) \Longrightarrow u(14)$? What is the value of $u^{-1}(false)$?

4.3.12 Let t be the following tuple of tuples of strings:

```
[["Tom", "Dick","Harry"],["Peter","Paul","Mary"],
 ["Mutt","Jeff"]]
```

Find t(2..3)(2..)(1)(2..)(1)(2).

4.3.13 Changing the value of a function: If a function, F, is represented by an ISETL func, smap, tuple, or string, then the following ISETL statement will change the value of the function at a single point:

$$F(x) := y;$$

Here, x is an expression whose value can be an element of the domain of F, that is, any ISETL value if F is a func or smap, any positive integer if F is a tuple, and any integer in [1..#F+1] if F is a string. Note that this can be used to introduce a new element in the domain of F or change the value for a domain element already present. Also, an element of the domain can be removed by setting the value of the function at that point equal to OM.

   a. Give the function in Exercise 4.2.1, part e. a name and then extend it, so that it has a reasonable value for $x = -2$.

   b. Restrict the function represented in Exercise 4.2.1, part n., so that it is only defined for integers.

   c. Change the function in Exercise 4.2.1, part o., so that it represents a circular list.

   d. Change the function in Exercise 4.2.1, part v., so that it represents a circular list.

   e. Restrict the domain of the function in Exercise 4.2.1, part s., so that it is equal to a constant function.

4.3.14 Write an ISETL func that will accept two smaps and determine if they are equal as functions.

4.3.15 Write an ISETL func that will accept two funcs and a set. Your func should determine whether the two functions represented by the funcs and restricted to the given set are equal. Run your func on at least one example in which the two given funcs are different as funcs but represent the same function.

4.3.16 Write an ISETL **func** that will accept a **func** that represents a function with a finite domain, the domain of the function, and an ISETL value. Your **func** should return the pre-image of the value. Make up and run test cases taken from Exercise 4.2.1, parts d., e., and f.

4.3.17 Write an ISETL **func** that will accept a **func** that represents a function with a finite domain, the domain of the function, and a finite **set**, S. Your **func** should return the image of S. Make up and run test cases taken from Exercise 4.2.1, parts d., e., and f.

4.3.18 Write an ISETL **func** that will accept a **func** that represents a function with a finite domain, the domain of the function, and a finite **set**, **S**. Your **func** should return the pre-image of S. Make up and run test cases taken from Exercise 4.2.1, parts d., e., and f.

4.3.19 Write an ISETL **func** that accepts an **smap**, a **tuple**, or a **string** that represents a function and an ISETL value. Your **func** should return the pre-image of the value. Make up and run test cases taken from Exercises 4.2.1, parts l.–y.

4.3.20 Write an ISETL **func** that accepts an **smap**, a **tuple**, or a **string** that represents a function and a finite **set**. Your **func** should return the image of the **set**. Make up and run test cases taken from Exercises 4.2.1, parts l.–y.

4.3.21 Write an ISETL **func** that accepts an **smap**, a **tuple**, or a **string** that represents a function and a finite **set**. Your **func** should return the pre-image of the **set**. Make up and run test cases taken from Exercises 4.2.1, parts l.–y.

4.3.22 Let F be the function of Exercise 4.2.2, part d. Find $F^{-1}(0)$, the image of the set of positive even integers less than 20, and the pre-image of the set of odd integers less than 20.

4.3.23 Let F be the function of Exercise 4.2.2, part f. Let S be the set of all squares in the domain of F whose vertices are labeled alphabetically in clockwise order. Find the image and pre-image of S.

4.3.24 Let F be the function of Exercise 4.2.2, part h. Let h be the ISETL **func** that always returns the value 0. Find $F^{-1}(h)$

4.3.25 Write an ISETL **func** called **is_one_to_one** that will accept an **smap**, a **tuple**, or a **string**, test if the function it represents is one-to-one, and return *true* or *false*. Test your **func** on the examples given in Exercises 4.2.1, parts l.–y.

4.3.26 For each of the following functions, decide if it is 1-1. Explain your answer.

a. The domain and range of F is the set of real numbers. Its process is given by

$$F(x) = \begin{cases} -x & \text{if } x \le 0 \\ x+1 & \text{otherwise} \end{cases}$$

b. The domain and range of G is the set of real numbers. Its process is given by

$$G(x) = \begin{cases} -x & \text{if } x \le 0 \\ x-1 & \text{otherwise} \end{cases}$$

c. The domain and range of H is the set of integers. Its process is given by

$$H(n) = n^3 + 1$$

d. The domain and range of K is the set of real numbers. Its process is given by

$$K(t) = t^2 + t$$

4.3.27 For each of the functions in Exercise 4.3.26, determine if it is onto. If it is, give a proof. If it is not, compute its image.

4.3.28 For each of the domain and range pairs given in Exercise 4.2.10, calculate the number of functions that can be constructed that are 1-1. Do the same for onto and bijection.

4.3.29 Draw a diagram that illustrates a function by showing its domain and range as finite sets of points (encircled by some boundary) and its process by lines with arrowheads at one end. Explain how the picture can be used to show that a function is not 1-1.

4.3.30 Write an ISETL func that will accept two funcs, smaps, or tuples that represent functions and returns a representation (using the same ISETL structure as the inputs) of a function that is the sum of the two functions represented by the inputs. Test your func on appropriate pairs taken from Exercise 4.2.1.

4.3.31 Write an ISETL func that will accept two funcs, smaps, or tuples that represent functions and returns a representation (using the same ISETL structure as the inputs) of a function that is the product of the two functions represented by the inputs. Test your func on appropriate pairs taken from Exercise 4.2.1.

4.3.32 Write an ISETL func that will accept two funcs, smaps, or tuples that represent functions and returns a representation (using the same ISETL structure as the inputs) of a function that is the quotient of the two functions represented by the inputs. Your func will have to do something about the possibility of dividing by 0. Test your func on appropriate pairs taken from Exercise 4.2.1.

4.3.33 Write an ISETL **func** that will accept two **funcs**, **smaps**, or **tuples** that represent functions and returns a representation (using the same ISETL structure as the inputs) of a function that is the composition of the two functions represented by the inputs. Test your **func** on appropriate pairs taken from Exercise 4.2.1. Try to find as many pairs as you can that will produce a reasonable composed function.

4.3.34 Express mathematically the product of the two functions represented in Exercises 4.2.1, parts d. and f.

4.3.35 Express the function that is the sum of the function represented in Exercise 4.2.1, part j., and given in Exercise 4.2.2, part d.

4.3.36 Let F, G, and H be three functions that satisfy the relation $H = F \circ G$. Suppose that G is the function given in Exercise 4.2.2, part c. and H is the function whose domain and range are both the set of all real numbers and whose process is given by

$$H(x) = \begin{cases} x & \text{if } x > 0 \\ 0 & \text{if } x = 0 \\ \text{sgn}(x) & \text{if } x < 0 \end{cases}$$

Express the function F mathematically.

4.3.37 Let F, G, and H be three functions that satisfy the relation $H = F \circ G$. Suppose that F is the function given in Exercise 4.2.2, part f., and H is the function given in Exercise 4.2.2, part g. Find the function G.

4.3.38 Let F, G, and H be three functions that satisfy the relation $H = F \circ G$. Suppose that the domain of F and of H is the set of all possible words—that is, finite strings of letters. Let the process of H consist of counting the number of words containing the string "TH", and let the process of F consist of counting the words containing the string "OW". Find G.

4.3.39 Let F, G, and H be three functions that satisfy the relation $H = F \circ G$. Suppose that the domain of G and of H is the set of all triples of integers. Let the process of H consist of adding the second and third integers and let G replace the second integer of the triple by the sum of the original three integers. Find F.

4.3.40 Suppose that each of the following three functions has as its domain the set of all positive real numbers. Find all possible ways of substituting A, B, and C for F, G, and H so that the relation $H = F \circ G$

holds.

$$
\begin{aligned}
A(x) &= 1/x^2 \\
B(x) &= \log(1+x) \\
C(x) &= \log(x^2+1) - \log(x^2)
\end{aligned}
$$

4.3.41 With A and C as in the previous exercise, what is the value of $(A \circ C)^{-1}(1)$?

4.3.42 With A and B as in Exercise 4.3.40, what is the pre-image under the function $A \circ B$ of the set $\{-2, -1, 0, 1, 2\}$?

4.3.43 Write an ISETL func that will accept an smap, test if it is a bijection, and, if so, return its inverse. Test your func with appropriate examples taken from Exercise 4.2.1.

4.3.44 Write an ISETL func that will accept a func that represents a function and a finite set that is a subset of the domain of that function. Your func should determine if the restriction of the function to that subset is a bijection and, if so, return the inverse of that restriction. Test your func with appropriate examples taken from Exercise 4.2.1.

4.3.45 Is the composition of two bijections a bijection? Show that it is true or give a counterexample.

4.3.46 Let F and G be two functions that are bijections. For each of the following two relations, show that it is true or provide a counterexample.

    a. $(F \circ G)^{-1} = F^{-1} \circ G^{-1}$

    b. $(F \circ G)^{-1} = G^{-1} \circ F^{-1}$

4.3.47 Let T be a tuple of sets. Explain, as best you can, the value of the following two expressions.

    a. `%+{%*{T(i) : i in [1..n]} : n in [1..#T]}`

    b. `%*{%+{T(i) : i in [1..n]} : n in [1..#T]}`

4.3.48 Simplify the expression at the end of Section 4.3.6.

4.3.49 Let X and Y be tuples of numbers each having length n. For each of the following equations, explain why it is always true or give a counterexample.

    a. `%+[X(i)+Y(i) : i in [1..n]] = %+X + %+Y`

    b. `%*[X(i)*Y(i) : i in [1..n]] = %*X * %*Y`

    c. `%+[X(i)*Y(i) : i in [1..n]] = %+X * %+Y`

     d. `%*[X(i)+Y(i) : i in [1..n]] = %*X + %*Y`

4.3.50 Expand each of the following expressions.

    a. $\displaystyle\sum_{i=1}^{n}(x_i + y_i)^2$

    b. $\displaystyle\left(\sum_{i=1}^{n} x_i \cdot y_i\right)^2$

    c. $\displaystyle\left(\sum_{i=1}^{n} x_i\right)^2$

4.3.51 Let T be the **tuple**

$$T := [F, [G, 23], -17];$$

where F and G are representations of the functions $F, G : \mathcal{R} \longrightarrow \mathcal{R}$ with domain and range equal to the set of real numbers and process defined by

$$F(x) = 2^x, \qquad G(x) = x^2 - 4$$

What value would ISETL return if given the following expression?

$$\texttt{T(3)*(T(1)(-3)-T(2)(1)(4))+T(2)(2)}$$

4.3.52 Let F, G, and H be three functions whose domains and ranges are the set of all real numbers. Briefly describe the meaning of each of the following:

    a. $F \circ (G + H)$

    b. $(F + G) \cdot H$

    c. $F \circ G - G \circ F$

    d. $F \circ G \circ H$

4.3.53 Assuming that all compositions are valid, which of the following statements are true and why?

    a. $F \circ (G \circ H) = (F \circ G) \circ H$

    b. $F \circ G = G \circ F$

4.3.54 Let x and y be two functions whose domains and ranges are the set of real numbers, and let F denote a real number. Which of the following statements make sense and which do not? Explain your answers.

    a. $(x \circ y)(F)$

   b. $x \circ (y(F))$

   c. $x(F) + y$

   d. $(x + y)(F)$

   e. $x + y(F)$

   f. $x(F) + y(F)$

   g. $(x(F)) \circ y$

4.3.55 Let F, G, and H be three functions such that all compositions are possible. For each of the following two expressions, determine if it implies that $G = H$ (ignoring the range). If it does, explain how you can be sure. If it does not, give additional conditions on F under which it would.

   a. $F \circ G = F \circ H$

   b. $G \circ F = H \circ F$

4.3.56 Let mapset be the set defined in the text on page 198.

   a. How many 1-1 functions are represented in mapset?

   b. Write an ISETL expression whose value will be the set of *onto* functions that are represented in mapset.

   c. How many onto functions are represented in mapset?

4.3.57 Referring to the weighted Newtonian gravity function, g, defined in Section 4.3.8, what is the value of the following expressions?

   a. g_2

   b. $g_x(1)$

   c. $curry(g)(0)$

   d. $curry(g)(0)(0)$

   e. $curry(g)(1)(2)$

4.3.58 Referring to the functions defined in Section 4.3.8, which of the following equations are correct? Explain your answer.

   a. $curry(g)(4)(5) = curry(g)(5)(4)$

   b. $curry(g)(\sqrt{3})(\sqrt{2}) = curry(g)(\sqrt{2})(\sqrt{3})$

   c. $q(1, 2, 3, 4, 5, 6) = pol(1, 2, 3, 4, 5)(6)$

   d. $q(0, \cdot, \cdot, \cdot, \cdot) = pol(0, \cdot, \cdot, \cdot, \cdot)$

   e. $r(-1) = pol(-1, \cdot, \cdot, \cdot, \cdot)$

4.3.59 Referring back to the notation of Chapter 1, page 33, as well as that of Section 4.3.8, explain the meaning of the following expressions:

    a. $curry(a6)(2)$

    b. $curry(m6)$

4.3.60 Let A and B be finite sets with $\#(A) = k$. If there are a total of n functions with domain A and range B, what is $\#(B)$?

4.3.61 Let A and B be finite sets with $\#(B) = k$. If there are a total of n functions with domain A and range B, what is $\#(A)$?

4.3.62 Let F be a function, A_1 and A_2 subsets of its domain, and B_1 and B_2 subsets of its range. For arbitrary sets A and B, denote by $F(A)$ the image under F of A and by $F^{-1}(B)$ the pre-image under F of B. Which of the following statements are true? If true, explain why, and if false, give a counterexample.

    a. $F(A_1 \cup A_2) = F(A_1) \cup F(A_2)$

    b. $F(A_1 \cap A_2) = F(A_1) \cap F(A_2)$

    c. $F^{-1}(B_1 \cup B_2) = F^{-1}(B_1) \cup F^{-1}(B_2)$

    d. $F^{-1}(B_1 \cap B_2) = F^{-1}(B_1) \cap F^{-1}(B_2)$

4.3.63 Let F, G, and H be functions that satisfy $H = F \circ G$. Which of the following statements are true? If true, explain why, and if false, give a counterexample.

    a. If F and G are 1-1, then so is H.

    b. If F and G are onto, then so is H.

    c. If F and H are 1-1, then so is G.

    d. If F and H are onto, then so is G.

    e. If H and G are 1-1, then so is F.

    f. If H and G are onto, then so is F.

4.3.64 Which of the functions in Exercises 4.2.1, parts l.–y., 4.2.2, parts a.–e., and 4.2.4, parts a.–d. have an inverse? (Define the range so as to give it the best possible chance.) Where the answer is yes, find the inverse, where it is no, explain why.

4.4   Applications of Functions

In the text and exercises for this section, we describe a number of situations in which the concept of function can be used as a powerful tool for description, analysis, and construction.

4.4.1 Data Bases

A *data base* is an organized collection of information. It includes a facility for storing this information and also the ability to answer questions about the data. One important issue is the amount of data that can be stored and retrieved efficiently with respect to space and time. A second issue with which we will be concerned in this section is the ease of formulating questions about the data that the system can answer. The point is that there is no hope of predicting beforehand all of the specific queries that one might desire to be put to a data base. Rather, one tries to structure the information so that relatively simple constructions can formulate very complicated questions. This is the kind of thing for which functions are good.

A data base is usually organized into individual sets of information or *records*. A record could correspond to a person, city, school, or whatever. Each record has the same structure corresponding to the categories of information that are to be stored. For example, you can think of a data base of students' academic performance in a university. There would be one record for each student. A record might contain the name of the student; personal information, such as age, sex, height, weight, address; and information about academic performance, such as a listing, for each course taken, of the title, the grade received, and when it was taken.

In general, there are two aspects of a record in a data base. First, the record must be distinguishable from all other records. This can be achieved by using a name, number, or any feature that will permit the system to tell one record from another. Whatever is used is called a *key*. The second aspect of a record is the set of *attributes* or categories of information. Each category (such as name, age, etc.) is a set of possible values, and the data in the data base consists of a value in each category for each record. The structure of the record is the way in which this information is organized.

One conceptually simple, yet powerful, way of constructing a data base begins by constructing a set, which we will call *keys*, of abstract points, one for every record. Then, for each category a function is defined. The domain of the function is the set *keys* and the range is the set of possible values for that category. The value of a function (i.e., attribute) at a point (i.e., record) is the data in that category for that record. Ignoring issues of efficiency of performance, one can make it fairly easy to express quite complicated queries in formal language by using these functions along with the predicate calculus and set formers. The resulting language is almost identical to ISETL syntax, and so a data base can be implemented without a great deal of difficulty.

To see how this works, let's look at an example. Consider a data base that contains information about students at a certain university and their academic performance. The set **keys** of abstract points would correspond to the students—one element of **keys** for each student. Then, using **keys** as the domain, we can construct the following four functions listed with their name and a description of the result of evaluating them at a particular point of keys.

> *name*: a **tuple** of 3 **strings** corresponding to the first, middle, and last name of the student

> *age*: a positive **integer** giving the age of the student

> *address*: a **tuple** whose 6 components are a positive **integer**, 3 **strings**, a 5-digit **integer**, and a 10-digit **integer**, corresponding, respectively, to the street number, street, city, state, zip code, and telephone number of the student

> *perf*: a function whose domain is a **set** of **strings** corresponding to course titles and whose range is a 3-**tuple** consisting of a character, a 4-digit **integer**, and a **string** corresponding to the grade received, the year taken, and the semester or quarter in which it was taken

Having set up this structure one can easily write expressions, such as

$$name(x), \qquad perf(x)(\text{``MA55''})(1)$$

The value of the first is the name of the student whose key is x and the second gives the grade that this student received in the course titled MA55.

You might think of using the student's name as the key. One reason why abstract points are better is that with names there is the difficulty that occurs if two students have the same name. This difficulty does not arise with course titles and so the function *perf*, which is really a "sub-data base" embedded in the larger one, does key on the titles.

Expressions for queries can be simplified by naming certain constants corresponding to tuple indices. For example, if you defined the constant *zip* to have the value 5, then

$$address(x)(zip)$$

gives the zipcode of the student whose key is x.

You can see how to implement all of this in ISETL. A **set keys** of atoms will represent the keys and four **smaps** will represent the categories. The ISETL structure of these **smaps** is completely specified by the descriptions of the four functions. You might make the following constant assignments to increase readability.

```
first := 1;    middle := 2;    last := 3;
number := 1;   street := 2;    city := 3;
state := 4;    zip := 5;       tel := 6;
grade := 1;    year := 2;      semester := 3;
```

You might also want to define other auxiliary structures, for instance, an **smap** called **points** that assigns 4 to the grade A, 3 to B, etc., or, if you don't have two students with the same name, an **smap** that is the inverse of **name**.

Now, you are ready to express queries in ISETL code. Here is ISETL code that calculates **GPA**, the grade-point average, of the student whose key is **x**:

```
total := %+{points(perf(x)(c)(grade)) : c in courses};
GPA := total/#(courses);
```

and here is an expression whose value is the **set** of names of students with at least three failing grades on their records:

```
{name(x)(first)+" "+name(x)(middle)+" "+name(x)(last)
   : x in keys
   | (#{c : c in domain(perf(x))
          | perf(x)(c)(grade)="F"} >= 3)};
```

4.4.2 Counting Onto Functions

There are some interesting problems that, once interpreted properly, can be solved just by counting how many onto maps there are with a given domain and range. Consider the following two problems.

1. Given a set with 25 elements, how many proper subsets does it have? Recall that a *proper* subset is one that is neither empty nor the whole set.

2. You are on a bus that has specific stops but will ignore them unless someone signals that they wish to get off. If you know that 32 people get off as the bus comes to three successive stops, what is the probability that the bus actually halts at all three of these stops?

Let's see how these problems can be first interpreted in terms of counting the number of elements in a set of functions and then solved. We begin with the first one.

If S is a set, then one way to represent a subset, $A \subseteq S$, is by a function χ_A, defined as follows:

$$\chi_A : S \longrightarrow \{0,1\} \qquad \text{by} \qquad \chi_A(x) = \begin{cases} 1 & \text{if } x \in A \\ 0 & \text{if } x \notin A \end{cases}$$

We call χ_A the *characteristic function of A with respect to S*.

Now, the number of subsets of S is exactly the number of characteristic functions defined on S. Moreover, a proper subset of S is precisely one for which the characteristic function neither always has the value 1 nor always has the value

0; in other words, it must take on both values 0 and 1, that is, it must be onto.

Thus, the first problem amounts to counting the number of onto functions from a set S to $\{0, 1\}$, where $\#S = 25$. Since all but 2 of the functions from S to $\{0, 1\}$ will be onto, it is easy to count all of the functions and then subtract 2.

Counting all of the functions corresponds to counting all of the ways in which a 0 or a 1 can be assigned to each element of S. We have 2 choices of our assignment to one element of S. For each of these choices, we have 2 choices for an assignment to another element of S, making 4 choices in all for assignments to two elements. For each of these 4 choices, there are 2 choices for an assignment to a third element, making 8 in all, and so on. You can see that if S has n elements, then taking them all into account, there are 2^n total ways of assigning 0 and 1 to them. Hence, there are $2^{\#S}$ functions from S to $\{0, 1\}$ and $2^{\#S} - 2$ of them are onto.

Note that this only works if $\#S > 1$; otherwise, if $\#S = 1$, then there is exactly one function and it is onto. In our case, $\#S = 25$, so S has $2^{25} - 2 = 33,554,430$ proper subsets.

Here is the solution to the second problem:

First, we interpret it in terms of counting sets of functions. To calculate a probability of this kind, you must count the total number of events and count the number of events in which the occurence you are interested in happens. Then, you divide the second number by the first. In this problem, it means that you have to count the total number of ways in which the 32 departing passengers can distribute themselves over the 3 stops. This means that you must count the number of functions from a set T with $\#T = 32$ to the set $\{1, 2, 3\}$. Then, you must count the number of ways in which the passengers can choose the stops in such a manner that each of the 3 stops is chosen at least once. This is just the number of functions from T to $\{1, 2, 3\}$ that are onto.

Using an argument analogous to what was done in the previous problem, you count the total number of functions by observing that for each element of T there are three choices for the value, and so, altogether, there are $3^{\#T} = 3^{32}$ functions.

To obtain the number of onto functions, you subtract the number of functions that are not onto. One of our functions can fail to be onto if its image is the set $\{1, 2\}$. We can use our solution to the first problem to determine that there are $2^{32} - 2$ of these and this is the same for functions whose image is $\{1,3\}$ or $\{2,3\}$. Thus, we must subtract the $3(2^{32} - 2)$ functions whose image is a two-element set. Finally, there are exactly 3 functions whose image is a one-element set and these must be subtracted as well. Hence, the total number of onto functions is

$$3^{32} - 3(2^{32} - 2) - 3 = 3^{32} - 3(2^{32}) + 3$$

To obtain the probability, we divide the number of onto functions, by the number of functions and so the solution to the problem is

$$\frac{3^{32} - 3(2^{32}) + 3}{3^{32}} = 1 - \frac{2^{32} - 1}{3^{31}}$$

which is a number that is pretty close to 1.

These two examples illustrate a more general analysis that leads to a formula for the number of onto functions between any two finite sets. For a complete and readable discussion on which our presentation was based, see *Discrete and Combinatorial Mathematics* by Ralph P. Grimaldi, Addison-Wesley, 1985.

4.4.3 Finite State Automata

People who work with computers often think about complex questions, such as: What exactly is it possible to do with a computer? How efficiently can it be done? How can you tell if two computers are "equivalent?" An important step in dealing with such questions is to construct a formal description of a computer. Several theoreticians have done this, and the product of their considerations has led to theoretical objects, such as the Turing machine, the Von Neumann machine, the Mealy machine, and the Moore machine. Investigations of theoretical questions about computers rely heavily on manipulations of representations of such formal descriptions.

The concept of function can be used as a tool for representing a formal description of a computer. In this section, we give a very brief initial exposure to how this is done.

You can think about a computer, frozen in time, as a set of *registers*, each of which has a particular setting. This can represent a humongous amount of information, but with our capacity for abstraction, we can encapsulate it all into a single datum, the *state*. We refer to the entire environment of a computer, with all the information that it contains at a particular point in time as the *state of the computer*. You can see that this idea of state includes the memory of the machine as the values held by some of the registers. A typical computer will have a very large number of states, but nevertheless, there are only finitely many of them. Also, the states of a computer change as it runs—automatically. That is why we refer to a computer as a *finite state automaton*.

Let's look at changes of state a little more closely. Actually, nothing will happen with a computer if you leave it alone. But if some input comes in, two things happen. The state of the computer changes, and there is some output. Here is an important point. These two results do not depend only on the value of the input. They also depend on the state of the machine when the input came in. Thus, a computer considers its state and its input and produces an output. This is a function of two variables. It also considers its state and its input to change to a new state. This is another function of two variables.

Let's see how it looks with an example. Suppose you have a computer whose input is either a 0 or 1, and its output is always a 0, unless the input was the last 0 in a 3-input sequence 010, in which case it outputs a 1. For example, if the input consists of the stream

$$00010111101011000101010001010101000$$

then the output will be

$$00001000000100000010101000101010100$$

We could construct such a machine (theoretically) with 3 states. First there would be a starting state, s_0, which will always give an output of 0. If the first input is 1, then this is not the start of a 010 sequence, so the machine stays in state s_0. If the input is a 0, then the machine goes to a different state, s_1. The interpretation of s_1 is that the machine is remembering that the last input was the first term in what could turn out to be a 010 string.

The output from state s_1 is still 0 no matter what the input is, since neither 01 nor 00 completes a 010 sequence. If the input is a 0, then the state returns to s_1 since the best that can be said is that this is the first term of a 010. If the input is a 1, however, then we have two terms and the state changes to s_2.

If the input to s_2 is a 0, then a 010 string has been completed, and the output is 1. Moreover, this could be the start of another 010, so the state returns to s_1. If the input when the machine is in s_2 is 1, then we have neither a completion of an 010 nor a part of one, so the state returns to s_0.

Do you agree that this description will do what we said? Here is a picture (a *state diagram*) of the machine.

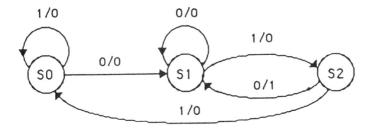

In this picture, the circles represent the states, the arrowed lines indicate the state transition, and the label x/y refers to the input x and the output y. Thus, for example, an arrow from state s_1 to s_2 with label $1/0$ means that if the machine is in state s_1 and receives the input 1, then it transfers to state s_2 and gives the output 0.

Now, the main point of this section is to see how functions can be used to describe such machines. First, there is a finite set $\mathcal{I}$ of symbols for the input. This is called the *input alphabet*, and in the above example, it was $\{0, 1\}$. Similarly, the *output alphabet* $\mathcal{O}$ was also $\{0, 1\}$ in the above example. Next, there is the set of states $\mathcal{S}$. This was $\{s_0, s_1, s_2\}$ in the above example. Finally, there are two functions. The *next state* function ν, which takes an input and a state and transforms it into a new state, can be symbolized as follows:

$$\nu : \mathcal{I} \times \mathcal{S} \longrightarrow \mathcal{S}$$

The output function ω, which takes an input and a state and transforms them into an output character, can be symbolized as follows.

$$\omega : \mathcal{I} \times \mathcal{S} \longrightarrow \mathcal{O}$$

For the above example, each of these functions has 6 elements in its domain, so we can describe them explicitly in the following table.

	ν		ω	
	0	1	0	1
s_0	s_1	s_0	0	0
s_1	s_1	s_2	0	0
s_2	s_1	s_0	1	0

Now, you can see how easy it is to implement all of this in ISETL. The finite sets $\mathcal{I}$, $\mathcal{O}$, and $\mathcal{S}$ can be represented as they are and the functions ν and ω can be represented as **smaps** since, really, a table is the same as an **smap**. You will have a chance to practice with this in the exercises.

Summary of Section 4.4

The point of this section was to illustrate how the function concept can be useful in dealing with a variety of complicated situations. In data bases, functions describe how categories are used to organize large amounts of meaningful data. Together with set operations and logical connectives, functions can make it easy to obtain information about this data. Difficult counting problems can be made tractable through the introduction of functions. As you study more Mathematics, you will learn a number of other ways to use abstract ideas in solving concrete problems. Finally, the dynamic process of the operation of a computer can be captured with functions. The study of automata theory consists largely of using the function concept to consider questions about what a computer can and cannot do—and how long it takes.

Exercises

4.4.1 Using the data base constructed in Section 4.4.1, write ISETL code that will answer the following queries.

   a. List the titles of all courses given in which there was a student older than 21.

   b. In which years was the total number of students taking both MA222 and MA346 at least 5?

   c. What is the average age of all students who live in a postal district having a zip code that starts with 100, took MA141 before 1985, and received a grade of B or better?

4.4.2 Make up your own data base that records all matches that took place during one season of a particular athletic league. The basic record should be the matches (or games) played. Categories should include the two (or more) teams (or individuals) that participated, score, winner, date, place, number of spectators, etc. Queries should include the participants, their final records, league standings on a particular date, average attendance for particular dates, teams, etc.

4.4.3 Derive a formula for the number of onto functions from a set with $n (\geq 3)$ elements to a set with 3 elements.

4.4.4 Let A and B be two sets and let χ_A and χ_B be their characteristic functions. In terms of these functions, write expressions for the characteristic functions of the following sets.

   a. $A \cup B$

   b. $A \cap B$

c. $A - B$

4.4.5 The following state diagram describes a finite state automaton.

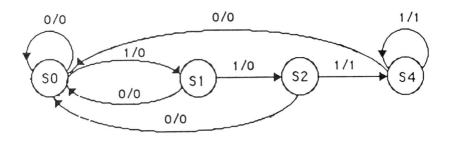

    a. Construct a table for the next-state and output functions.

    b. Explain what the machine does.

4.4.6 The following table gives both the next-state function ν and the output function ω of a finite state automaton.

	ν		ω	
	0	1	0	1
s_0	s_0	s_1	0	0
s_1	s_2	s_1	0	0
s_2	s_0	s_1	0	1

    a. Construct a state diagram for this machine.

    b. Explain what the machine does.

4.4.7 Construct the state diagram and tables for the next-state and output functions for a finite state automaton that gives an output of 1 everytime it sees a 1 that is the third of a string of three consecutive 1's and an output of 0 otherwise.

4.4.8 Construct the state diagram and tables for the next-state and output functions for a finite state automaton that gives an output of 0 if the number of 1's it has seen is even and an output of 1 if this number is odd.

4.4.9 Write an ISETL **func** that will simulate an arbitrary finite state automaton.

Chapter 5

Predicate Calculus

To the Instructor

In this chapter, our method of using ISETL to learn Mathematics begins to produce benefits. Predicate calculus refers to the study of quantified logical statements and various operations that can be performed on them. This topic is critical for understanding most of the important concepts in Mathematics. For example, it is hard to imagine how anyone could master the notions of limit and continuity in calculus, linear independence in algebra, or compactness in topology without a solid foundation in working with quantifications.

On the other hand, predicate calculus is a very difficult subject and students usually have a lot of trouble with it. Often, it is not even considered part of the undergraduate curriculum. Because it is so essential to so many topics that are part of that curriculum, however, and because our experience has been that students who study quantification according to the methods presented in this chapter *can* succeed in mastering the basic principles, we have decided to include it as a major theme in this book. Indeed, your students have already done some work with predicate calculus in previous chapters, and we will be reminding them of their past activities as the occasion arises.

No major new ISETL features will be introduced in this chapter. Rather, computer experiences will rely heavily on previous work with sets and Boolean expressions, including the ISETL constructs, `forall` and `exists`.

We assume, as a starting point, that the student has a mental image of a set as a collection of values and can imagine iterating through this collection in a `for` loop or evaluating quantified expressions, such as

```
forall x in S | P(x)              exists y in T | Q(y)
```

where P and Q are propositions with a parameter, as described in Chapter 2, that is, functions whose range is the set {*true*, *false*}, as discussed in

Chapter 4. We will call these *proposition valued functions.* We expect that the student knows what such a statement says, can express appropriate English statements in this syntax, can negate such statements, and can answer simple questions about these expressions, such as "What do we know about the truth value of the statement if we know that the set is empty?" All of this *single-level quantification* will be reviewed in the beginning of the chapter.

The next step is critical. It consists of coordinating two of these statements to obtain the familiar

for every x in S, there is a y in T such that $P(x, y)$

or

there exists an x in S such that for every y in T, $P(x, y)$

which we will refer to as *two-level quantification.* Finally, we will introduce *three-level quantification.* To see how important this is in Mathematics, just think about how many major concepts are expressed using these forms.

Our first goal for the student in this chapter is to acquire the ability to translate single-level quantifications back and forth between English and formal notation. It is crucial that the notation should be more than a string of symbols for the student. He or she should imagine the operation of the computer as it iterates a variable through a set, checks a proposition, and, depending on whether the quantification is *universal* or *existential*, returns the appropriate Boolean value. Thus, the notation should be a convenient description for the student of the mental manipulations that he or she understands and imagines performing. We achieve this goal by having the student read and write ISETL expressions. The student should be reminded as often as possible to think about how the computer might evaluate these expressions.

Here is how we handle the passage from *single-level* to *two-level* quantification, which is our next goal. We begin by having the student write a **func** that accepts a proposition valued function and a finite set that is to be its domain. The **func** returns a Boolean value obtained by "quantifying the variable out." (This is a particular case of the considerations in Section 4.3.8 of Chapter 4.) There are, of course, two such **funcs** corresponding to existential and universal quantification.

Next, these **funcs** are extended to accept a proposition depending on two variables. One of the variables is quantified out and the **func** returns a proposition depending on a single variable. Make no mistake about it, this is a major step for the student. Procedures and calculations are expected to return numbers, and when they return a function, the student will have some difficulty. Again, this was considered in Chapter 4. Most students will probably have to confront this notion more than twice before it is fully understood. The analogy with multiple integrals is exploited, although it is not obvious which way the help is going.

At this point, it is an easy programming step to compose two (or more) of these constructions to obtain two-level (and higher) quantifications. The student who has written such programs and can think about how the computer might evaluate them should develop a dynamic mental image of quantified logical propositions. It can be helpful at this point for the student to try to represent the computer operations as (nested) **for** loops.

Throughout all of this, at each level, we give the student a lot of practice with translating back and forth between English statements, ISETL syntax, and mathematical notation. Since the last two are so close, we will often blur the distinction so that the student can use the former to understand the latter. The question of negation is handled in three ways: encapsulating all but the first quantification so that negation consists of negating a single-level statement followed by negating the rest, which is another quantified statement; formal application of a rule; and negating directly from the meaning of the statement. The idea of a "generalized De-Morgan's rule" is helpful here. The student is asked to reason about various quantified statements, using the image of the computer as an aid. Some of the questions that must be dealt with can be extremely difficult and go far beyond what undergraduates usually learn to do. Even a small measure of success here may be considered to be a major achievement for the student. Finally, a number of applications of quantification are discussed.

5.1 Preview

In this chapter, you will put together some of the things that you studied in the previous three chapters and learn to handle very complicated logical statements. Don't worry about being able to manage this stuff. The material is hard, but you are ready for it. Your progress may be slow and painful (like most growth), but it will be steady. As a result, you will have acquired some skills that will be extremely useful in any Mathematics courses that you take in the future. Here are some examples of the statements that you will be analyzing.

1. For every tire in the library, there is a car in the parking lot such that if the tire fits the car, then the car is red.

2. Among all the fish flying around the gymnasium, there is one for which there exists, in every Computer Science class, a Physics major who knows how much the fish weighs.

3. For every distance in miles, there is an academic subject at Podunk U. such that any two Podunk students who are presently majoring in this subject have their respective cities of birth within that distance of each other.

4. For every positive number, ϵ, there is a positive number, δ, such that for every x in the domain of the function F, if x is within δ of c, then the value of the function at x is within ϵ of the value of the function at c.

Here are some examples of things you will learn to do as a result of studying this chapter.

- You will have a picture in your mind of the situation described by each of the above statements.

- You will be able to express these statements accurately using formal notation (ISETL syntax or mathematical notation).

- You will be able to figure out the negation of these statements.

- You will be able to answer a number of "thought" questions about these statements. For example:

    a. What do you know about the first statement, if there are no red cars in the parking lot? Suppose all of them are red?

    b. What do you know about the second statement, if there are no Computer Science classes? Suppose there are Computer Science classes but no Physics majors in any of them?

    c. Under what conditions is the third statement *true* if you know that all students at Podunk were born in the same city? Suppose you know that there is only one subject in which it is possible to major at Podunk?

    d. What do you know about the fourth statement if the function F is a constant?

- You will begin to learn about how these ideas relate to other ideas in Mathematics, such as dependence relations between variables and calculus concepts of limits and continuity.

5.2   Single-Level Quantification

In Chapters 2 and 3, you worked with statements such as:

(1) Every one of the integers b_1, b_2, b_3, and b_4 is less than 10.

(2) One of the integers c_1, c_2, or c_3 is greater than 0.

You learned to express them in ISETL syntax as follows:

```
forall b in {b1, b2, b3, b4} | b < 10;
exists c in {c1, c2, c3} | c > 0;
```

or, more compactly you could define two sets:

$$S \; := \; \{\texttt{b1, b2, b3, b4}\};$$
$$T \; := \; \{\texttt{c1, c2, c3}\};$$

so that the two expressions could be written as

```
forall b in S | b < 10;
exists c in T | c > 0;
```

Here is the way such statements are usually expressed in Mathematics:

$$\forall b \in S, \; b < 10$$

$$\exists c \in T \; \ni \; c > 0$$

There is still another way of using ISETL syntax to express these statements. It can be done by constructing a set of propositions and then using the compound operator %. Thus, for the first, we could write

```
B := {b < 10 :   b in S};
%and B;
```

and for the second,

```
C := {c > 0 :   c in T};
%or C;
```

No matter which way you choose to represent the expression, the important thing to keep in mind is the action that the syntax represents. Thus, for the first statement, which is a single-level *universal* quantification, the variable b is iterated through all of the values in the set S, and for each value of b the proposition $b < 10$ is evaluated. If the result is *true* for *every* one of these tests, then the entire expression has the value *true*. Thus if the set of numbers S is $\{3, -5, 7, 9\}$, then (1) has the value *true*, but if S has the value $\{3, 19, -34, 12\}$ then (1) is *false*. As you think of how the computer might be performing these calculations, you should try to develop a mental image of the operations that you can use for thinking about computations to be done on paper or in your head.

The same idea holds for the second statement, which is a single-level *existential* quantification, except that the entire expression has the value *true* if the test succeeds *at least once* and *false* if it fails every time. Thus, if T has the value $\{3, -5, -7\}$, then (2) is *true*, whereas if T is $\{-3, -5, -7\}$, then (2) is *false*.

These two expressions have a general form that is important to keep in mind. The two forms in ISETL and mathematical notation are

```
forall b in S | P(b);                ∀b ∈ S,  P(b)
```

which you can read as "for all values of b in S, it is the case that P evaluated at b is *true*," and

```
exists c in T | Q(c);
```
$$\exists c \in T \ni Q(c)$$

which you can read as "there exists at least one value of c in T such that Q evaluated at c is *true*."

In these forms, P and Q represent Boolean expressions, such as $b < 10$ or $c > 0$, which depend on one or more parameters; S and T can be any sets; and b and c represent variables that iterate through S and T, respectively, and provide actual values for the expressions P and Q.

In this kind of situation b and c are referred to as *bound variables* or *dummy variables*. It is exactly the same as the variable i in

$$\sum_{i=1}^{N} i^2$$

or t in

$$\int_0^{\pi} \sin(2t)\, dt$$

One important comment is that the Boolean expressions P and Q are functions, so, as we saw in Chapter 4, they could be represented as ISETL **funcs**. The only restriction is that the range of the function is {*true, false*}, so the **func** must return a **Boolean** value. For example, we could define the following **func**:

```
POL_INEQ := func(n);
                return 2**n > 2*(n**2) - n - 8;
            end;
```

Then, we could write ISETL expressions, such as

```
exists x in [1..10] | POL_INEQ(x);
forall i in {1,2,3} | POL_INEQ(i);
not exists n in [9..186] | not POL_INEQ(n);
```

Can you read these statements? How would you express in English the negation of each statement, that is, the assertion that it is *false*?

A Boolean function can also be represented as a **tuple** provided the domain is a finite set of integers, that is, the bound variable runs through a finite set of integers. More generally, it can be represented as an ISETL *smap* as long as the domain is some finite set. There were many examples of this in Chapter 4.

Quantification is a very powerful structure that can be useful (even in the single-level case) to express a wide variety of situations. Recall, for example, the chemical factory problem introduced in Chapter 2 (page 95). In that problem, three **strings** representing materials were chosen, and one of the requirements was: "For at least one of the materials chosen, both of the following conditions must hold." If M1, M2, and M3 represent the materials and COND1 and COND2 are the names of the conditions that have been expressed as **smaps** or **funcs**, then this requirement can be written

```
exists M in {M1, M2, M3} | COND1(M) and COND2(M);
```

You might think about how to express this statement if there were a lot more conditions. We will see at the end of this chapter how the structures of ISETL can be used to solve the chemical factory problem with a very short, elegant, and, yet, totally understandable program.

Here is another application of quantification in a different context. Suppose that F is a function whose domain is the set of real numbers. Let c be a number in the domain of F and ϵ and δ given positive numbers. Suppose you wanted to express the statement that any number in the domain of F that is close to c (as measured by δ) has its corresponding value of the function close to $F(c)$ (as measured by ϵ). You could express this very compactly as

$$\forall x \in dom\_F, \quad |x - c| < \delta \Longrightarrow |F(x) - F(c)| < \epsilon$$

Here, dom_F stands for the domain of F. This expression reads, *for every x in the domain of F, if the absolute value of x minus c is less than δ, then the absolute value of F of x minus F of c is less than ϵ.*

Now, let's take a look at negation. Recall that in Chapter 2 we made use of the fact that an expression like **exists x in S | P(x)** could be considered as a list of statements—one for each **x** in the finite set **S**, all connected by the operator **or**. Then, using DeMorgan's law, the expression is negated by negating each of the statements in the list and replacing each operator **or** by the operator **and**. In other words, the quantifier **exists** means that at least one of the statements is *true*; thus, the negation is simply the proposition that *all* of the statements are *false*.

For example, consider the statement

```
exists C in {C1, C2, C3} | C > 0;
```

It can be considered as the following connected list of statements:

```
(C1 > 0) or (C2 > 0) or (C3 > 0);
```

so its negation is

```
(C1 <= 0) and (C2 <= 0) and (C3 <= 0);
```

or, in more compact notation,

```
forall C in {C1, C2, C3} | C <= 0;
```

This method of negation is completely general. For example, if **Lot** is the set of all cars in the parking lot; **Tires** is a set of tires; **Fits** is a representation (as a **func** or **smap**) of a function whose domain is the set of all pairs [**tire, car**], where **tire in Tires**, **car in Lot**, and **Fits** tells whether the **tire** fits the **car**; and **Red** is a function whose domain is **Lot** that tells if the **car** is red, then the statement

> There is a car in the parking lot such that if the tire fits the car, then the car is red.

can be represented in ISETL as

> `exists car in Lot | Fits(tire, car) impl Red(car);`

and its negation in ISETL is

> `forall car in Lot | Fits(tire, car) and not Red(car);`

or, in English,

> Every car in the parking lot is fit by the tire and is not red.

The universal quantifiers work the same way. For example (recall the definition of POL_INEQ on page 230), the statement that the inequality

$$2^n > 2n^2 - n - 8$$

is *true* for $n = 1, 2, 3$ is represented in ISETL as

> `forall n in {1, 2, 3} | POL_INEQ(n);`

so its negation is

> `exists n in {1, 2, 3} | not POL_INEQ(n);`

or, in English, the original statement was *the inequality holds for each value of n*, so its negation is the assertion *the inequality is false for at least one of these values*—that is,

$$2^n \leq 2n^2 - n - 8$$

for at least one value of $n = 1, 2, 3$. (Compare this to Exercise 3.5.3.)

Now, we come to the consideration of *extreme* cases. Suppose, in expressions such as `forall x in S | P(x)` or `exists x in S | P(x)`, the set S is empty? It is probably fairly obvious to you that the second statement, the *existential* quantification, should be *false*. We would like to convince you that it makes sense for us to agree that, in this situation, the first statement, the *universal* quantification, is *true*—it comes from the negation. As we just saw, the negation of

> `forall x in S | P(x);`

is

> `exists x in S | not P(x);`

which, if S is empty, is *false*. Hence, the original statement is *true*. Notice here that if S is empty, the value of P(x) does not matter.

In general, the expression

$$\texttt{forall x in S | P(x)}$$

is equivalent (has the same truth value for any **set** S and function P) as

$$\{\texttt{x : x in S | P(x) } \} \texttt{ = S}$$

and

$$\texttt{exists x in S | P(x)}$$

is equivalent to

$$\{\texttt{x : x in S | P(x) } \} \texttt{ /= \{ \}}$$

Moreover, this all works when S is the empty set.

It might help for you to think about these quantifications in terms of ISETL **for** loops. Here is how a universal quantification would look:

```
G := { };
for x in S do
    if P(x) then G := G with x; end;
end;
G = S;
```

This code is identical to

$$\texttt{forall x in S | P(x);}$$

and you can see explicitly in the former that if S is empty, the loop will be omitted and *true* will be written.

Similarly, for existential quantification, it is only necessary to change the last line of this code from

$$\texttt{G = S;}$$

to

$$\texttt{G /= \{ \};}$$

To see how this works in an example, observe that if there are no cars in the parking lot then it is not the case that *there is a car in the parking lot such that if the tire fits the car then the car is red,* so that, if there are any tires in the library, the statement

> For every tire in the library, there is a car in the parking lot
> such that if the tire fits the car, then the car is red.

from the Preview is *false;* whereas, if there are no tires in the library, then it is *true* whether there are cars or not.

Of course, propositions that are quantified statements can be operated on by the standard operators, such as **and**, **or**, and **impl**, just like any propositions. Thus, in the above notation, the statement

If every b in $\{b_1, b_2, b_3, b_4\}$ is less than 10, then there exists a
c in $\{c_1, c_2, c_3\}$ is positive

can be expressed in ISETL as

```
(forall b in {b1, b2, b3, b4} | b < 10) impl
                  (exists c in {c1, c2, c3} | c > 0);
```

A single-level quantification can work on more than one variable running
through the same or different sets. Thus, statements such as

There are numbers x and y in the set $\{b_1, b_2, b_3, b_4\}$ such that
x is more than twice y.

For every x in the set $\{b_1, b_2, b_3, b_4\}$ and y and z in the set
$\{c_1, c_2, c_3\}$, x is less than the sum of y and z.

can be expressed in ISETL as

```
exists x,y in {b1, b2, b3, b4} | x > 2*y;
forall x in {b1, b2, b3, b4}, y,z in {c1, c2, c3} | x<y+z;
```

and in mathematical notation,

$$\exists x, y \in \{b_1, b_2, b_3, b_4\} \; \ni \; x > 2y$$

$$\forall x \in \{b_1, b_2, b_3, b_4\}, \; y, z \in \{c_1, c_2, c_3\}, \; x < y + z$$

Summary of Section 5.2

In this section we studied propositions that were single quantifications—
either existential (there exists) or universal (for all)—over sets. A quantification can be expressed in English, ISETL syntax, or mathematical
notation. It is important to be able to translate back and forth between
these representations. The statement has the same meaning whichever form
is used, and you should always make sure you understand this meaning.
Whenever possible, enter the ISETL statements that occur in the following
exercises, assign appropriate values, and evaluate the expressions. It helps
to think about what operations the computer might be doing as it performs
the evaluation.

Quantifications can be used to describe complicated situations. We
discussed applications to a manufacturing situation and to Mathematics.

A quantification over a finite set can be considered as a list of propositions connected by conjunctions (and) for univerals and disjunctions (or)
for existential quantifiers. The statement can then be negated by using
DeMorgan's laws. In this way, you can develop the ability to negate a
quantified statement formally and then learn how to do it directly from
the meaning.

Finally, this section contains a first introduction to reasoning about
quantified statements—to determine the truth or falsity of the statement
in light of auxiliary information.

Exercises

5.2.1 Translating   back and forth between English, ISETL syntax, and mathematical notation.

   a. Translate each of the following ISETL expressions into as simple and clear English as you can.

   i) `exists x in dom_F |`
            `(abs(x-c) < d) and (abs(F(x)-F(c)) > e);`
   (Here, `F` is a function, `dom_F` is the domain of this function, `c` is an element of `dom_F`, and `d`, `e` are fixed positive `floating-point` numbers.)

   ii) `exists ch in "banana" |`
            `ch in { "a", "e", "i", "o", "u"};`

   iii) `forall s,t in {7,9..33} | (s-t) > 1;`

   iv) `forall wd in stringset | ch in wd;`
   (Here, `stringset` is a set of character **strings** and `ch` is a character.)

   v) `forall N in [1..50] | (2**N > 2*N**2 - N - 2)`
            `impl (2**(N+1) > 2*(N+1)**2 - N - 1);`

   vi) `exists N in [3..12], M in [-7,-11..-20]`
            `| N+M+16 < 0;`

   vii) `forall k in [1..100] |`
            `%+[i**2:i in [1..k]] = k*(k+1)*(2*k+1)/6;`

   viii) `exists s,t in [1..u] | 3*s + 5*t = u;`
   (Here, `u` is an integer.)

   ix) `forall i in [P..Q] | (i mod 2 = 0) or`
            `((i+1) mod 2 = 0);`
   (Here, `P` and `Q` are positive integers.)

   x) `forall r in {-1..b} | 2**(r+1) > %*[1..(r+2)];`
   (Here, `b` is an integer.)

   b. Express each of the statements from part a. in mathematical notation.

   c. Translate each of the following mathematical expressions into as simple and clear English as you can.

   i) $\forall x \in \{2, 4, ..., 10\}, \quad 4 \le P(x) \le 100$
   (Here, P is a function that assigns to an integer x, the integer x^2.)

   ii) $\exists y \in S \ni y > 2 \implies y$ odd
   (Here, S is a set of integers.)

   iii) $\exists c \in st \ni c$ is a vowel
   (Here, st is a character **string**.)

    iv) $\forall c \in \text{dom\_G}, \quad G(c) > 0$
    (Here, G is a function whose domain is dom_G.)

    v) $\forall x \in S, \quad x > 3.7$
    (Here, S is a set of floating-point numbers.)

d. Express each of the statements from part c. in ISETL syntax. Where necessary, specify auxiliary objects.

e. Express each of the following statements in both mathematical notation and ISETL syntax. If any auxiliary objects (sets, functions, propositions, etc.) are involved, describe them or construct them explicitly.

    i) Among all the teams in the league, one must finish last in the standings. (Use $<$ to express the relative ranking of teams.)

    ii) Every character in the **string** appears twice.

    iii) Every integer in the set is either less than 10 or odd.

    iv) One of you is lying.

    v) School is canceled every time it snows.

    vi) There is a mayor of a city in Massachusetts and a governor of a state west of the Mississippi who have the same last name.

    vii) Any star in the heaven that moves back and forth is a planet.

    viii) None of my students this semester had trouble with quantification.

    ix) All of the socks in my drawer are of different colors.

    x) The binary operation has an identity.

5.2.2 Negating single-level quantifications. The negation of a quantification is expanded by replacing the universal (existential) quantification with an existential (universal) quantification and negating the rest. This is easy to do if the statement is expressed in ISETL syntax or mathematical notation. If it is given in English, then it can be translated into a formal expression, negated, and translated back. After some practice, you should begin to develop the ability to negate directly from the English version.

a. Negate each statement in Exercise 5.2.1, part a., and translate the result into English.

b. For each of the following statements, translate it into ISETL syntax or mathematical notation, negate it, and then translate the negation back into English.

    i) All the cranes are flying.

ii) The equation $x^2 - x - 6 = 0$ has an integer solution.

iii) Every solution of the equation $x^2 - x - 6 = 0$ is an integer.

iv) Everyone holding a first class ticket or assigned to a seat in rows 21 to 40 boards the airplane.

c. Negate each statement in Exercise 5.2.1, part e., directly from English without using its formal expression.

d. Negate each of the following statements directly from English.

i) There is an integer x between 0 and 11 such that $(x + 3)$ mod $12 = 0$.

ii) Whenever the pair $[a, b]$ is in the set, the pair $[b, a]$ is in the set as well.

iii) Given a particular car, there is a tire in the library such that if the car is red, then the tire fits the car.

iv) Every student in the school and every teacher on the faculty looks forward to winter break.

v) One day next year there will be an eclipse of the sun.

vi) If any book in the library has been defaced, the library will be closed.

vii) There is a rational number that cannot be expressed in ISETL.

5.2.3 Reasoning about propositions. It may be helpful to imagine using a **for** loop to evaluate a quantification.

a. Each of the following exercises refers to one of the propositions given in a previous exercise in this section. In each case you are given additional information about the situation. Your task is to decide if, based on this information, you can be sure of the truth value of the statement. Explain your answer as fully as you can.

i) In Exercise 5.2.1, part a. iv), **stringset** is the null **string**.

ii) In Exercise 5.2.1, part a. viii), **u=0**.

iii) In Exercise 5.2.1, part a. x), **b=-2**.

iv) In Exercise 5.2.1, part c. ii), $\forall y \in S,\ y \leq 2$.

v) In Exercise 5.2.1, part c. ii), $\forall y \in S,\ y > 2$.

vi) In Exercise 5.2.1, part c. ii), $\forall y \in S,\ y$ mod $2 = 0$.

vii) In Exercise 5.2.1, part c. ii), $\forall y \in S,\ y$ mod $2 \neq 0$.

viii) In Exercise 5.2.2, part d. iv), there are no students in the school.

b. Give a proof or counterexample for each of the following statements. You may assume that neither S nor T is empty.

i) $(\forall x \in S, \ P(x)) \Longrightarrow (\exists y \in T \ni Q(y))$
   $\quad = \exists x \in S, y \in T \ni \neg P(x) \vee \neg Q(y)$

ii) $\neg(\forall x \in S, \ P(x)) \Longrightarrow (\exists y \in T \ni Q(y))$
   $\quad = \forall x \in S, \ y \in T, \ (P(x) \wedge \neg Q(y))$

iii) $(\exists x \in S \ni \forall y \in T, \ P(x,y))$
   $\quad \Longrightarrow (\forall y \in T, \ \exists x \in S \ni P(x,y))$

iv) $\forall x \in S, ((\forall y \in T, \ P(x,y))$
   $\quad = \forall y \in T, ((\forall x \in S, \ P(x,y))$

v) $\exists x \in S \ni (\exists y \in T \ni P(x,y))$
   $\quad = \exists y \in T \ni (\exists x \in S \ni P(x,y))$

5.2.4 Representing a single-level quantification as a **func**.

    a. Write two ISETL **funcs** that will each take a set S and a representation P of a proposition valued function (whose domain contains S) as parameters and return the truth value of the corresponding universal or existential quantifier over S.

    b. Run your two **funcs** for as many of the quantifications that appear in these exercises as you can.

    c. What would happen if you ran one of your **funcs** in a situation where S is not a subset of the domain of P?

5.2.5 Forming sets. Write ISETL one-liners to form the following sets.

    a. The set of those characters in the **string** that are followed later in the **string** by a vowel.

    b. The set of characters in the **string** such that all preceding characters are consonants.

    c. The set of **smaps** in S that represent functions that have at least one point in their domain for which the value of the function at that point is 0.

    d. The set of values in D that are zeros for every function in the set F. (Here, F is a set of functions whose domain and range is the set of real numbers and D is a finite set of real numbers.)

    e. The set of elements in H that, in the operation $\oplus$, commutes with every element of H. (Here, H is a finite set with a binary operation $\oplus$.)

5.3 Two-Level Quantification

The operation of a single-level quantification, for example, an existential quantifier, can be considered as a function, *EXI*, that takes as input a set S and a function P and returns a Boolean value. Here, P is a function that takes any element of S and returns a Boolean value. Thus, if we consider the statement

There exists a car in the parking lot that is red.

then we could take $S = Lot$ to be the set of all cars in the parking lot and P to be the function that takes one of these cars and returns *true* if it is red, *false* otherwise. The value of $EXI(S, P)$ would then be the truth or falsity of the above statement.

Suppose, however, that P depended on more than one variable. In the last section, for example, we had the statement, `Fits(tire, car) impl Red(car)` in which the proposition also depended on a tire. Thus, for a given *tire* and *car*, $P(tire, car)$ would have for its value the truth or falsity of this statement. We could consider *EXI* now to describe the following statement:

```
exists car in Lot | Fits(tire, car) impl Red(car);
```

Thus, $EXI(S, P)$ still takes a set S and a function P as input, but the function P now depends on *two* variables. Moreover, the value of $EXI(S, P)$ is not a Boolean value, but a Boolean *function* that, for a given *tire*, evaluates the truth or falsity of this new statement. In thinking about this, you should try to recall the examples that you studied in Chapter 4, Section 4.3.8, where the range of a function was a set of functions.

To distinguish between the case in which *EXI* returns a Boolean value and when it returns a function, let us call this new function *EXI_2* and the previous one *EXI_1*. The 1 and 2 refer to the number of variables on which the function P is to depend. We can represent each of them as an ISETL func.

```
$P is a function of one variable and S subset domain(P)
EXI_1 := func(S,P);
            return exists x in S | P(x);
        end;


$P is a function of two variables and T subset domain(P(x,·))
EXI_2 := func(T,P);
            return func(x);
                    exists y in T | P(x,y);
                end;
        end;
```

A similar construction can be done with a universal quantifier, and we will discuss this in the exercises. The resulting functions will be called *UNI_1 and UNI_2*, respectively.

Now, we can see how to put two of these statements together to construct a two-level quantification. Consider statement 1 from the Preview, which we parse with parentheses.

For every tire in the library, (there is a car in the parking lot such that if the tire fits the car, then the car is red).

Look at the part inside the parentheses first. Taken as a whole, it is a function that takes a tire and returns a Boolean value. It also makes use of auxiliary functions *Fits* and *Red*. We can describe it with *EXI_2*. If *Lot* is the set of cars in the parking lot and *P* is the function of two variables defined by

$$P(tire, car) = \text{if the tire fits the car, then the car is red}$$

or, in more mathematical notation,

$$P(tire, car) = (Fits(tire, car) \implies Red(car))$$

then the part of statement 1 inside the parentheses is described by

$$EXI\_2(Lot, P)$$

which is now a Boolean function of *one variable*. We can apply *UNI_1* to it and the set *Tires* of all tires in the library to obtain the expression

$$UNI\_1(Tires, EXI\_2(Lot, P))$$

If everything has been properly defined in ISETL, then the entire statement can be evaluated in ISETL by the code

```
UNI_1(Tires,EXI_2(Lot,P))
```

It can also be evaluated directly in ISETL by the code

```
forall tire in Tires |
        exists car in Lot | Fits(tire, car) impl Red(car);
```

or in mathematical notation as

$$\forall t \in T, \ \exists c \in L \ \ni \ Fits(t, c) \implies Red(c)$$

where $T = Tires$ and $L = Lot$. It might be useful, in working with nested quantifications, to think about a nested loop in ISETL. Thus you can imagine ISETL evaluating the above statement by running through the following loop (cf. page 233):

```
result1 := true;
for tire in Tires do
      result2 := false;
      for car in Lot do
            result2 := result2
                        or (Fits(tire,car) impl Red(car));
      end;
      result1 := result1 and result2;
end;
result1;
```

This code is a lot less convenient than what you get by using `forall` and `exists`, and the latter will also be helpful when you think about these constructions in more complicated situations. For a long time, however, you might want to run through nested loops in your mind in order to see clearly what is going on. When you are ready to dispense with this aid, it will disappear from your mind, naturally.

As another example, but this time with the order of universal and existential quantification reversed, try to see if you can figure out what the following mathematical expression says:

$$\exists x \in \{-4, -3, \ldots, 16\} \; \ni \; \forall y \in \{0, -1, \ldots, -4\}, \; x = y^2$$

The same thing can be expressed with the following three lines of ISETL code.

```
set1 := {-4..16};
set2 := {0,-1..- 4};
exists x in set1 | (forall y in set2 | x = y**2);
```

Notice, by the way, that the order in which the quantifiers are written makes a difference. Consider the statement we have just been discussing together with what we obtain by reversing this order.

$$\exists x \in \{-4, -3, \ldots, 16\} \; \ni \; \forall y \in \{0, -1, \ldots, -4\}, \; x = y^2$$

$$\forall y \in \{0, -1, \ldots, -4\}, \; \exists x \in \{-4, -3, \ldots, 16\} \; \ni \; x = y^2$$

The two expressions have very different meanings. The first says that there is a number in the set $\{-4, \ldots, 16\}$ that is equal to the square of *every* number in the set $\{0, -1, \ldots, 4\}$, whereas the second says that the square of each number in the set $\{0, -1, \ldots, 4\}$ is contained in the set $\{-4, \ldots, 16\}$ You can see that the first statement is *false*, while the second is *true*.

Two-level quantifications are similar in many ways to the double integrals we discussed briefly in Section 4.3.8 of Chapter 4. Suppose that F is a function of two variables, for example,

$$F(x, y) = \frac{x^3 + y^3}{x + y}$$

where x is to range over the interval $[1, 2]$ and y over the interval $[0, 1]$. Then, we can "integrate out" the variable y to obtain the function G of one variable,

$$G(x) = \int_0^1 F(x, y) \, dy$$

and then integrate G to obtain

$$\int_1^2 G(x) \, dx$$

which is a number. The whole process is often written as an iterated integral.

$$\int_1^2 \int_0^1 F(x,y)\,dy\,dx$$

You might think that this analogy breaks down when it comes to reversing the order. In fact, however, although it is true for most reasonable functions that reversing the order of integration does not change the value, in more advanced Mathematics courses, you will see examples in which this is not the case.

There is another kind of situation that can be handled with a two-level quantification. Consider, for example, the chemical factory problem introduced in Chapter 2 (page 95) and discussed earlier in this chapter (page 230). The second requirement in that problem was the following:

> For at least one of the three materials chosen, both of the following conditions must hold.

Let us denote, as before, by M1, M2, and M3 the materials chosen, and the two conditions which must hold by COND1 and COND2. As before, these are functions (represented by smaps or funcs or even tuples) that accept a material and return a Boolean value. Suppose that CHOICES is a set of choices of the three materials, that is, each element of CHOICES is a set of three materials. It is very easy now to express in ISETL the statement that the above requirement holds for every choice in CHOICES. It looks like this:

```
forall ch in CHOICES |
        (exists m in ch | (COND1(m) and COND2(m)));
```

In this case, the proposition COND1(m) and COND2(m) depends on only one variable m, but the set ch through which m varies also varies under control of the outer quantification. Thus, the expression exists m in ch | (COND1(m) and COND2(m)) is again a proposition valued function whose independent variable is ch. It is then possible to apply a second quantification to this proposition valued function, which is how the full statement is obtained.

We can express this situation generally in mathematical notation as follows:

$$\forall x \in S,\ \exists y \in T(x)\ \ni\ P(y)$$

Here, the expression $\exists y \in T(x)\ \ni\ P(y)$ represents a proposition valued function (with domain variable x), which we may call Q, so that the entire expression has the form

$$\forall x \in S,\ Q(x)$$

Of course, it is still possible that P depends on both x and y, which would give us the form

$$\forall x \in S,\ \exists y \in T(x)\ \ni\ P(x,y)$$

and nothing else is changed. In the exercises, you will have the opportunity to describe this situation with ISETL **funcs** analogous to our *UNI* and *EXI* **funcs**.

Now, we consider the problem of negating a two-level quantification. It is not difficult to negate such a statement if you think of it as an iteration of two quantifications, each of which you learned how to negate in the previous section. Consider, for example, the general form

$$UNI\_1(S, EXI\_2(T, P))$$

where P is a Boolean valued function of two variables, one of which ranges through the set S and the other ranges through the set S. Since the value of $EXI\_2(T, P)$ is, as we have seen, a Boolean valued function of one variable, we can give it a name, Q, and express the original proposition as

$$UNI\_1(S, Q) \qquad \text{where} \qquad Q(x) = EXI\_2(T, P)(x), \quad x \in S$$

The negation of this statement, according to what we learned in the previous section is

$$EXI\_1(S, \text{ not } Q)$$

That is, the original statement says that Q is *true* for every value of its parameter in the set S so the negation is the statement that Q is *false* for at least one of these values.

It remains to negate Q for a value of its parameter in S. For convenience, let us call this parameter x and denote the other parameter that ranges through T by y. Now, $Q(x)$ is the statement that there is a value of y in T for which $P(x, y)$ is *true*, so its negation is the statement that, for this value of x, $P(x, y)$ is *false* for every value of y in T, that is,

$$\text{not } Q = UNI\_2(T, \text{ not } P)$$

Substituting this in the negation of the entire statement, we obtain

$$EXI\_1(S, \text{not } Q) = EXI\_1(S, UNI\_2(T, \text{ not } P))$$

The whole operation can be expressed as a simple rule. Once the statement is expressed in terms of *UNI* and *EXI*, it can be negated by replacing every occurrence of *UNI* by *EXI* and every occurrence of *EXI* by *UNI* and negating the innermost proposition valued function.

Let's see how it looks in a particular example, such as (once again) statement 1 in the Preview:

> For every tire in the library, there is a car in the parking lot
> such that if the tire fits the car, then the car is red.

Proceeding as above, we can reason that the negation of this statement is

> There is a tire in the library for which the rest of the statement is *false*.

The *rest of the statement* is

> there is a car in the parking lot such that if the tire fits the car, then the car is red

and its negation is

> for every car in the parking lot the rest of the statement is *false*

Now, the *rest of the statement* becomes

> if the tire fits the car, then the car is red

and its negation is

> the tire fits the car, but the car is not red

Putting the last two together we obtain

> for every car in the parking lot, the tire fits the car but the car is not red

Substituting this back in our first negation we obtain as the negation of our original statement

> There is a tire in the library such that for every car in the parking lot, the tire fits the car, but the car is not red.

Thus, you see that a two-level quantification is considered as a nested quantification. It is negated by first considering the inner quantification as a single proposition valued function. Then, the outer quantification is negated and finally the inner one is negated. The same way of thinking works when the set in the inner quantification depends on the variable of the outer quantification. Consider again the statement from the chemical factory problem, which we expressed (see page 242) in ISETL as follows:

```
forall ch in CHOICES |
        (exists m in ch | (COND1(m) and COND2(m)));
```

It has the general form

$$\forall x \in S, \; Q(x) \qquad \text{where} \qquad Q(x) \; = \; \exists y \in T(x) \; \ni \; P(y)$$

Hence, its negation has the general form

$$\exists x \in S \; \ni \; \neg Q(x) \qquad \text{where} \qquad \neg Q(x) \; = \; \forall y \in T(x), \; \neg P(y)$$

and we obtain for the negation of the original statement

```
exists ch in CHOICES |
        (forall m in ch | (not COND1(m) or not COND2(m)));
```

or, in English,

> There is at least one choice of the three materials such that
> for each material chosen one or both of the conditions fails.

To summarize, our method for negating a two-level quantification is to replace the first quantification by an existential if it is universal or replace it by a universal if it is existential and then negate the rest of the statement— which may, of course, involve other quantifications. If you practice thinking about negation in this way, when you work on the exercises, you won't have much trouble extending this to three-level quantifications and beyond.

At this point, you have worked with two-level quantifications enough so that you can begin to think about the meaning of such statements and to figure out what happens under various special circumstances. Consider for example, statement 1 of the Preview:

> For every tire in the library, there is a car in the parking lot
> such that if the tire fits the car, then the car is red.

What is the value of the statement if there are no tires in the library? It is *true* because, as we saw in the previous section, a universal quantification over an empty set is always *true*. Now, supposing that there are some tires in the library, what is the situation if there are no cars in the parking lot? Then, the entire statement is *false* because an existential quantification over the empty set will always return *false*.

Here is a little harder one. You might want to think about nested loops in order to take it apart in your mind. Assuming that there are some tires in the library and there are some cars in the parking lot, suppose that none of the cars in the parking lot are red but at least one tire fits every car? Then, the implication will be *false* at least once for every tire, so the statement will be *false*. Similarly, if all of the cars in the parking lot are red or if no tire fits any car, then the implication is always *true*, so the statement will be *true*.

We close this section with one last comment on how one might think about a two-level quantification. Consider the two forms in mathematical notation:

$$\forall u \in S, \ \exists v \in T \ni P(u, v)$$

$$\exists u \in S \ni \forall v \in T, \ P(u, v)$$

If you think about the process of evaluating these statements, then, in the first one, you (or the computer) would run through all of the values of u in S trying to find a v in T for which the rest of the statement is *true*. If you are successful, that is, the two-level statement is *true*, then one by-product of determining this fact would be that for each u in S a particular v in

T was selected. This selection process can be considered to establish a dependency relation in which the value of v (that is, the choice) depends on the value of u.

Consider, for example, the following expression where F is a real valued function of a real variable, c is an element of the domain of F, and ϵ is a positive number:

$$\forall \delta > 0, \ \exists x \in \text{domain}(F) \ni |x - c| < \delta \wedge |F(x) - F(c)| \geq \epsilon$$

To say that this statement is *true* means that it is possible to set up a relationship between δ and x in which the value of x depends on the value of δ in such a way that the rest of the statement is *true*.

Looking at the second mathematical form, $\exists u \in S \ni \forall v \in T, P(u, v)$, the meaning of this statement is that it is possible to find a single value of u in S for which many statements are *true*, that is, each statement obtained by specifying a value of v in T. Thus, if we negate the above statement to obtain

$$\exists \delta > 0 \ni \forall x \in \text{domain}(F), \ |x - c| < \delta \Longrightarrow |F(x) - F(c)| < \epsilon$$

this asserts that it is possible to find a single positive number δ for which the implication is *true* for every possible choice of x in the domain of F.

Summary of Section 5.3

If the result of a quantification of a proposition valued function over a set is another proposition valued function (which can happen if the original function and/or its domain depends on a parameter), then a second "outer" quantification can be applied. In this section, we considered examples of such statements and explained how to negate them.

Aside from manipulating two-level statements formally, it is important to develop a mental image of the meaning of the statements and to be able to answer questions about them. Thinking about nested loops can help you construct such an image. It is also important to realize that the order in which the quantifications are performed will usually affect the value of the statement.

Exercises

5.3.1 Translating back and forth between English, ISETL syntax and mathematical notation.

     a. Translate each of the following statements into as simple and clear English as you can.

    i) `forall s in {1..4} |`
               `(exists t in {1..4} | (s*t) mod 5 = 1);`

  ii) $\exists t \in \{1, \ldots, 4\} \ni \forall s \in \{1, \ldots 4\}, \ s \cdot t \bmod 5 = 1$

 iii) $\forall N \in S, \ \exists k \in \{2, 3, \ldots, N-1\} \ni N \bmod k = 0$
     (Here, S is a set of positive integers.)

  iv) `exists M in Mapset |`
              `(forall x in domain(M) | [x,x] in M)`
     (Here, `Mapset` is a set of smaps.)

   v) `forall s in C`
     `| (exists y in domain(A)*image(B)`
      `|([s(1),y] in B and [y,s(2)] in A))`
     (Here, `A`, `B`, and `C` are smaps.)

  vi) $\forall s \in Stmts, \ T \in Time, \ \exists x \in Input, \ t \in Time \ni$
     $t > T \wedge P(x, t, s)$
     (Here, *Stmts* is the set of statements in a program, *Time*
     is a set of integers, *Input* is the set of all inputs to the
     program, and the value of $P(x, t, s)$ is the truth or falsity
     of the statement *given input x, the program is at statement
     s after t steps.*)

 vii) The same as vi) with P replaced by Q, where the value of
     $Q(x, t, s)$ is the truth or falsity of the statement *if s is not
     a conditional statement, then P(x,t,s) is true.*

b. Use the **EXI funcs** and analogously defined **UNI funcs** to express each of the statements in part a.

c. Translate each of the following statements into ISETL syntax. If any auxiliary objects are required, make sure to specify them.

    i) There is a vowel in every word of the text.

   ii) The five polynomials have a common zero.

  iii) Last year it rained at least one day in every week.

  iv) The binary operation *bop* on the set G has the property that there is an element z of G that, when combined with any other element of G, gives the answer z, no matter which order is used.

   v) To each even integer i from 2 to 30 there corresponds an integer j between 11 and 40 such that if j divides i, then i is larger than 14.

  vi) Every point on the graph can be connected to at least one other point on the graph.

 vii) There is a point on the graph that is connected to every other point on the graph by a path that either has length greater than 5 or passes through the point P.

viii) Every positive integer x in the set has a positive even divisor that is less than $x - 5$.

d. Translate each of the statements of part c. into mathematical notation. Make sure that in each case the sets and the variables that run through them are specified.

e. Write ISETL funcs analogous to the UNI, EXI funcs that will handle two-level quantifications of the following forms.

i) $\forall x \in S,\ \exists y \in T(x)\ \ni\ P(x,y)$

ii) $\exists x \in S\ \ni\ \forall y \in T(x),\ P(x,y)$

f. For each of your funcs in part e., make up three English statements, translate them to ISETL syntax, and run your funcs on them.

5.3.2 The order in which the quantifications are performed is important.

a. Explain why the two statements in part a. of Exercise 5.3.1, parts i) and ii), are not equal.

b. Replace the proposition valued function in part a. of Exercise 5.3.1, parts i) and ii), by a proposition valued function P such that the two statements will be equal.

c. Give three examples of pairs of sets S and T such that

$$(\exists x \in S\ \ni\ \forall y \in T,\ P(x,y))$$
$$= (\forall y \in T,\ \exists x \in S\ \ni\ P(x,y))$$

no matter what the proposition valued function P is.

d. For each of the following conditions, draw three curves in the plane representing functions F_1, F_2, and F_3 such that the conditions are satisfied.

i) $\exists x > 0\ \ni\ \forall i \in \{1,2,3\},\ F_i(x) = 1$

ii) $\forall i \in \{1,2,3\},\ \exists x > 0\ \ni\ F_i(x) = 1$

5.3.3 Negation of two-level quantifications.

a. Negate each of the statements in Exercise 5.3.1, part a., and translate the result into reasonable English.

b. Negate each of the statements in Exercise 5.3.1, part c. Try to do it directly from the English without translating into ISETL syntax or mathematical notation.

c. Negate each of the following statements directly from the English.

i) In all the classes that I have taught, there is one in which every student got an A.

ii) Every dog has his day.

iii) The program has a specific running time for each value of its input.

   iv) At least one of the six functions has a derivative that never
       vanishes.
   v) For a given $\epsilon > 0$, it is the case that for every $\delta > 0$ there
      is an x in the domain of the function such that $|x| < \delta$ but
      $|F(x)| > \epsilon$.

5.3.4 **Reasoning**   about two-level quantifications. Think about nested
loops if it helps.

   a. What do you know about the truth value of the statement in
      Exercise 5.3.1, part a. v), if you know that the domain of **A** and
      the image of **B** have no values in common? Does your answer
      in this case depend on any property of **C**?

   b. What do you know about the truth value of the statement in
      Exercise 5.3.1, part c. viii), if you know that

      i) every integer in the set is odd?
      ii) the set is $\{1,2,3,4\}$?

   c. What do you know about the truth value of the statement in
      Exercise 5.3.3, part c. v), if you know that the domain of the
      function is the set of real numbers and

      i) $F(x) = 0$ for each element of its domain and $\epsilon = 0.001$?
      ii) $\epsilon = -1$?
      iii) $F(x) = x^2$ for each element of its domain and $\epsilon > 0$?

5.3.5 The $\epsilon - \delta$ simulator: first version. The object of this exercise is to
write a **func** that will accept a positive number ϵ, a function F, and
a number c in the domain of F. The **func** is to return the truth
value of the following statement:

$$\exists \delta > 0 \ni \forall x \in \mathrm{domain}(F), \ |x - c| < \delta \Longrightarrow |F(x) - F(c)| < \epsilon$$

The problem is that domain(F) and the values of δ form infinite sets.
It is possible to "approximate" the situation by replacing $\delta > 0$ by
the condition (cf. Exercise 1.8.3)

```
delta in {n*t :   n in [1..50]}
```

where **t** is an appropriately chosen small positive number. In a
similar manner, one can approximate domain(F). (Note: When
approximating the domain of F, choose a smaller value for t than
the value used to approximate the domain of δ.)

Write such a **func** and run it for various choices of **c** with the functions defined in Exercise 1.8.4.

5.3.6 Set formers. Write ISETL "one-liners" to form the following sets.

   a. The set of integers in the set S for which there is a second
      integer in S such that the sum of the two integers is not divisible
      by any element of S. (Here, S is a finite set of positive integers.)

   b. The set of characters, ch, in the **string** such that for each
      vowel in the **string** that comes after ch, there is a character
      that comes before ch and occurs in the alphabet after the vowel.

5.4   Three-Level Quantification

What was done in the previous section to coordinate two quantifications
can be iterated for three and even more quantifications. Whatever number
of quantifications the statement has, the innermost proposition will depend
on that number of variables, each of which has a set through which it is
to vary. Consider, for example, statement 3 of the Preview, which we can
parse to show the different levels of quantification.

> For every distance in miles, (there is an academic subject
> at Podunk U. such that (any two Podunk students who are
> presently majoring in this subject have their respective city
> of birth within that distance of each other)).

If we want to express this in ISETL, then we might let **Miles** be a finite
set of nonnegative integers, **Subjects** the set of all academic subjects at
Podunk U., **Students** the set of all students at Podunk U., **Major** a function
(represented by a **func** or **smap**) that assigns to each student the subject
in which he or she is majoring, **Distance** a function that assigns to each
pair of cities their distance in miles, and **Birth** a function that assigns to
each student the city of her or his birth. Then, the statement is expressed
by the following ISETL code.

```
forall mi in Miles | exists subj in Subjects |
   forall stud1, stud2 in Students |
      (Major(stud1)=subj and Major(stud2)=subj)
         impl (Distance(Birth(stud1), Birth(stud2)) <= mi);
```

The same statement can be expressed in Mathematics (using abbrevia-
tions for names) as follows:

$$\forall m \in Mi, \quad \exists s \in Su \ni \forall s1, s2 \in St,$$
$$((Ma(s1) = s) \wedge (Ma(s2) = s)) \implies (D(B(s1), B(s2)) \leq m)$$

There are several ways in which one can go about   negating such a
statement. The most straightforward is to work directly with the parsed
English version. Thus, we can negate successively as follows.

There exists a distance in miles such that the rest of the statement is *false*.

There exists a distance in miles such that for every academic subject at Podunk U., the rest of the statement is *false*.

There exists a distance in miles such that for every academic subject at Podunk U., there exists two Podunk students such that the rest of the statement is *false*.

There exists a distance in miles such that for every academic subject at Podunk U., there exists two Podunk students such that the students are majoring in this subject, but their respective cities of birth are not within that distance of each other.

So you see that once the form of the statement is well understood, the negation can proceed in an orderly, step-by-step manner. It might be helpful in organizing your thinking to use the *UNI* and *EXI* constructs that we described in the previous section. This will result in the following ISETL expression:

```
UNI_1(Miles, EXI_2(Subjects, UNI_3(Students, P)));
```

Here, P represents a function of the three variables: Miles (for miles), Subjects (for academic subjects), and [stud1, stud2] (for pairs of students). The result of applying the function to a particular set of values of these three variables is the truth or falsity of the statement

> if the two students are majoring in the subject, then their cities of birth are within that distance in miles from each other

The negation of the entire statement can then be obtained by applying the somewhat mechanical rule of interchanging UNIs and EXIs and negating the innermost proposition. Thus, we obtain

```
EXI_1(Miles, UNI_2(Subjects, EXI_3(Students, not P)));
```

which translates directly to the same English statement of the negation that we obtained above.

The same rule can be applied if the original statement is expressed in mathematical notation to obtain the negation as follows.

$$\exists m \in Mi \ \ni \ \forall s \in Su, \ \exists s1, s2 \in St \ \ni$$
$$(Ma(s1) = s) \wedge (Ma(s2) = s) \wedge (D(B(s1), B(s2)) > m)$$

Leaving the negation and returning to the original statement, we can use the formal notation to help answer questions about it. Suppose we know that all students at Podunk were born in the same city. Then, the condition

$$D(B(s1), B(s2)) \leq m$$

is always *true*, since the left-hand side is 0 and the right-hand side is never negative. This would mean that the statement is always *true except* if the set Su of subjects is empty, in which case the statement is *false*. (Note that the set Mi of distances is not empty).

This example, hard as it may have seemed, was made easier by writing the original English version in a form that was very close to the logical structure of the statement so that the translation was fairly simple. Sometimes, the necessity of writing relatively clear English can force the logical structure to be somewhat hidden, and then it is more difficult to express the statement formally. Nevertheless, operations such as negation and other kinds of reasoning about the statement (such as consideration of extreme cases) can become hopelessly complicated unless the statement is expressed correctly in formal notation. Essentially, what you must do is use the English version to understand the situation, and then use your ability to think logically to analyze what is being described. Finally, this description is expressed formally, either in ISETL syntax or mathematical notation. Once again, you may find that thinking about nested loops in ISETL can be helpful in resolving some of the difficulties.

Consider now statement 2 in the Preview.

> Among all the fish flying around the gymnasium, there is one for which there exists, in every Computer Science class, a Physics major who knows how much the fish weighs.

This is an example in which the logical structure is not immediately obvious. You might think at first that the sequence of quantification is universal, existential, existential, universal. However, if you read the statement carefully and think about what is being described, then you realize that a simple translation of words such as *all* and *every* to universal quantifiers and words such as *there is* and *exists* to existential quantifiers does not work. In this statement, the first thing that is being said is that *there is a fish*. The initial phrase of the sentence is not a quantifier but tells you what set the fish comes from. The next thing that appears in the sentence seems to be an existential quantification. The order of the statement in English does not, however, correspond to the logical order in which the quantifications are to be performed. The meaning is that something is about to be said regarding *every Computer Science class*. It is only afterward that the statement asserts that *there exists a Physics major* for which something is *true*.

Thus, the statement can be expressed, for example, in mathematical notation, as follows:

$$\exists f \in G \; \ni \; \forall cl \in Cs, \; \exists m \in Ph \; \ni \; K(f, cl, m)$$

where G is the set of fish flying around the gymnasium, Cs is the set of Computer Science classes, Ph is the set of Physics majors, and K is a function that tells us if a Physics major m is in the class cl and knows how much the fish f weighs.

You should now be able to negate this statement formally and express that negation in English. Also, you should be able to answer questions about it, such as: What can you conclude if you know that there are no Computer Science classes? Suppose there are Computer Science classes, but no Physics majors in any of them; then what can you conclude?

Summary of Section 5.4

In this section, the considerations of Section 5.3 were extended to three-level quantifications. The usual topics of translation between various representations, negation, and reasoning about statements were discussed.

Exercises

5.4.1 Translating back and forth between English, ISETL syntax, and mathematical notation.

   a. Translate each of the following statements into as simple and clear English as you can.

   i) `exists i in {0..17} | (forall x in {0..17} |`
      `(exists y in {0..17} | (x*y) mod 18 = i));`

   ii) $\forall c \in \mathrm{domain}(F),\ \epsilon > 0,\ \exists \delta > 0\ \ni\ \forall x \in \mathrm{domain}(F),$
      $|x - c| < \delta \Longrightarrow |F(x) - F(c)| < \epsilon$

   iii) $\forall \epsilon > 0,\ \exists \delta > 0\ \ni\ \forall x, c \in \mathrm{domain}(F),$
      $|x - c| < \delta \Longrightarrow |F(x) - F(c)| < \epsilon$

   iv) `forall t1 in T | (exists t2 in T |`
      `(forall i in [1..#t1] | t1(i) < t2(i)));`
      (Here, `T` is a set of `tuples`.)

   v) $\exists x \in \mathcal{R}\ \ni\ \forall F \in \mathcal{S},\ \exists y \in \mathcal{R}\ \ni\ F(x) > F(y)$
      (Here, $\mathcal{R}$ is the set of real numbers and $\mathcal{S}$ is a set of functions whose domains are $\mathcal{R}$.)

   b. Translate each of the following statements into ISETL syntax. Specify any auxiliary objects that are used.

   i) For every `set` in $\mathcal{S}$, there is a `tuple` in $\mathcal{T}$ such that each integer in the `set` divides every component of the `tuple`. (Here, $\mathcal{S}$ is a set of `sets` of integers and $\mathcal{T}$ is a set of `tuples` of integers.)

ii) There is an **smap** in $\mathcal{M}$ whose image has an element in common with the image of every other **smap** in $\mathcal{M}$. (Here, $\mathcal{M}$ is a set of **smaps**.)

iii) There is a real number that is larger than at least one solution of every equation in the system.

c. Translate each of the statements in part b. into mathematical notation.

5.4.2 Negation   of three-level quantifications.

a. Negate each of the statements in Exercise 5.4.1, part a., and translate the result into good English.

b. Negate each of the statements in Exercise 5.4.1, part b. Try to do it directly from the English without translating into ISETL syntax or mathematical notation.

c. Negate each of the following statements directly from the English.

i) Every function in the set has a point at which its maximum occurs.

ii) For every interval of real numbers centered at the point 1.2, there is a term of the sequence beyond which every term of the sequence is in this interval.

iii) Among all the fish flying around the gymnasium, there is one for which there exists, in every Computer Science class, a Physics major who knows how much the fish weighs.

5.4.3 Reasoning   about three-level quantifications.

a. In statement 3 of the Preview (page 227), what can you conclude if you know that there is only one subject in which it is possible to major at Podunk U.?

b. In statement 3 of the Preview (page 227), what can you conclude if you know that all students at Podunk U. were born within 100 miles of Podunk?

c. In statement 2 of the Preview (page 227), what can you conclude if you know that there are no Computer Science classes? There are classes, but no Physics major in any of them?

d. Let P and Q represent the statements in Exercise 5.4.1, part a. ii) and iii), respectively. Let F be some function. Show that $Q \Longrightarrow P$.

e. Explain why statements ii) and iii) in part a. of Exercise 5.4.1 are not equal.

f. What can you conclude in Exercise 5.4.1, part a. v), if you know that $\mathcal{S}$ is a nonempty set of constant functions?

5.4.4 The $\epsilon - \delta$ simulator: full version. The object of this exercise is to write a func that will accept a function F and a number c in the domain of F. The func is to return the truth value of the following statement:

$$\forall \epsilon > 0, \quad \exists \delta > 0 \ \ni$$
$$\forall x \in \text{domain}(F), \ |x - c| < \delta \Longrightarrow |F(x) - F(c)| < \epsilon$$

The problem is that domain(F) and the sets of values of ϵ and δ form infinite sets. It is possible to "approximate" the situation by replacing $\epsilon > 0$ by the condition (cf. Exercise 1.8.3).

```
epsilon in {n*t :n in [1..50]}
```

and similarly for δ, where t is an appropriately chosen small positive number. In a similar manner, using a smaller value for t, one can approximate domain(F).

Write such a func and run it for various choices of c with the functions defined in Exercise 1.8.4.

5.5 Quantification in Mathematics and Computer Science

In the beginning of this chapter, we said that quantification is a key concept that is very important for you to acquire. Now that you have developed a little expertise (or at least some experience) in working with this idea, it is time to consider specific situations in which quantification arises naturally and where the skills that you have acquired while studying this chapter will be useful in solving actual problems. We consider four concrete examples: two in Mathematics, one in Computer Science, and one application.

5.5.1 Continuity of Functions

Consider the following statement:

$$\forall \epsilon > 0, \ \exists \delta > 0 \ \ni \ \forall x \in \text{domain}(F), \ |x - c| < \delta \Longrightarrow |F(x) - F(c)| < \epsilon$$

Here, F is a function whose domain and range are sets of real numbers and c is a number in the domain of F. This statement is, in fact, the definition of F *is continuous at* c. It says that in order for this to be the case, there must be, first of all, a dependency relation whereby for each ϵ it is possible to choose a certain δ. This δ must have the property that an infinite number of statements must be *true*, namely, for each value of x in the domain of F, the implication must hold.

The situation can be described in the following picture. The choice of ϵ establishes a vertical strip within which the values of the function must be contained. In order to arrange this, the choice of δ reduces the area of consideration to values of x that keep the points of the graph within a vertical strip. The intersection of these two strips produces a box and the graph of the function must remain in this box. In the picture, the value of δ_1 is not a good choice for ϵ, but δ_2 does work. If a smaller value of ϵ were chosen, then it might be necessary to choose a smaller value of δ.

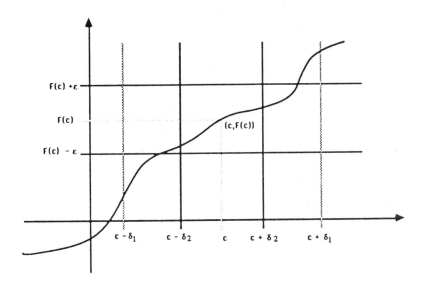

As a simple example, consider the function F whose domain is the set of all real numbers and whose values are given by

$$F(x) = \begin{cases} 2x + 1 & \text{if } x \geq 0 \\ 2x - 1 & \text{if } x < 0 \end{cases}$$

Suppose we considered, for instance, $c = 0.01$ so that $F(c) = 1.02$. Take, for example, $\epsilon = 0.5$. If we chose a value of $\delta = 0.1$, then the condition would not be satisfied because, for example, $x = -0.05$ would be in the horizontal strip since

$$|x - c| = |-0.05 - 0.01| = 0.06 < \delta$$

but $F(-0.05) = -1.1$, so that

$$|F(x) - F(c)| = |F(-0.05) - F(0.01)| = |-1.1 - 1.02| = 2.12 > 0.5 = \epsilon$$

It is necessary to take a smaller value for δ, say $\delta = 0.005$. Then, as long as $x \in [c - \delta, c + \delta] = [0.005, 0.015]$, since x is always positive, we have

$$|F(x) - F(c)| = |2x + 1 - 1.02| = |2x - 0.02| \leq 0.01 < 0.5 = \epsilon$$

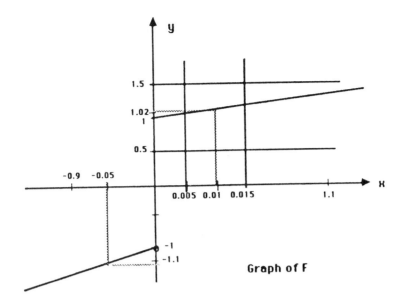

Graph of F

In fact, you can see that as long as $c > 0$, any choice of δ that is small enough to force x to always be positive and to differ from c in absolute value by less than $\epsilon/2$ will work, that is, if $\delta < c$ and $\delta < \epsilon/2$, then whenever $|x - c| < \delta$,

$$|F(x) - F(c)| = |(2x + 1) - (2c + 1)| = 2|x - c| < 2\delta < \epsilon$$

The point is that the choice of δ depends on the value of ϵ. A similar situation exists for any $c < 0$. Thus, this function is continuous at any $c \neq 0$.

Consider now the possibility that a function is *not* continuous at a particular point. For this, we must look at the negation of the definition

$$\exists \epsilon > 0 \; \ni \; \forall \delta > 0, \; \exists x \in \text{domain}(F) \; \ni \; |x - c| < \delta, \; |F(x) - F(c)| \geq \epsilon$$

It says that a single ϵ must be chosen such that the rest of the statement holds. That is, given this choice of ϵ, it must be possible to set up a dependency relation whereby for each positive δ it is possible to find an x in the domain of F within δ of c for which the value of the function is not within ϵ of the value of $F(c)$.

Returning to our example, we can see that this is the case if $c = 0$ and we choose, say, $\epsilon = 1$. Then, $F(c) = F(0) = 1$ and so, no matter what value of δ is chosen, there will always be at least one permissible value of x *that is negative*, and for this value, $F(x) < -1$, so $|F(x) - F(c)| > 2$, which is larger than ϵ. Thus, this function is not continuous at $c = 0$.

Once you understand how to handle quantifications, many difficult ideas in Mathematics become much easier. For example, consider the following theorem from calculus. *Every constant function is continuous at every point in its domain.* To see that this is *true*, we consider once again the definition of continuity given at the beginning of this section.

$$\forall \epsilon > 0, \; \exists \delta > 0 \; \ni \; \forall x \in dom\_F, \; |x - c| < \delta \Longrightarrow |F(x) - F(c)| < \epsilon$$

Now, the meaning of the theorem's hypothesis that the function F has a constant value is that, for any $c, x \in \text{domain}(F)$, $F(x) = F(c)$, or $|F(x) - F(c)| = 0 < \epsilon$. Thus, the proposition valued function in the definition, that is, the implication, is always *true* because the conclusion is always *true*. Hence, the statement holds, so the function is continuous for any choice of c in its domain.

5.5.2   The Chemical Factory Problem

We will now show how this problem (see page 95 and page 230) can be solved with a very simple ISETL program. First, we describe the data structures that must be established to record the information given in the statement of the problem. All of this information is contained in the three tables at the end of Chapter 2, page 95. For the first table, Table 2.5, we construct a **set** Inad, which will consist of two-element **sets** of two-character **strings** (those listed in Table 2.5). Thus, the fact that a pair is inadmissible is recorded by placing it as a **set** (since the order does not matter) in Inad. Next, for Table 2.6, we establish three smaps, Chang, Oper, and Mat, which represent three functions giving, respectively, for each material the changeover cost, operating cost ratio, and material cost ratio. Finally, we record the quality control information (Table 2.7) in three **sets**, Qual1, Qual2, and Qual3. A material satisfies a quality control condition iff it is in the appropriate one of these three sets.

Here is the ISETL code that determines whether a choice M1, M2, or M3 of three materials satisfies the three conditions. We omit the code that actually constructs the data structures and the code that checks the validity of the input. This may be long, but it is completely straightforward. Checking the conditions, which is the heart of the problem, can be done with a single ISETL statement consisting of three Boolean expressions connected by the operation and:

```
(forall m1, m2 in {M1, M2, M3} | not({m1, m2} in Inad))
                          and
(exists m in {M1, M2, M3} |
    (Chang(m) < 10000 and (Oper(m) > 110 impl Mat(m) <= 115)))
                          and
(exists m1, m2 in {M1, M2, M3} |
    (m1 /= m2) and (m1 in Qual1) and (m2 in Qual2) and
              (((m1 in Qual1) and (m2 in Qual2)) or
              ((m1 in Qual3) and (m2 in Qual3))));
```

This is a very compact program, but it is easy to understand because the syntax follows the logical structure of the conditions and is close to the English version of the statement of the problem.

5.5.3 Abstract Algebra

In Mathematics, there are many definitions that are very difficult to understand because their logical structure is full of quantifications. Because of your ability to understand these constructs, you will find many ideas much easier to understand in more advanced Mathematics courses. In this section, we describe a very simple example that occurs extremely often.

In Chapter 1, we introduced modular arithmetic. We will consider here, as an example, *addition mod 12* and *multiplication mod 12*. That is, you take two integers, perform the arithmetic operation, divide the result by 12, and take the remainder as your answer. For example,

$$(6 + 8) \bmod 12 = 2;$$
$$14*20 \bmod 12 = 6;$$

which return the values *true* and *false*, respectively. These expressions are often written in Mathematics as follows:

$$6 + 8 \equiv 2 \pmod{12}$$
$$14 \cdot 20 \equiv 4 \pmod{12}$$

Since all of the answers to these mod operations are between 0 and 11, it makes sense to restrict the domain and range of the operators to this set. Then, we can define two **funcs** to implement the binary operations:

```
add_mod12 := func (x,y);
                if x,y in {0..11} then
                        return (x + y) mod 12;
                end;
           end;

mult_mod12 := func(x,y);
                if x,y in {0..11} then
                        return (x * y) mod 12;
                end;
           end;
```

Now, we can discuss identities and inverses. The general situation is that we have a set and a binary operation on that set, as in the two **funcs** we have just defined. An *identity* in such a situation is a member e of the set that has the property that if you perform the operation with e and *any* other element x of the set, in either order, you get x back as the answer.

Thus, if $\bowtie$ stands for the operation and S denotes the set, the requirement is that

$$x \bowtie e = e \bowtie x = x \qquad \text{for every } x \in S$$

It is easy to see that for the binary operations defined above, 0 is an identity for add_mod12 and 1 is an identity for mult_mod12. Here is another interesting example. If we consider the union of two sets to represent a binary operation, then the empty set is an identity since the union of any set and the empty set gives back the original set. If U is a *universal* set in the sense that all sets in a particular discussion are to be considered as subsets of U, then the intersection of sets is a binary operation and the set U is an identity for it.

Once there is an identity, it is possible to talk about inverses. An element x of the set has an *inverse* if there is an element y in the set such that performing the operation on x and y, in either order, gives the identity, e. An important property of a binary operation is the existence of inverses for all elements of the set. The clearest way to state this is to use quantification, as follows:

$$\forall x \in S, \ \exists y \in S \ \ni \ x \bowtie y = y \bowtie x = e$$

It is even possible to include both the statement that the operation has an identity and that every element has an inverse in one large quantification.

$$\exists e \in S \ni \forall x \in S,$$
$$((x \bowtie e = e \bowtie x = x) \wedge (\exists y \in S \ni x \bowtie y = y \bowtie x = e))$$

It is not hard to see, with 0 as the identity for the operation add_mod12, that every element has an inverse. On the other hand, for mult_mod12, although 1 is an identity, several elements fail to have inverses.

5.5.4   Statements About Programs

A very complicated statement about a program can often be expressed clearly, compactly, and accurately in the language of quantification. Not only does this help in understanding the statement, but, using a language like ISETL, which implements quantification directly, it becomes much more reasonable to write programs that analyze other programs.

For example, suppose you were working with some programming language that used procedures, variables that were local to those procedures, and also global variables that could be accessed by any procedure. You might want to ask questions like the following about a program written in this language.

Is there a global variable such that for each procedure there is a (possibly the same) global variable that does not depend

on any local (to this procedure) variable that depends on the
first global variable?

The first thing you would have to decide is what it means for one variable **x** to *depend on* another variable **y**. The simplest version would be that there is an assignment statement that has **x** on the left-hand side and **y** on the right-hand side. You could make it more complicated by allowing a sequence of assignment statements connected by the property that every variable on the right-hand side of an assignment statement appears on the left-hand side of the next one. Then, if **x** was on the left-hand side of the first statement and **y** was on the right-hand side of the last one, you could say that **x** *depends on* **y**. We will see in Chapter 8 how to use the idea of *transitive closure* to go from the simple version to the more complicated one.

In any case, all information about variables using other variables in a program could be recorded in an ISETL map **Uses**, which consists of pairs [x,y], where **x** *depends on* **y**. Thus, the ISETL expression **Uses**{x} would have as its value the set of all variables that **x** depends on.

What else would you have to represent in order to ask the above question in an ISETL program? Of course, there would have to be the set **Glob** of all global variables. Then, there would have to be the set **Proc** of all procedures and a function **Loc_vars** (represented by an **smap** or **func**) that gives, for each procedure, the set of variables that are local to it. The following ISETL expression would then evaluate to *true* or *false* according to whether the answer to the above question is positive or negative.

```
exists u in Glob | (forall p in Proc |
     (exists v in Glob | (forall w in Loc_vars(p) |
          (([w,u] in Uses) impl ([v,w] notin Uses)))));
```

Summary of Section 5.5

In this section, we consider specific situations where quantification is used to solve actual problems. We consider what it means for a function to be continuous, expressing the definition as a three-level quantification. We show how the chemical factory problem can be solved with a very simple ISETL program when you use quantified statements. We consider an application to abstract algebra and discuss how you can write quantified statements in ISETL to show, for a given binary operation, the existence of an identity and its associated inverses. Finally, we show how quantification can be used to express a complicated statement about a program.

Exercises

5.5.1 Functions continuous at a point.

a. Show that if F is a function given by $F(x) = x$ and c is a point in the domain of F, then F is continuous at c.

b. Let F be a function and c a point in its domain such that F is continuous at c. Let G be a function with the same domain as F and with values given by

$$G(x) = 3F(x), \qquad x \in \text{domain}(F)$$

Show that G is continuous at c.

5.5.2 Functions continuous everywhere and uniformly continuous functions. Let F be a function and let D be its domain. We say that F is *continuous everywhere in D* or just *continuous everywhere* if (cf. Exercise 5.4.1, part a. ii))

$$\forall c \in \text{domain}(F),\, \epsilon > 0,\, \exists \delta > 0 \ \ni \ \forall x \in \text{domain}(F),$$
$$|x - c| < \delta \Longrightarrow |F(x) - F(c)| < \epsilon$$

On the other hand, we say that F is *uniformly continuous on D* (cf. Exercise 5.4.1, part a. iii)) if

$$\forall \epsilon > 0,\, \exists \delta > 0 \ \ni \ \forall x, c \in \text{domain}(F),$$
$$|x - c| < \delta \Longrightarrow |F(x) - F(c)| < \epsilon$$

a. Show that a uniformly continuous function is continuous everywhere.

b. Show that if F is constant then it is both continuous everywhere and uniformly continuous on its domain.

c. Show that if the domain of F is the interval (0,1) of real numbers (the points 0 and 1 are *not* in the domain) and the values of F are given by

$$F(x) = 1/x, \qquad x \in (0, 1)$$

then F is continuous everywhere.

d. Show that the function F of part c. is not uniformly continuous.

5.5.3 The chemical factory problem.

a. Write an ISETL program that will find all choices of materials that satisfy the conditions of the problem.

b. Write an ISETL program that will find all choices of materials that satisfy the first and second conditions of the problem but fail the third.

    c. Write an ISETL program that will find all choices of materials that satisfy the first and third conditions of the problem but fail the second.

    d. Write an ISETL program that will find all choices of materials that satisfy the second and third conditions of the problem but fail the first.

5.5.4 Commuting elements. Let S be a set and $\odot$ a binary operation on S. We say that elements x and y of S *commute* with each other if

$$x \odot y = y \odot x$$

    a. Express in formal language the assertion that there is at least one element in S that commutes with every element of S.

    b. Express in formal language the statement that a subset C of S has the property that every element of C commutes with every element of S.

    c. For $x \in S$ and $H \subseteq S$, write

$$xH = \{x \odot y : y \in H\} \qquad Hx = \{y \odot x : y \in H\}$$

      i) Use formal language to describe the situation when the following equation holds for every $x \in S$:

$$xH = Hx$$

      ii) Is the condition of i) the same as the condition on C in part b.? Explain your answer.

5.5.5 Write an ISETL expression that uses an appropriate proposition valued function P to express the assertion that every statement in a given program will be executed for some input, provided you are willing to wait long enough.

Chapter 6

Combinatorics, Matrices, Determinants

To the Instructor

This chapter is a potpourri of topics for which **sets** and **tuples** are the unifying factors. By this point, students should view **sets** and **tuples** as objects to which certain operations can be applied. The chapter helps to strengthen these notions, in addition to introducing them to some important applications.

We discuss several counting ideas by first introducing the multiplication and addition principles. To illustrate the concepts being presented, we look carefully at examples similar to those the students have previously solved intuitively. We emphasize the use of n-**tuples** and **sets** to describe the possibilities. The distinction between **tuples** and **sets** in ISETL helps the student to understand the difference between permutations and combinations.

A matrix is represented in ISETL by a **tuple** of **tuples**. In developing the vector and matrix operations, the compound operator **%** turns out to be a very powerful tool, since **%+** and **%*** correspond to $\sum$ and $\prod$, respectively. For instance, it allows students to write a one-line ISETL statement to evaluate the product of two given matrices.

Finally, we develop the mathematical definition of determinant by encouraging the student to think about the definition in terms of ISETL **funcs** and operations.

6.1 Preview

The purpose of this chapter is to introduce you to a few fundamental mathematical ideas: counting, matrices, matrix operations, and determinants.

Development of each of these ideas relies heavily on you constructing, manipulating, and using `tuples` and `sets`. By this time, however, you should feel very comfortable doing this.

The first section of the chapter discusses different ways of determining "how many." Obviously, counting is one mathematical concept that you use all the time! We asked you in previous chapters to think about these types of things, and we encouraged you to try to answer the questions we posed by looking at examples and by using your intuition. Do not stop using this approach—frequently, it is the best way to count the number of possible outcomes. The goal of this section, however, is to help you think about counting questions in a more organized fashion and to help you see how many of the questions you've thought about previously are related.

The second section explores matrices and their operations. A *matrix* is sometimes referred to as a table or a two-dimensional array. It can be represented in ISETL by a `tuple` of `tuples`, where each component of the outer `tuple` is a `tuple` of the same length. Of course, as with many of the other mathematical objects you have studied so far in this book, such as integers, sets, or propositions, there is a collection of operations associated with matrices. You will learn how to add two matrices together and how to multiply a matrix by a scalar or by another matrix, and all the time you can be thinking about manipulating `tuples`. An interesting thing about matrices is that they do not behave like most of the other mathematical objects to which you have been exposed. For example, although real multiplication, the disjunction of two propositions, and set union are commutative, that is,

$$
\begin{aligned}
a \cdot b &= b \cdot a & &\text{for all real numbers } a \text{ and } b \\
P \wedge Q &= Q \wedge P & &\text{for all propositions } P \text{ and } Q \\
S \cap T &= T \cap S & &\text{for all sets } S \text{ and } T
\end{aligned}
$$

it is not true that

$$
A \cdot B = B \cdot A \qquad \text{for all matrices } A \text{ and } B
$$

In fact, $B \cdot A$ may not even be defined, even though $A \cdot B$ has a value.

The last section has to do with finding a particular number associated with "square" matrices, called the *determinant* of a matrix. Determinants are useful for solving a system of n linear equations in n unknowns or when determining whether or not a matrix has an inverse. The mathematical definition of determinant is fairly complex, and consequently, many texts just show you how to find the number for matrices of several different sizes, and then ask you to mimic their examples. We are convinced that by using the ISETL approach, you can develop an understanding of the statement of the definition, and that you will be able to use the definition to establish some important properties of determinants.

6.2 Combinatorics

6.2.1 The Multiplication and Addition Principles

A typical counting question is How many (distinct) ISETL identifiers can be formed by concatenating a single letter and a single digit? A useful way to try to answer this question is to construct a **set** of 2-**tuples** to describe the possibilities. First, let **letters** be the **set** of all upper- and lowercase letters and let **digits** be the **set** of all single digits. Then, the *Cartesian product* of **letters** and **digits**, *letters* × *digits*, is the **set** of 2-**tuples**

```
Names_2 := {[l,d] :  l in letters, d in digits};
```

whose cardinality is precisely the number of ISETL variable names that consist of a letter followed by a digit. To answer the question, all you need to do is find the number of distinct **tuples** in **Names_2**. Now, for *each* value of the first component of a **tuple** in **Names_2**, there are 10 possible choices for the second component. For instance, if the value of **l** is "a", then **Names_2** contains

```
["a",0], ["a",1], ["a",2], ["a",3], ["a",4]
["a",5], ["a",6], ["a",7], ["a",8], ["a",9]
```

Therefore, since there are 52 possible ways to choose a value for the first component (26 lowercase and 26 uppercase letters), the cardinality of **Names_2**, and hence the number of variable names meeting the specified requirements, is $52 \cdot 10$ or 520. Notice that 520 is the product of the cardinalities of the two sets, **letters** and **digits**.

We solved this counting problem by finding the cardinality of the Cartesian product of two sets. This is a very helpful approach to determining *how many*; in fact it is used so often that it has a name, the *multiplication principle*. Formally, the principle states that if A and B are finite sets, then

$$\#(A \times B) = \#\{[x, y] : x \in A, y \in B\} = \#A \cdot \#B$$

(As in ISETL, the mathematical symbol # stands for the cardinality operator.) The rule follows from the fact that there are $\#A$ possible values for the first component of a 2-**tuple** in the Cartesian product, and for each of these values, there are $\#B$ possibilities for the second component. Therefore, there are $\#A \cdot \#B$ possible outcomes. You probably have used the multiplication principle before, even though you were not aware of it. Let's look at a few familiar examples, where the rule is applicable.

In Exercise 3.4.2 in Chapter 3, we asked you to determine how many iterations are needed to construct a set such as

```
S := {x*y :  x in {1, "counting is fun", 2.0, -16},
             y in {-2..10} |
                   is_integer(x) and (y mod 2 = 0)}};
```

where counting the number of iterations needed to construct S is equivalent to counting the number of times ISETL executes the statements in the body of the nested **for** loop:

```
S := {};
for x in {1, "counting is fun", 2.0, -16} do
    for y in {-2..10} do
        if is_integer(x) and (y mod 2 = 0) then
            S := S with x*y;
        end;
    end;
end;
```

The body of the inner loop is executed once for each value of the pair [x,y] as x and y iterate through their respective sets. So, to count the number of iterations needed to construct S, it suffices to find the cardinality of the Cartesian product:

```
{[x,y] :   x in {1, "counting is fun", 2.0, -16},
           y in {-2..10}};
```

Since there are 4 ways to choose a value for x and 13 for y, the 2-tuple [x,y] has 4·13 possible values. Thus, 52 iterations are required to construct S.

The multiplication principle for counting the number of possible 2-**tuples** can be extended to a rule for counting n-**tuples**. The extended rule can be applied, for example, when counting the number of elements in the power set of a given finite set. Recall that in Chapter 4, Section 4.4.2, we showed that

$$\#(pow(S)) = 2^n \qquad \text{whenever } \#S = n$$

We did this by noting that each subset, $A \subseteq S$, defines a unique characteristic function, $\chi_A : \longrightarrow \{0,1\}$ where

$$\chi_A(x) = \left\{ \begin{array}{ll} 1 & \text{if } x \in A \\ 0 & \text{if } x \notin A \end{array} \right.$$

Therefore, counting the number of subsets of S is equivalent to counting the number of distinct characteristic functions associated with S. Each characteristic function, χ_A, however, can be represented by an n-**tuple** of 0's and 1's, T_A, whose ith component is given by

$$T_A(i) = \left\{ \begin{array}{ll} 1 & \text{if } \chi_A(x_i) = 1 \\ 0 & \text{if } \chi_A(x_i) = 0 \end{array} \right.$$

where we are assuming that

$$S = \{x_1, x_2, \ldots, x_n\}$$

So, counting the number of subsets of a set with cardinality n corresponds to counting the number of n-tuples that can be constructed whose components are 0 or 1, that is, to finding the cardinality of the set

$$\{[t_1, t_2, \ldots, t_n] : t_1, t_2, \ldots, t_n \in \{0, 1\}\}$$

Each n-tuple in this set has two possible values for its first component, 0 or 1; two possibilities for its second component; and so on. Therefore, there are

$$\underbrace{2 \cdot 2 \cdots \cdots 2}_{n \text{ factors}}$$

distinct n-tuples whose components are 0 or 1, and the cardinality of the power set is 2^n as desired.

The *extended multiplication principle* says that the cardinality of the Cartesian product of n given finite sets, $S_1, \ldots, S_n$—which is exactly the number of distinct n-tuples that can be formed from the given sets—equals the product of the cardinalities of each of the sets. That is,

$$\#(S_1 \times \cdots \times S_n) = \#\{[t_1, \ldots, t_n] : t_1 \in S_1, \ldots, t_n \in S_n\} = \prod_{i=1}^{n} \#S_i$$

Another counting task you have encountered is to determine the number of possible functions with a given domain and range. Consider a specific example, such as how many functions map the set $\{$"a","b","c"$\}$ to the set $\{$1..4$\}$? Since there are three objects in the domain, each possible function can be represented by a 3-tuple, whose first component contains the value of a function at one member of the domain, say at "a", the second contains the function's value at another member of the domain, "b", and the third contains the value at "c". Therefore, counting the number of functions is equivalent to finding the cardinality of

$$\{[x, y, z] : x, y, z \text{ in } \{1..4\}\};$$

Since the range of the function is $\{$1..4$\}$, there are four possible values for each component and, hence, $4 \cdot 4 \cdot 4$ or 64 possible functions.

How does your approach change if you restrict your search to one-to-one functions? The difference is that although you still have four ways of assigning a value to the first member of the domain, there are only three possible values for the image of the next element in the domain since the requirement that the function be one-to-one means that you cannot assign two different members of the domain to the same item in the range. Applying the same line of argument, there are only two possible values for the last item in the domain. So, to determine the number of one-to-one functions, you need to find the size of the set

$$\{[x, y, z] : x \text{ in } \{1..4\}, y \text{ in } \{1..4\} - \{x\},$$
$$z \text{ in } \{1..4\} - \{x, y\}\};$$

which is $4 \cdot 3 \cdot 2$ or 24.

Obviously, not all counting problems can be solved using the multiplication principle. Suppose you wanted to determine how many iterations are performed by ISETL when constructing the set

```
Pairs := {{i,j} :  i in {1..10}, j in {i+1..10}};
```

If you think about constructing Pairs using a nested for loop, then the number of iterations equals the cardinality of the set of 2-tuples,

```
{[i,j] :  i in {1..10}, j in {i+1..10}}
```

The problem is that this set is not the Cartesian product of two sets since the number of possible values of j depends on the particular value of i. So, you cannot use the multiplication principle to find its cardinality. Observe, however, that you can express this set of 2-tuples as the union of 10 subsets:

```
{[1,j] :  j in {2..10}} + {[2,j] :  j in {3..10}}
       + ··· + {[10,j] :  j in {11..10}}
```

If you examine any two of these subsets, you can see that no pair has any elements in common—so they are said to be *pairwise disjoint*—and hence, you can find the number of elements in their union and, thus, the cardinality of Pairs by finding the size of each subset and adding the sizes together. Therefore,

$$9 + 8 + 7 + 6 + 5 + 4 + 3 + 2 + 1 + 0$$

is the total number of iterations needed to construct Pairs.

The method we used to count the number of iterations performed by ISETL when constructing Pairs illustrates the idea underlying the *addition principle*. While you can think of the multiplication principle in terms of the cardinality of the Cartesian product of two sets A and B—that is, counting the number of 2-tuples determined by A and B—the addition principle corresponds to finding the cardinality of the union of A and B, where A and B have *no* elements in common. Formally, the addition principle states that

$$\#(A \cup B) = \#A + \#B \qquad \text{whenever } A \cap B = \emptyset$$

That is, to find the cardinality of the union of two disjoint sets, you can count the number in A, count the number in B, and add the results together. In the last example, we applied an extended version of the addition principle, which says that for any collection of pairwise disjoint sets, the cardinality of their union equals the sum of their cardinalities, that is,

$$\# \left(\bigcup_{i=1}^{n} S_i \right) = \sum_{i=1}^{n} \#S_i$$

6.2.2 Permutations

Consider the **set**

$$notes := \{\texttt{"do"}, \texttt{"re"}, \texttt{"mi"}\};$$

Since **sets** are unordered, a reasonable question to ask yourself is "How many different ways might ISETL output the value of **notes**?" That is, if you entered

$$\texttt{print notes;}$$

what is the total number of possible outcomes?[1] A similar question is How many different orders might ISETL evaluate the **for** loop

```
for x in notes do
        Statement list
end;
```

In both of these cases, you are interested in determining the number of *arrangements* or *orderings* that can be formed from the given nonempty collection of objects, **notes**. You want to find the cardinality of the **set**

$$\{\texttt{[x, y, z]} : \texttt{x in notes, y in notes} - \{\texttt{x}\},$$
$$\texttt{z in notes} - \{\texttt{x, y}\}\};$$

which, using ideas from the previous section, is $3 \cdot 2 \cdot 1$ or $3!$.[2] ISETL defines a particular ordering of the **set** by choosing a value for the first component of the **tuple** denoting the ordering, let's assume it picks **"do"**, a value for the second component, possibly **"mi"**, and finally a value for the last component, which must be **"re"** since this is the only item in **notes** that has not already been chosen. Let's call this arrangement **Perm1**, so

$$\texttt{Perm1 := ["do", "mi", "re"];}$$

Take a moment to list the other five possible orderings of the items in **notes**.

An arrangement of the items in a given set, S, is called a *permutation* and can be denoted by a **tuple**, whose length equals the size of the set. But, from our discussions in Chapter 4, Section 4.2.3, you know that an ISETL **tuple** can be used to represent a function. For example, **Perm1** represents a function that can be described mathematically as follows:

$$Perm1 : \{1..\#(notes)\} \longrightarrow notes$$

where

$$Perm1(1) = \text{``do''}, \ Perm1(2) = \text{``mi''}, \ \text{and} \ Perm1(3) = \text{``re''}$$

---

[1] Because of the way ISETL works, not all permutations are equally likely. In fact, some permutations of **notes** will never appear using **print notes;**.

[2] This *will* generate all $3!$ possibilities, however.

Thus, **Perm1** represents a one-to-one, onto function *Perm*1 with domain {1..3} and image the set represented by **notes**.

In fact, every *n*-**tuple** corresponding to a permutation of the items in a set S, where $\#S = n$, represents a one-to-one, onto function from {1..*n*} to S. Therefore, determining the number of possible arrangements or permutations of the items in the given set is equivalent to counting the number of one-to-one, onto functions, with domain {1..*n*} and image S. Since permutations on S are represented by *n*-**tuples**, where there are n possible values for the first component, $n - 1$ for the second, and so on, there are

$$n \cdot (n - 1) \cdot (n - 2) \cdot \cdots \cdot 1 \quad \text{or} \quad n!$$

distinct orderings of the members of a set with cardinality n. (We read the notation $n!$ as "n factorial.")

Suppose instead of ordering *all* the items in a set S, you are only interested in arrangements that contain r items, where $r \leq n = \#S$. As in the previous discussion, you are still thinking about arrangements and **tuples**. The difference is that the **tuple** representation of a particular ordering defines a one-to-one function from {1..*r*} to S that is not necessarily onto. As an example, suppose you wanted to list (in order) three members of the set

```
Friends := {"Alice", "Amy", "Alex", "Allen", "Alicia"};
```

You can accomplish this with the ISETL loop

```
Friends3 := [ ];
for i in {1..3} do
    x := random(Friends);
    Friends3 := Friends3 with x;
    Friends := Friends less x;
end for;
print Friends3;
```

That is, you can use ISETL to construct a **tuple** of length 3, **Friends3**, by initializing **Friends3** to be empty and then repeating the following steps three times: choose an arbitrary member of **Friends**; add the choice to the end of the new **tuple**; and delete the item from **Friends** (so you don't accidentally choose the same object twice). This process is probably very similar to the process you go through when you construct the ordered list in your mind.

We call a particular value of **Friends3** an *r-permutation* or a *permutation of size r*, where in this case $r = 3$. Observe that **Friends3** represents a one-to-one function from {1..3} to **Friends**. In general, an r-permutation of a set S is a one-to-one function with domain {1..*r*} and range S that can be represented by an r-**tuple**.

Look again at the ISETL code defining `Friends3`. How many outputs might ISETL return? That is, how many different values might `Friends3` have? A more general question is How many 3-tuples or 3-permutations can be formed from a set containing 5 items? Well, the first time through the `for` loop there are 5 possible values for x and hence for `Friends3(1)`, while during the next pass through the loop there are 4 possible ways to assign a value to `Friends3(2)` and, in the last pass, 3 possible values for `Friends3(3)`. Therefore, the number of different ways to construct `Friends3` is

$$5 \cdot 4 \cdot 3 = 5 \cdot 4 \cdot 3 \cdot \frac{2 \cdot 1}{2 \cdot 1} = \frac{5!}{(5-3)!}$$

Consequently, if we use $P(n, r)$ to denote the number of possible r-permutations, that is, the number of different ordered choices of r members of a set S with cardinality n, then finding the value of $P(n, r)$ is the same as determining how many r-tuples can be constructed from the elements in S, or counting the number of one-to-one functions that exist with domain $\{1..r\}$ and range S. Since there are n possibilities for the first component, $(n-1)$ possibilities for the second, and finally $(n-r+1)$ possible values for the rth component, we have

$$
\begin{aligned}
P(n, r) &= n \cdot (n-1) \cdot \cdots \cdot (n-r+1) \\
&= n \cdot (n-1) \cdot \cdots \cdot (n-r+1) \cdot \frac{(n-r) \cdot \cdots \cdot 1}{(n-r) \cdot \cdots \cdot 1}
\end{aligned}
$$

Therefore,

$$P(n, r) = \frac{n!}{(n-r)!}$$

Up to this point, we have only considered permutations of distinct objects. Let's shift our focus from counting the number of possible ways to arrange the members of a set (all of which are distinct) to counting arrangements of the elements in a `tuple` (some of which may be repeated). That is, we are interested in counting the number of n-tuples that can be constructed from a given n-tuple. This occurs, for instance, when trying to determine the number of permutations of the letters in the word "PIZZA". One way to solve this counting question is to find how many 5-tuples can be constructed from the characters in the ISETL `string`

```
T := "PIZZA";
```

Since there are five characters in `T`, your immediate guess might be 5!, but this is too many. The reason is that the `tuples` formed by interchanging the values of the two Z's both represent the same arrangement of the elements in the given `string`. For instance, P1 and P2 both represent `T` itself, where

```
P1 := [T(1), T(2), T(3), T(4), T(5)];
P2 := [T(1), T(2), T(4), T(3), T(5)];
```

while

```
P3 := [T(3), T(4), T(1), T(2), T(5)];
P4 := [T(4), T(3), T(1), T(2), T(5)];
```

both correspond to the arrangement "ZZPIA". As a matter of fact, for each
5-tuple there is an equivalent 5-tuple that is constructed by interchanging
the two Z's. Thus, the total number of distinct 5-tuples is 5!/2 or 5!/2!.

As another example, how many arrangements are there of the characters
in the ISETL string

$$W := "EERIE";$$

If the characters in "EERIE" were all distinct, then there would be 5! possible orderings, but this is not the case since E occurs three times. So, we
need to determine how many tuples with the same value can be formed by
interchanging the positions of the three E's. Well, there are 3! or 6 tuples
representing "EERIE", namely,

```
R1 := [W(1), W(2), W(3), W(4), W(5)];
R2 := [W(1), W(5), W(3), W(4), W(2)];
R3 := [W(2), W(1), W(3), W(4), W(5)];
R4 := [W(2), W(5), W(3), W(4), W(1)];
R5 := [W(5), W(1), W(3), W(4), W(2)];
R6 := [W(5), W(2), W(3), W(4), W(1)];
```

Moreover, each arrangement of characters in "EERIE" is represented by
3! distinct 5-tuples, so the total number of distinct arrangements of the
characters in "EERIE" is 5!/3!. On the other hand, there are 6!/(3! · 2!)
different orderings of the 6 characters in the string "EERIER", since the
character E appears three times and R occurs two times.

6.2.3 Combinations

A *combination* is a selection where order is *not* taken into consideration.
When you think about r-permutations, you are interested in the number
of r-tuples that can be constructed from the elements of a given set S.
In this case, the order of the items is important since obviously different
orderings define different r-tuples. If you eliminate the idea of *order* from
the discussion, then you would be thinking about constructing *subsets* of S
that contain r items, instead of r-tuples. A subset containing r items from
a given set is called an *r-combination*. Observe that since r-combinations
are represented by sets they do not permit repetitions.

You might have thought about r-combinations earlier in the book without actually using the term combination, when you considered the ISETL
func npow (see Exercises 3.2.7 and 3.4.7).

```
npow := func(S,r);
        if is_set(S) and is_integer(r) and r>=0 then
            return {X : X in pow(S) | #(X) = r};
        end if;
    end func;
```

This **func** accepts a **set** S and a nonnegative **integer** r and returns the collection of all subsets of S with cardinality r. Consequently, npow(S,r) is the **set** of all the possible r-combinations of items from S.

Of course, not only are we interested in constructing r-combinations, we also want to count the number of distinct possibilities or find the cardinality of npow(S,r). We know that there are $n!/(n-r)!$ r-permutations whenever $\#S = n$. Therefore, to count the number of r-combinations, all we need to do is determine how many repetitions result from the fact that we are now interested in **sets** and not **tuples**.

Let's examine some of the differences between permutations and combinations by looking at a specific example. Consider again the set **Friends**

```
Friends = {"Alice", "Amy", "Alex", "Allen", "Alicia"};
```

If you choose three items from **Friends**, say the strings "Alice", "Amy", and "Alicia", then there is only one subset or 3-combination of **Friends** that contains these strings, namely,

$$\{"Amy", "Alicia", "Alice"\};$$

On the other hand, there are 6 or 3! different 3-**tuples**, or 3-permutations, having the **strings** as values for their components. In fact each collection of three items from **Friends** leads to the formation of 3! distinct 3-**tuples** or 3-permutations, while you can construct only *one* subset or 3-combination. Thus, the number of 3-permutations is 3! times the number of possible 3-combinations, or

$$P(5,3) = 3! \cdot C(5,3)$$

where we have used the notation $C(n,r)$ to denote the number of r-combinations that can be formed from the set whose cardinality is n. Thus,

$$C(5,3) = \frac{P(5,3)}{3!} = \frac{5!}{(5-3)!\,3!}$$

Generally, if S is a set containing n items, then corresponding to each subset of S with cardinality r there are $r!$ distinct r-**tuples**. So,

$$P(n,r) = r! \cdot C(n,r)$$

and

$$C(n,r) = \frac{P(n,r)}{r!} = \frac{n!}{(n-r)!\,r!}$$

Summary of Section 6.2

This section introduces you to a number of important counting concepts. We restated each problem that we wished to solve in terms of counting the number of n-tuples in a set of tuples or the number of elements in the union of a collection of pairwise disjoint sets. We then applied the multiplication and addition principles to help find the solutions. With these ideas in mind, we discussed permutations with and without repetitions, r-permutations, and r-combinations.

This section is different from many of the others in that it contains several formulas to help you determine "how many." Try to develop the formulas for yourself by thinking in terms of the familiar concepts, sets and tuples—try to *understand* what they mean rather than memorize their statements.

Exercises

6.2.1 Some important ISETL funcs.

   a. Define a func that accepts a nonnegative integer n and returns $n!$. (Note that 0! equals 1.) Test your func on several values of n, noting that $n!$ grows very fast.

   b. Define a func P that accepts two nonnegative integers n and r and returns $P(n, r)$ whenever $0 \leq r \leq n$.

   c. Define a func that accepts a set S and an integer r and returns a tuple representing an r-permutation of S, where $0 \leq r \leq \#S$.

6.2.2 ISETL and counting.

   a. Determine the number of iterations performed by ISETL when constructing the following sets:

      i) T1 := {[r,s,t] : t,s,r in {-2..2}};

      ii) T2 := {x*y + z : x in {2..6}, y in {-3,-5..-16},
                            z in {"a", 0, "b", 1} |
                                is_integer(z) impl (x < y)};

      iii) T3 := {p - q : p in {5,4..1}, q in {1..p}};

   b. Determine the number of ways ISETL might output the list of values in the following sets:

      i) {1..10};

      ii) {x : x in {1..10} | even(x)};

   c. How many different values might ISETL assign to the tuple

      [w : w in {2,4..100}];

    d. How many different `tuples` can ISETL construct from the characters in the `string "POPPA"`?

    e. You have 10 words that you wish to have ISETL alphabetize.

       i) How many different possible ways can you give the original list as input if each item on the list is distinct?

      ii) How many different ways can you supply the list as input if two items each appear on the list twice and one item appears three times?

    f. In how many different ways (orders) might ISETL evaluate the following code fragments?

       i)
```
for count in {-10,-7..26} do
      ⋮
end;
```

      ii)
```
for r in {1..6} do
    for s in {1..10} do
          ⋮
    end;
end;
```

     iii)
```
for i in {1..10} do
      ⋮
end;
for j in {1..5} do
      ⋮
end;
```

    g. Suppose you have a list of n distinct names and you write ISETL code to randomly choose three names from the list.

       i) How many possible `sets` might ISETL return?

      ii) How many possible `tuples` might ISETL return?

     iii) How are your answers to parts i) and ii) related?

6.2.3 Let $\sigma = [2,1,3,5,4]$ and $\tau = [3,5,2,1,4]$ be two permutations of $\{1..5\}$.

    a. Since a permutation is a one-to-one, onto function it makes sense to find the *composition of two permutations*. Write an ISETL `func` called `comp` that accepts two permutations of $\{1..n\}$ represented as `tuples` and returns the `tuple` representing their composition. Use your `func` to find

       i) $\sigma \circ \sigma$

      ii) $\sigma \circ \tau$

iii) $\tau \circ \sigma$

iv) $\tau \circ \tau$

v) $\sigma \circ (\tau \circ \sigma)$

vi) $(\sigma \circ \tau) \circ \sigma$

b. Consider the following two definitions:

    1. The permutation $\iota = [1..n]$ is called the *identity* permutation on $\{1..n\}$.

    2. The permutation β is the *inverse* of the permutation α if

$$\alpha \circ \beta = \beta \circ \alpha = \iota$$

in which case, we write $\beta = \alpha^{-1}$.

With these definitions in mind:

i) Show that

$$\pi \circ \iota = \iota \circ \pi = \pi$$

for every permutation π of $\{1..n\}$ by tracing through the code of your ISETL **func comp**.

ii) Write an ISETL **func** called **inverse** that accepts a permutation and returns its inverse. Use your **func** to find σ^{-1} and τ^{-1}.

iii) Use the **funcs comp** and **inverse** to show that $\sigma \circ \sigma^{-1} = \iota$ and $\tau \circ \tau^{-1} = \iota$.

c. Write an ISETL **func** called **perm_pow** that accepts a permutation π and a positive integer r and returns π^r, where π^r is π composed with itself r times. Use your **func** to find

i) σ^r for $r = 2$.

ii) τ^r for $r = 2, 3, 4, 5$.

6.2.4 Permutations of **tuples** with repeated items.

a. How many arrangements (permutations) are there of the characters in the strings

i) EXISTENTIAL

ii) UNIVERSAL

iii) ONE-TO-ONE

b. Given an n-**tuple**, or a **string** of length n, that contains n_1 copies of one object, n_2 copies of a second object, ..., and n_k copies of a kth object, find a general formula for the number of distinct arrangements that can be constructed from the given **tuple** or **string**.

6.2.5 The *inclusion-exclusion principle* states that for any finite sets A and B

$$\#(A \cup B) = \#A + \#B - \#(A \cap B)$$

   a. Explain why the inclusion-exclusion principle holds.

   b. Explain how you can use the rule to solve the following counting problems.

   i) Given sets S and T, assume

   $$\#S = 75, \ \#T = 30, \ \text{and} \ \#(S \cup T) = 100$$

   How many objects are in both S and T?

   ii) Suppose out of 100 freshmen students 50 are taking Discrete Math, 30 are taking Calculus, and 20 are taking both. How many freshmen are taking at least one of these courses? How many are taking neither course?

   c. Give a statement of the inclusion-exclusion principle for a collection of three sets. Use Venn diagrams (see Chapter 3, Section 3.3) to illustrate the extension of the rule.

6.2.6 Let $\Sigma = \{a, b, c, d, e, f, g\}$ be a set of characters, or *letters*, called an *alphabet*, and let Σ^* be the set of all strings, or *words*, that can be formed from the members of Σ (including the empty string, " ").

   a. How many words have length 7?

   b. How many words can be constructed from the given alphabet (including the empty word)?

   c. How many words are there of length 3 where no letter appears in a word more than one time?

   d. How many words of length 6 contain an a in the first position?

   e. How many words of length 6 contain at least one c?

   f. How many words of length 5 contain exactly one b?

   g. How many 4-letter words can be constructed from the letters a, c, e, f?

   h. How many words of length 5 have 3 a's and 2 b's?

   i. How many words are there that have an even length, less than or equal to 8, and whose odd components (1st, 3rd, etc., characters) are f?

   j. How many 5-letter words begin with c and end with d?

   k. How many 5-letter words begin with c or end with d?

6.2.7 The *pigeonhole principle* says that if m objects are to be placed into n pigeonholes, where $m > n$, then at least one pigeonhole will receive more than one object.

Use this rule to explain why the following statements are true.

    a. If $\mathcal{F}$ is a function from set A to set B, where $\#A > \#B$, then $\mathcal{F}$ is not one-to-one.

    b. If $\mathcal{G} : S \longrightarrow T$ is a function, where the number of elements in its domain is greater than the number in its range, then there exists a $y \in T$ such that $\mathcal{G}^{-1}\{y\}$ contains more than one element of S.

    c. If a bag contains five marbles that are black, white, or pink, then two of the marbles must be the same color.

6.2.8 Let $A = \{$"apples", "peaches", "oranges", "pears"$\}$ and $B = \{8..10\}$.

    a. How many functions are there from A to B? B to A?

    b. How many onto functions are there from A to B? B to A?

    c. How many one-to-one functions are there from A to B? B to A?

6.2.9 The integers $C(n, r)$ are also called *binomial coefficients* because they are the coefficients of the terms when $(x + y)^n$ is expanded. In particular,

$$(x + y)^n = C(n, 0)x^n y^0 + C(n, 1)x^{n-1}y^1 + \cdots + C(n, n)x^0 y^n$$
$$= \sum_{r=0}^{n} C(n, r)x^{n-r}y^r$$

This is called the *binomial theorem*.

    a. Show that the binomial theorem holds for $n = 0, 1, 2, 3, 4$.

    b. By choosing appropriate values for x and y in the binomial theorem, show that

        i) $\sum_{r=0}^{n}(-1)^r C(n, r) = 0$

        ii) $\sum_{r=0}^{n} C(n, r) = 2^n$

6.2.10 Use the definition of $C(n, r)$ to prove the following *combinatorial identities*.

    a. $C(n, 0) = 1$

    b. $C(n, n) = 1$

    b. $C(n, 1) = n$

    c. $C(n, n - 1) = n$

    d. $C(n, r) = C(n, n - r)$

    e. $C(n, r) = C(n - 1, r - 1) + C(n - 1, r)$

6.2.11 For each $n = 0, 1, 2, \ldots$, the values of

$$C(n, r) \qquad \text{where } 0 \le r \le n$$

can be displayed in a triangular form, called *Pascal's triangle* as in Table 6.1.

$$
\begin{array}{ll}
n = 0 & C(0, 0) \\[2mm]
n = 1 & C(1, 0) \quad C(1, 1) \\[2mm]
n = 2 & C(2, 0) \quad C(2, 1) \quad C(2, 2) \\[2mm]
n = 3 & C(3, 0) \quad C(3, 1) \quad C(3, 2) \quad C(3, 3) \\[2mm]
& \vdots
\end{array}
$$

Table 6.1: Pascal's Triangle

Moreover, using the combinatorial identities in Exercise 6.2.10, it follows that the first and last entries in each row are 1, while the remaining entries in the row can be calculated from the entries in the previous row. That is,

$$
C(n, r) = \begin{cases} 1 & \text{if } r = 0 \text{ or } r = n \\ C(n - 1, r - 1) + C(n - 1, r) & \text{if } 0 < r < n \end{cases}
$$

where $n = 0, 1, 2, \ldots$ and $0 \le r \le n$. For instance, assuming that the first row in the triangle is row 0, and hence, the $n + 1$st row is row n, the first and last entries in row 2, $C(2, 0)$ and $C(2, 2)$, equal 1, while

$$C(2, 1) = C(1, 0) + C(1, 1) = 1 + 1 = 2$$

The values for $C(n, r)$ can be displayed in the triangle shown in Table 6.2.

    a. Calculate the entries in rows 5, 6, 7, and 8 of Pascal's triangle.

    b. Using the ideas underlying the formation of Pascal's triangle, define a *recursive* ISETL func—that is, a func that "calls itself"—that accepts nonnegative integers n and r, where $0 \le r \le n$, and returns the value of $C(n, r)$.

6.2.12 Write a recursive func that accepts a nonnegative integer n and returns the set of all permutations of $\{1..n\}$.

$$\begin{array}{ll}
n = 0 & 1 \\
n = 1 & 1 \quad 1 \\
n = 2 & 1 \quad 2 \quad 1 \\
n = 3 & 1 \quad 3 \quad 3 \quad 1 \\
n = 4 & 1 \quad 4 \quad 6 \quad 4 \quad 1 \\
& \vdots
\end{array}$$

Table 6.2: Values in Pascal's Triangle

6.3 Matrices

6.3.1 Representation of Matrices

We've discussed in earlier chapters the idea of using `tuples` to represent an ordered list, which computer scientists usually call a *one-dimensional array* and mathematicians refer to as a *finite sequence* or a *vector*. For example, you might be interested in the alphabetically ordered list of all the students in Discrete Math:

```
Student_Names := ["Deb", "Jed", "Meg", "Sue", "Tom", "Zac"];
```

But, what if you wanted to store the students' scores on two quizzes, an in-class exam, a final, and their average homework grades? To do this, you need to store information in a table, such as Table 6.3, instead of in a list.

	Deb	Jed	Meg	Sue	Tom	Zac
Quiz1	90	81	92	88	85	75
Quiz2	76	80	95	78	67	60
Homework	78	75	99	85	72	67
Exam1	79	83	92	80	66	77
Final	80	82	90	85	75	70

Table 6.3: Student grades

One way to think about a table is as a list of lists, since the first row in the table is a list (which can be represented by a `tuple`), the second row is also a list (which again can be represented by a `tuple`), and so on. So, we can represent a table in ISETL as a `tuple` where each component of the `tuple` is itself a `tuple`. For example, Table 6.3 can be represented by

```
Grades := [[90, 81, 92, 88, 85, 75],
           [76, 80, 95, 78, 67, 60],
           [78, 75, 99, 85, 72, 67],
           [79, 83, 92, 80, 66, 77 ],
           [80, 82, 90, 85, 75, 70]];
```

This **tuple** of **tuples** is called a *matrix* or a *two-dimensional array*.

You refer to an item in a **tuple** or list by giving the index of its position in the list. For example, the **string "Meg"** is stored in the third component of the **tuple** Student_Names and is accessed as follows:

> Student_Names(3);
"Meg";

You can do the same thing with the matrix **Grades**, but in this case entering

Grades(3);

gives you the third *row* in the matrix, which contains all the students' homework grades, namely,

[78, 75, 99, 85, 72, 67];

instead of a single item. If you only wanted to know how Tom did on his homework, then you need to look in the fifth *column* of the third row of the matrix, or the fifth component of Grades(3), namely,

Grades(3)(5);

Generally, Grades(i)(j) is the value of the item in the ith row and jth column of the matrix **Grades**. What values does ISETL return if you enter

Grades(2)(4); Grades(4)(2); Grades(6)(1);

Well, the first statement tells ISETL to look in the second row, fourth column of **Grades** for Sue's Quiz2 grade, which is 78. The value of the next expression is 83, which is Jed's in-class exam grade that is stored in the fourth row, second column of the matrix. The last expression returns an error message since Grades(6) is OM, so it makes no sense to try to find its value at 1.

If a matrix has m rows and n columns, we say that the *size* of the matrix is $m \times n$, which we read "m by n." Thus, **Grades** is a 5×6 matrix. Notice that the number of rows in **Grades** equals the number of components in the **tuple**, which is #Grades, or 5, while the number of columns in the matrix equals the number of items in any one of its components (all of which must contain the same number of items), #(Grades(1)) or 6.

In general, if A is an $m \times n$ matrix in ISETL, then A has m rows and n columns, where

$$m = \#A \qquad \text{and} \qquad n = \#A(1)$$

The element in the ith row, jth column, or the (ij)-entry, is given by

$$A(i)(j)$$

while the ith row,

```
[A(i)(j) :  j in [1..n]]
```

is obtained by letting j iterate through the column values, `[1..n]`. Therefore, to represent the entire matrix in ISETL, you let i iterate through the possible row values, `[1..m]`, which gives

$$\overbrace{[[\ \underbrace{\texttt{A(i)(j)}}_{\textit{ith } \texttt{row}}}^{(ij)-\texttt{entry}}\ :\ \texttt{j in [1..n]]}\ :\ \texttt{i in [1..m]]};$$

A matrix can be interpreted as a function of two variables, where the first variable iterates through the values of rows of the matrix and the second through the values of the columns. Let's examine this statement more closely, keeping in mind the discussion we had in Chapter 4, Section 4.3.8, concerning functions of several variables. An $m \times n$ matrix can be represented by an m-`tuple`, A. But, every `tuple` represents a function. In particular, you can think of A as

$$A : \{1..m\} \longrightarrow \text{ the set of all } n\text{-}\texttt{tuples}$$

where $A(i)$, $1 \le i \le m$, contains the entries in the ith row of the matrix. Furthermore, since each $A(i)$ is itself a `tuple`, namely, an n-`tuple`, it represents a function whose domain is $\{1..n\}$ and range is the collection of all ISETL objects, **U**. That is,

$$A(i) : \{1..n\} \longrightarrow \mathbf{U}$$

where for a fixed value of i, $A(i)(j), 1 \le j \le n$, contains the (ij)-entry of the given matrix. Therefore, you can think of a matrix as a function of two variables

$$A : \{1..m\} \times \{1..n\} \longrightarrow \mathbf{U}$$

where

$$A(i, j) = A(i)(j)$$

Mathematics notation uses double *subscripts* to denote an element in a matrix, with the first subscript specifying the row and the second the column of the desired component. Thus, if A is an $m \times n$ matrix, then a_{ij} denotes the entry in row i, column j, or the (ij)-entry, and we write

$$A = \begin{bmatrix} a_{11} & a_{12} & \cdots & a_{1n} \\ a_{21} & a_{22} & \cdots & a_{2n} \\ \vdots & & & \vdots \\ a_{m1} & a_{m2} & \cdots & a_{mn} \end{bmatrix}$$

or use the shorthand representation,

$$A = (a_{ij})$$

Finally, a $1 \times n$ matrix contains exactly one row, so it is called a *row matrix* or a *row vector*. In ISETL, it is represented by a `1-tuple` whose single component is an n-`tuple`, such as

$$[[2, 4, 6, 8]];$$

or it can be represented by an n-`tuple`, such as

$$[2, 4, 6, 8];$$

which is then referred to as a *vector*. An $m \times 1$ matrix, on the other hand, contains exactly one column and thus is called a *column matrix* or a *column vector*. The ISETL representation of a $m \times 1$ column vector is an m-`tuple`, all of whose components are singleton `tuples`.

6.3.2 Operations on Matrices

You know how to add, multiply, or divide two real numbers to get another real number, and you can apply the operators **and** ($\wedge$), **or** ($\vee$), or **impl** ($\Longrightarrow$) to two propositions to get another proposition, but what operations can we perform on matrices? In this section, we'll examine some matrix operations, such as what it means to add two matrices together and what it means to multiply a matrix by a given number or *scalar*, by a row or a column vector, and finally by another matrix. You'll have to be careful about two things. First, when talking about matrix operations, the size of the matrices involved is very important. The second thing is the fact that, algebraically, matrices do not always act like other mathematical objects, such as the real numbers, with which you are familiar. Earlier in the book, we discussed various operations on `tuples`. Let's recall some of these ideas, and then extend them to operations on matrices.

In Chapter 4, page 174, we defined the *coordinate-wise sum* of two sequences represented by the `tuples` X and Y to be the `tuple`

$$[X(j) + Y(j) : j \text{ in } [1..\min(\#X, \#Y)]];$$

If the two `tuples` have the same length, this operation is called *vector addition*. For example, using the notation $[x, y]$ to denote the *vector* from the origin $[0, 0]$ to the point with coordinates $[x, y]$ in the real plane, suppose

$$\vec{a} = [3, 1] \text{ and } \vec{b} = [2, 5]$$

Then, the coordinate-wise sum of these two vectors is

$$\vec{a} + \vec{b} = [3 + 2, 1 + 5] = [5, 6]$$

The sum is itself a vector that can be interpreted geometrically in $\mathcal{R} \times \mathcal{R}$ as the diagonal of the parallelogram formed by $\vec{a}$ and $\vec{b}$ as shown in Figure 6.1.

We also discussed in Chapter 4 (see page 174) the idea of multiplying every component in a given sequence, X, by a fixed number or scalar, r:

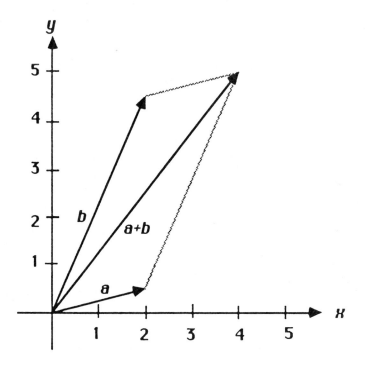

Figure 6.1: Sum of two vectors in $\mathcal{R} \times \mathcal{R}$

```
[r * X(j)  :   j in [1..#X]];
```

Notice that applying this operation to a given **tuple** returns another **tuple** or vector that is called the *scalar product* of r and $\vec{x}$. It is denoted by $r\vec{x}$ in Mathematics. Therefore, assuming the vectors $\vec{a}$ and $\vec{b}$ are defined as before,

$$2\vec{a} = [2 \cdot 3, 2 \cdot 1] = [6, 2]$$

$$-4\vec{b} = [-4 \cdot 2, -4 \cdot 5] = [-8, -20]$$

and

$$0.5\vec{b} = [0.5 \cdot 2, 0.5 \cdot 5] = [1.0, 2.5]$$

What happens graphically when you multiply a given vector $\vec{v}$ by a scalar in the real plane? More specifically, why do the terms change direction, shrink, and stretch describe the relationship between the vectors $\vec{v}$ and $r\vec{v}$ when $r < 0$, $0 \le r \le 1$, and $r > 1$, respectively? To help answer these questions, choose a value for $\vec{v}$ and locate the vector in the real plane, then find the value of $r\vec{v}$ for various values of r and locate these vectors in the plane. What is happening here?

Of course, vectors in the real plane can be represented in ISETL by 2-tuples. One word of caution, however, should be mentioned concerning hopping back and forth between the mathematical notation for these operations and the ISETL notation: using the operators "+" or "*" with tuples in ISETL indicates concatenation and replication, and not vector addition and scalar product! As an example, look at what happens in the following ISETL session:

```
>  a := [3,1]; b := [2,5];
>  2 * a;
[3, 1, 3, 1];
>  a + b;
[3, 1, 2, 5];
```

But, you can easily define ISETL funcs for these operations using the ideas from Chapter 4. The func TupPlus, for example, returns the sum of two tuples with the same length:

```
TupPlus := func(V,W);
              local n;
              if is_tuple(V) and is_tuple(W)
                        and #V = #W then
                  n := #V;
                  return [V(j) + W(j) :  j in [1..n]];
              end;
          end;
```

So, we would have

```
>  [3,1] .TupPlus [2,5];
[5,6];
```

whereas

```
>  [1,2,3] .TupPlus [-1,0];
OM;
```

These vector and tuple operations extend easily to operations for matrices. Just as you find the sum of two vectors (with the same length) by adding their corresponding components, you find the sum of two matrices (with the *same* size) by adding their corresponding entries. Similarly, just as you find the product of a scalar and a tuple by multiplying each component in the tuple by the scalar, you find the scalar product of a scalar and a matrix by multiplying every entry in the matrix by the scalar.

First, let's think about the operation of matrix addition. Recall that a matrix can be represented by a `tuple`, where each component of the `tuple` is itself a `tuple` that contains the entries in a particular row of the matrix. So, if two matrices have the same number of rows (their `tuple` representations have the same lengths), it makes sense to think about adding the matrices together by adding together the corresponding rows. As long as the rows have the same length, you can do this using the operator `TupPlus`. That is, you can find the first row in the matrix corresponding to the sum of the matrices `A` and `B` by finding

$$A(1) \text{ .TupPlus } B(1)$$

and the second by evaluating

$$A(2) \text{ .TupPlus } B(2)$$

and so on. Expressing the sum as a `tuple` former, where `m = #(A)`, we have

$$[A(i) \text{ .TupPlus } B(i) : \quad i \text{ in } [1..m]];$$

Notice that to construct this `tuple` representing the sum not only do `A` and `B` have to have the same number of rows (that is, `#A = #B`) but they must also have the same number of columns (that is, `#A(1) = #B(1)`, ..., `#A(m) = #B(m)`) in order to add the corresponding rows together. Therefore, for the sum of two matrices to be defined, they must have the same size. Let's go one step further. To find the `tuple` `A(i) .TupPlus B(i)`, you add the first component of `A(i)`, `A(i)(1)`, to the first component of `B(i)`, `B(i)(1)`, and so on, for the rest of the entries in the two rows. Thus, if the two matrices have the same size, where

$$m := \#A; \qquad n := \#A(1);$$

then the `tuple` representing their sum is

$$[[\underbrace{A(i)(j) + B(i)(j)}_{i\text{th row}} : j \text{ in } [1..n]] : i \text{ in } [1..m]];$$

where $A(i)(j) + B(i)(j)$ is labeled "(ij)−entry".

Here is an example using the mathematical notation for matrices:

$$\begin{bmatrix} 14 & 0 & -8 \\ 3 & 11 & -16 \end{bmatrix} + \begin{bmatrix} -9 & 1 & 0 \\ 2 & 0 & 21 \end{bmatrix} = \begin{bmatrix} 5 & 1 & -8 \\ 5 & 11 & 5 \end{bmatrix}$$

As you evaluate this sum, think about how ISETL evaluates the expression — it adds corresponding rows to corresponding rows by adding corresponding entries in each row to corresponding entries.

Using the mathematical notation for matrices, if $A = (a_{ij})$ and $B = (b_{ij})$ are two $m \times n$ matrices and $C = (c_{ij})$ is their sum, then

$$C = A + B = (a_{ij}) + (b_{ij}) = (a_{ij} + b_{ij})$$

and

$$c_{ij} = a_{ij} + b_{ij}$$

Look once more at the expression A(i)(j) + B(i)(j) in the ISETL tuple former used to construct the sum of the matrices A and B and notice how close this is to the mathematical notation for the sum.

Matrix addition satisfies a number of properties that are listed in Exercise 6.3.5. For instance, matrix addition is commutative. That is, for any $m \times n$ matrices, $A = (a_{ij})$ and $B = (b_{ij})$,

$$A + B = B + A$$

This follows from the fact that addition of real numbers is commutative, and hence,

$$a_{ij} + b_{ij} = b_{ij} + a_{ij}, \qquad \text{for } 1 \le i \le m, \ 1 \le j \le n$$

So we can add two matrices together, but what about the operation of multiplication on matrices? First, let's think about matrices and scalar multiplication. You know how to multiply a tuple by a scalar by multiplying each component in the tuple by the scalar. Similarly, to multiply a matrix by a scalar, you multiply each component in the matrix by the scalar. Therefore, if Q is an $m \times n$ matrix and r is a scalar, then the tuple that represents their scalar product is

$$\overbrace{[[\underbrace{\texttt{r * Q(i)(j)}}_{\text{ith row}}}^{(ij)-\text{entry}} \ : \ \texttt{j in [1..n]]} \ : \ \texttt{i in [1..m]];}$$

In Mathematics, just as we write $3\vec{x}$ to denote the product of 3 and the value of the variable $\vec{x}$, we write $3A$ for the scalar product of 3 and the matrix A. So, if $Q = (q_{ij})$ is an $m \times n$ matrix, then the scalar product of r and Q is

$$rQ = r(q_{ij}) = (r \cdot q_{ij})$$

Again, notice the similarity to the expression for the (ij)-entry in the ISETL tuple former.

As an example, suppose you want to find $2F - G$, where

$$F = \begin{bmatrix} 2 & -7 & 0 \\ 0 & 1 & -6 \end{bmatrix} \qquad G = \begin{bmatrix} 0 & -11 & 2 \\ 1 & 2 & 3 \end{bmatrix}$$

Then, noting that

$$2F - G = 2F + (-1)G$$

and that scalar multiplication has precedence over matrix addition (as usual, you multiply before you add),

$$2F - G = 2 \cdot \begin{bmatrix} 2 & -7 & 0 \\ 0 & 1 & -6 \end{bmatrix} + (-1) \cdot \begin{bmatrix} 0 & -11 & 2 \\ 1 & 2 & 3 \end{bmatrix}$$

$$= \begin{bmatrix} 2 \cdot 2 & 2 \cdot (-7) & 2 \cdot 0 \\ 2 \cdot 0 & 2 \cdot 1 & 2 \cdot (-6) \end{bmatrix}$$

$$+ \begin{bmatrix} (-1) \cdot 0 & (-1) \cdot (-11) & (-1) \cdot 2 \\ (-1) \cdot 1 & (-1) \cdot 2 & (-1) \cdot 3 \end{bmatrix}$$

$$= \begin{bmatrix} 4 & -3 & -2 \\ -1 & 0 & -15 \end{bmatrix}$$

Now, let's examine how you might multiply a matrix by another `tuple` or even another matrix. We'll first investigate multiplication by a `tuple`. Consider the following problem: You are the grader for the Discrete Math class whose grades are stored in Table 6.3, page 282, and in the matrix `Grades`, where

```
Grades := [[90, 81, 92, 88, 85, 75],
           [76, 80, 95, 78, 67, 60],
           [78, 75, 99, 85, 72, 67],
           [79, 83, 92, 80, 66, 77],
           [80, 82, 90, 85, 75, 70]];
```

Suppose that for each student you want to find their weighted grade average where each quiz counts half as much as the in-class exam, homework counts one and a half times as much, and the final counts twice as much. To accomplish this, you need to do two things. First, find the weighted sum of each student's grades. That is, multiply the entries in the first two rows of a given column (containing a particular student's grades for Quiz1 and Quiz2) by 0.5, the entry in the third row (Homework) by 1.5, the entry in the next row (Exam1) by 1, and finally the entry in the last row of the column (Final) by 2, and then find the sum of these products. For instance, for the first column in the matrix `Grades`, you need to evaluate

$$0.5 \cdot 90 + 0.5 \cdot 76 + 1.5 \cdot 78 + 1 \cdot 79 + 2 \cdot 80$$

Then, the second step is to divide each of the results by the sum of the weights, namely, $5.5 = 0.5 + 0.5 + 1.5 + 1 + 2$.

The first step can be done by finding the product of the `tuple` containing the weights,

$$W := [0.5, 0.5, 1.5, 1, 2];$$

and the matrix Grades *in that order*, which results in

```
WG := [0.5*90 + 0.5*76 + 1.5*78 + 1*79 + 2*80,
       0.5*81 + 0.5*80 + 1.5*75 + 1*83 + 2*82,
       0.5*92 + 0.5*95 + 1.5*99 + 1*92 + 2*90,
       0.5*88 + 0.5*78 + 1.5*85 + 1*80 + 2*85,
       0.5*85 + 0.5*67 + 1.5*72 + 1*66 + 2*75,
       0.5*75 + 0.5*60 + 1.5*67 + 1*77 + 2*70];
```

Carrying out the calculations,

$$WG = [439, 440, 514, 460.5, 400, 385]$$

Then, to complete the problem, you find the scalar product of 1/5.5 and WG, which is

$$[79.81, 79.99, 93.45, 83.72, 69.99]$$

So, for instance, the student whose grades are stored in the first column has the average 79.81.

How do the entries in WG relate to the entries in W and Grades? Let's look again at the sum of the desired products for a particular student, say the student whose grades are stored in the second column of the matrix Grades. But, this time let's write the sum in terms of the two given matrices, using G to denote the matrix Grades:

```
W(1)*G(1)(2) + W(2)*G(2)(2) + W(3)*G(3)(2)
    + W(4)*G(4)(2) + W(5)*G(5)(2)
```

This gives the value of the entry in the second column of the product WG. Can you identify the pattern? To find the (1,2)-entry in WG, you add together the products of corresponding components in row 1 of W and column 2 of Grades or G. Notice that in order for this to be possible, the number of entries in W must equal the number of entries in a column of G. Furthermore, if you look carefully at the other entries in the product, you will see that to find the value of the entry in the jth column of the row vector WG, where $1 \le j \le 6$, you add together the products of the corresponding components of the first row of W and jth column of G. Hence, the $(i1)$-entry in the product is given by

```
W(1)*G(1)(j) + W(2)*G(2)(j) + W(3)*G(3)(j)
    + W(4)*G(4)(j) + W(5)*G(5)(j)
```

where j in [1..6]. This can be expressed using the compound operator % as follows

```
%+[W(k)*G(k)(j) :  k in [1..5]]
```

where j in [1..6]. Using sigma notation, the entry in the jth column of the row vector containing the product is

$$\sum_{k=1}^{5} w_k \cdot g_{kj}, \qquad \text{for } 1 \leq j \leq 6$$

where $W = (w_k)$ and $G = (g_{kj})$. Therefore, the product WG of W and G is given by

```
[ %+[W(k)*G(k)(j) :   k in [1..5] :   j in [1..6]]];
```

In general, given a $1 \times p$ row vector $R = (r_k)$ and a $p \times n$ matrix $T = (t_{kj})$, the product $RT = P = (p_j)$ is a $1 \times n$ row vector, whose jth component is given by

$$\sum_{k=1}^{p} r_k \cdot t_{kj}, \qquad \text{for } 1 \leq j \leq n$$

In order for the product to be defined, R must have the same number of columns as T has rows. Their product then has the same number of rows as R and the same number of columns as T. That is,

$$\underbrace{1 \times \overbrace{p \qquad p}^{\text{equal}} \times n}_{\text{size of product}}$$

In our example, W is 1×5 and G is 5×6, so we have

$$\underbrace{1 \times \overbrace{5 \qquad 5}^{\text{equal}} \times 6}_{\text{size of product}}$$

Look again at what happened in the example, and make sure you can explain why this will always be the case.

This finally leads us to thinking about how to find the product of two matrices. Suppose you've defined an ISETL func called TupProd that accepts a row vector (represented as a tuple) and a matrix and returns the row vector (again represented as a tuple) that is their product, as long as the two inputs have the appropriate sizes. That is, using the Boolean func is_matrix, which is described in Exercise 6.1.1, assume

```
TupProd := func(R,T);
            local p, n;
            if is_tuple(R) and is_matrix(T)
                            and #R = #T then
                p := #T; n := #T(1);
                return [%+[R(k)*T(k)(j) :  k in [1..p]] :
                                    j in [1..n]];
            end if;
        end func;
```

How can you use TupProd to find the product of two matrices, A and B, in that order? Now, the rows in A are represented by the tuples, namely, A(1), A(2), ..., A(m), where #A = m. So, for instance,

$$A(1) \ .TupProd \ B$$

will return a tuple provided the number of entries in A(1) is the same as the number of rows in B, which is the case whenever A is $m \times p$ and B is $p \times n$. In fact, if A has the same number of columns as B has rows, then for *each* row of A—that is, for each i in [1..m]—you can find the tuple

$$A(i) \ .TupProd \ B$$

We take this tuple to be the ith row in the product of A and B *in that order*. So, the product of A and B is the matrix

$$[A(i) \ .TupProd \ B \ : \ i \ in \ [1..m]];$$

Replacing the call to TupProd with its definition and setting

$$m \ := \ \#A; \qquad p \ := \ \#B; \qquad n \ := \ \#B(1);$$

the product AB is the $m \times n$ matrix

$$\overbrace{[[\underbrace{[\% + [A(i)(k) * B(k)(j) \ : \ k \ in \ [1..p]]}_{i\text{th row of product}} \ : \ j \ in \ [1..n]]}^{(ij)-\text{entry}} \ : \ i \ in \ [1..m]];$$

So, if $C = AB$, the process of calculating the (ij)-entry in $C = (c_{ij})$ can be expressed in mathematical notation by

$$c_{ij} = \sum_{k=1}^{p} a_{ik} \cdot b_{kj} \qquad \text{where } 1 \le i \le m, \ 1 \le j \le n$$

Let's summarize what we've done. You can multiply two matrices, A and B, as long as the number of columns in A is the same as the number of rows in B. The number of rows in the resulting matrix is the number of rows in A, while the number of columns in the product equals the number of columns in B. That is,

$$\underbrace{m \times \overbrace{p \qquad p}^{\text{equal}} \times n}_{\text{size of product}}$$

To find the entry in the ith row and the jth column of AB, ISETL iterates (simultaneously) across the entries in the ith row of A and down through the entries in the jth column of B, finding the sum of the products of the corresponding entries. This can also be expressed using a series of nested for loops (assuming A and B are given matrices and m, p, and n are defined as above) as follows:

```
Prod:= [ ];
for i in [1..n] do
    Next_Row := [ ];
    for j in [1..n] do
        Next_Entry := %+[A(i)(k)*B(k)(j) :  k in [1..p]];
        Next_Row := Next_Row with Next_Entry;
    end;
    Prod := Prod with Next_Row;
end;
```

For example,

$$\begin{bmatrix} 1 & -7 & 0 \\ 0 & 1 & -6 \end{bmatrix} \begin{bmatrix} 0 & -11 & 2 & 8 \\ 1 & 2 & 3 & 7 \\ 0 & 2 & 2 & 5 \end{bmatrix} = \begin{bmatrix} -7 & -25 & -19 & -41 \\ 1 & -10 & -9 & -23 \end{bmatrix}$$

To find the (1,3)-entry in the product, you find the sum of the products of the corresponding entries in row 1 of the first matrix and column 3 of the second, or

$$1 \cdot 2 + (-7) \cdot 3 + 0 \cdot 2$$

and to find the value in the (2,4)-entry of the product, you consider the entries in row 2 of the matrix on the left and column 4 of the matrix on the right, evaluating

$$0 \cdot 8 + 1 \cdot 7 + (-6) \cdot 5$$

How do you find the (2,2)-entry? The (3,3)-entry?

As an application of matrix multiplication, we return to the concept of a permutation. In Section 6.2.3, we represented a given permutation,

$$\pi : \{1..n\} \longrightarrow \{1..n\}$$

by an n-tuple. You can also represent a permutation by an $n \times n$ matrix, $P = (p_{ij})$, where

$$p_{ij} = \begin{cases} 1 & \text{if } \pi(i) = j \\ 0 & \text{otherwise} \end{cases}$$

for $1 \le i, j \le n$. Here, P is called the *permutation matrix*, where each row and each column of P has exactly one entry whose value is 1, while all the remaining entries are 0. For example, if $\tau = [3, 1, 2]$, $\sigma = [2, 1, 3]$, and T and S are their respective permutation matrices, then

$$T = \begin{bmatrix} 0 & 0 & 1 \\ 1 & 0 & 0 \\ 0 & 1 & 0 \end{bmatrix}, \qquad S = \begin{bmatrix} 0 & 1 & 0 \\ 1 & 0 & 0 \\ 0 & 0 & 1 \end{bmatrix}$$

One interpretation of the product ST is as the permutation matrix representing the composition of τ and σ, $\tau \circ \sigma$ (notice the *order* of the composition with respect to the *order* of the product). Verify this for the given example by (1) finding the permutation matrix for $\tau \circ \sigma$ by first representing the composition as a `tuple` and then finding the `tuple`'s associated matrix, (2) finding the product ST, and then (3) comparing the two matrices. (They should be the same!) Let's look at the (ij)-entry in the permutation matrix of $\tau \circ \sigma$. If

$$\sigma(i) = r \qquad \text{and} \qquad \tau(r) = j$$

then

$$\tau \circ \sigma(i) = \tau(r) = j$$

So, the (ij)-entry in the permutation matrix of the composition is 1, while the remaining entries in the ith row and in the jth column are 0. Now, let's look at the (ij)-entry in ST, and show that it is also 1. Observe that if $S = (s_{ik})$ and $T = (t_{kj})$ then

$$s_{ik} = t_{kj} = \begin{cases} 1 & \text{if } k = r \\ 0 & \text{if } k \neq r \end{cases}$$

So,

$$\sum_{k=1}^{n} s_{ik} \cdot t_{kj} = s_{ir} \cdot t_{rj} = 1$$

while all the other entries in the ith row and jth column of ST are equal to 0. Therefore, the permutation matrix $\tau \circ \sigma$ and the product of the product of the permutation matrices of σ and τ (in that order) are equal.

6.3.3   Matrix Inversion

The number 1 is said to be the *multiplicative identity* for the real numbers because, given any number r, you can multiply r on the right and on the left by 1 and the result is r. That is,

$$r \cdot 1 = 1 \cdot r = r$$

Furthermore, every nonzero real number x has a *multiplicative inverse*, $1/x$. For example, $1/4$ is the multiplicative inverse of 4, since

$$\frac{1}{4} \cdot 4 = 4 \cdot \frac{1}{4} = 1$$

The set of all $n \times n$ matrices also has a multiplicative identity, namely, the matrix that has 1 in the (ii)-position, for all $1 \leq i \leq n$, and 0 everywhere else. This is called the *identity matrix* of size n. To construct this matrix in ISETL, let's first define a `func`, say `id` that returns the entries in the identity matrix when given the row and column values.

```
                    id := func(i,j);
                            if i=j then return 1;
                            else return 0;
                            end;
                    end;
```

Then, the ith row of the identity matrix is represented by the **tuple** of **tuples**

$$[id(i,j) : j \text{ in } [1..n]];$$

and a **func**, **I**, that accepts an integer n and returns the identity matrix of size n is given by

```
I := func(n);
        return [[ id(i,j) : j in [1..n] ] : i in [1..n]];
      end;
```

So, **I(4)** has the value

```
                    [[1, 0, 0, 0],
                     [0, 1, 0, 0],
                     [0, 0, 1, 0],
                     [0, 0, 0, 1]];
```

Of course, the mathematics notation for the identity matrix is very similar. Namely,

$$I_n = (\delta_{ij}) \qquad \text{where} \qquad \delta_{ij} = \begin{cases} 1 & \text{if } i = j \\ 0 & \text{if } i \neq j \end{cases}$$

for $1 \leq i, j \leq n$. We leave it as an exercise for you to show that the identity matrix of size n actually is a multiplicative identity for the set of all $n \times n$ matrices. That is,

$$I_n A = A I_n = A$$

whenever A is an $n \times n$ matrix. (See Exercise 6.3.5.)

Since the set of all $n \times n$ matrices with real entries has a multiplicative identity, does every nonzero matrix have a multiplicative inverse? That is, given any nonzero matrix A, does there exist a matrix B such that

$$AB = BA = I_n$$

Let's look at a few examples. First, consider the matrix

$$A = \begin{bmatrix} 1 & 2 \\ 3 & 4 \end{bmatrix}$$

Does it have an inverse? If so, what is it? That is, do there exist values for w, x, y, and z such that

$$\begin{bmatrix} 1 & 2 \\ 3 & 4 \end{bmatrix} \begin{bmatrix} w & x \\ y & z \end{bmatrix} = \begin{bmatrix} 1 & 0 \\ 0 & 1 \end{bmatrix}$$

Multiplying the matrices on the left side of the expression gives

$$\begin{bmatrix} 1 \cdot w + 2 \cdot y & 1 \cdot x + 2 \cdot z \\ 3 \cdot w + 4 \cdot y & 3 \cdot x + 4 \cdot z \end{bmatrix} = \begin{bmatrix} 1 & 0 \\ 0 & 1 \end{bmatrix}$$

and equating corresponding entries yields the system of equations

$$
\begin{aligned}
1 \cdot w + 2 \cdot y &= 1 \\
3 \cdot w + 4 \cdot y &= 0 \\
1 \cdot x + 2 \cdot z &= 0 \\
3 \cdot x + 4 \cdot z &= 1
\end{aligned}
$$

Solving the first two equations for w and y and the last two for x and z gives the matrix

$$B = \begin{bmatrix} -2 & 1 \\ 3/2 & -1/2 \end{bmatrix} = -1/2 \begin{bmatrix} 4 & -2 \\ -3 & 1 \end{bmatrix}$$

as our candidate for the inverse of A. It remains to show by multiplying out that $BA = I_2$, which it does. So, we say that A is *invertible*.

But, what happens when you try to find the inverse of the matrix

$$X = \begin{bmatrix} 2 & 1 \\ 6 & 3 \end{bmatrix}$$

If you try to solve for the inverse by looking at the associated system of 4 equations with 4 unknowns, namely,

$$
\begin{aligned}
2 \cdot w + 1 \cdot y &= 1 \\
6 \cdot w + 3 \cdot y &= 0 \\
2 \cdot x + 1 \cdot z &= 0 \\
6 \cdot x + 3 \cdot z &= 1
\end{aligned}
$$

you'll discover that there is no solution to the system, and consequently X does not have an inverse, that is, X is not invertible.

Therefore, unlike the real numbers, some nonzero matrices have inverses, and some don't. If A is invertible, then we denote its inverse by A^{-1}.

In the exercises, you will work with a number of properties involving inverses. Some can be proved by tracing through an ISETL program, others can be proved by directly applying the definition. For instance, how might you prove the fact that for any two invertible matrices with the same size, the inverse of their product equals the product of their inverses in *reverse* order? That is,

$$(PQ)^{-1} = Q^{-1}P^{-1}$$

To prove this, you need to show that $Q^{-1}P^{-1}$ is the inverse of PQ or that

$$(PQ)(Q^{-1}P^{-1}) = (Q^{-1}P^{-1})(PQ) = I_n$$

Accepting the facts that matrix multiplication is associative and that I_n is the multiplicative identity (see Exercise 6.3.5), then

$$
\begin{aligned}
(PQ)(Q^{-1}P^{-1}) &= P(QQ^{-1})P^{-1} \\
&= PI_nP^{-1} \\
&= PP^{-1} \\
&= I_n
\end{aligned}
$$

Similarly, it can be shown that when you multiply PQ on the left by $Q^{-1}P^{-1}$ you also get I_n. Hence, PQ is invertible and $Q^{-1}P^{-1}$ is its inverse.

Summary of Section 6.3

An $m \times n$ matrix can be represented by an *m*-tuple of *n*-tuples. There are a number of definitions associated with matrices, such as what it means for a matrix to be *square* or *symmetric* or for two matrices to be *equal*. We include definitions of these terms in the exercises and ask you to write ISETL property testers for each one. We also introduce the *diagonal* and *trace* of a square matrix, the *transpose* of a given matrix, and a special matrix called the *zero* matrix.

In this section, we discussed what it means to find the sum of two matrices and what it means to multiply a matrix by both a scalar and another matrix whenever all the entries in the matrix are numbers. While doing the exercises, we encourage you to "think ISETL"—that is, to represent matrices and their operations in ISETL and then think about how ISETL evaluates matrix expressions. When applying the operations, the sizes of the matrices involved is an important consideration—be sure to keep this in mind.

Just as there are rules for sets and equivalence laws for propositions, there are rules for matrix operations. Matrices, however, don't always behave like real numbers. We pointed out some of the differences in this section; others are indicated in the exercises.

We discussed using a matrix to represent a permutation and showed the relationship between the composition of two permutations and the product of their associated permutation matrices. In Chapter 8, we'll use matrices to represent relations and graphs and discuss additional applications for the product.

The section ends with a discussion of what it means for a matrix to be invertible. In the next section on determinants, we'll see the relationship between the value of the determinant of a matrix and the existence of an inverse.

Exercises

6.3.1 Some useful ISETL funcs.

  a. Define an ISETL matrix tester is_matrix that accepts an ISETL object M and returns *true* whenever M is a matrix and *false* otherwise.

  b. Define an ISETL func called size that accepts a matrix M and returns the size of M as a 2-tuple whose first component is the number of rows in M and whose second component is the number of columns.

  c. Define a func called row that accepts a matrix M and an integer i and returns the ith row in M.

  d. Define a func called column that accepts a matrix M and an integer j and uses a tuple former to return the jth column in M.

6.3.2 Some important definitions.

  1. A matrix A is a *square matrix* if it has the same number of rows as columns.

  2. The *main diagonal* of a square $(n \times n)$ matrix, $A = (a_{ij})$, consists of all the entries in the matrix that have the same row and column numbers, namely, $a_{11}, a_{22}, \ldots, a_{nn}$.

  3. The *trace* of a square matrix is the product of all the entries along the main diagonal.

  4. An *upper triangular matrix* is a square matrix in which all the entries below the main diagonal are equal to 0.

  5. A *diagonal matrix* is a square matrix in which every entry that is not on the main diagonal of the matrix is 0.

  6. The *transpose* of an $m \times n$ matrix A is the $n \times m$ matrix whose jth column is the jth row of A. The mathematical notation for the transpose of A is A^T.

  7. A matrix A is said to be *symmetric* whenever $A = A^T$.

  8. A square matrix I is called an *identity matrix* if it is a diagonal matrix and every item on the main diagonal is 1.

  9. A square matrix Z is called the *zero matrix* if every entry in Z is 0.

  10. Two matrices A and B are said to be *equal* when A and B have the same size and the corresponding entries in A and B are equal.

With these definitions in mind, do the following:

    a. Give an example of a 3 × 3 matrix that illustrates each of the definitions in the list.

    b. Define the following ISETL funcs. Run your funcs on a variety of inputs.

        i) Funcs called is_square, is_upper_triangular, is_diagonal, is_symmetric and is_identity that accept a matrix A and return *true* if A is a square matrix, an upper triangular matrix, a diagonal matrix, a symmetric matrix, or the identity matrix respectively. Otherwise they return *false*. Try to write your property testers in one line and make your funcs as simple as possible by using previously defined funcs.

       ii) A func called transpose that accepts a matrix and returns its transpose.

      iii) A func called diagonal that accepts a square matrix and returns a tuple containing its diagonal entries.

      iv) A func called zero that accepts two positive integers m and n and returns the $m \times n$ zero matrix.

       v) A func called trace that accepts a square matrix and returns its trace.

      vi) A func called equal that accepts two matrices and returns *true* if they are equal and *false* otherwise.

6.3.3 Define ISETL funcs for each of the following matrix operations. In each case, be sure and check that the input is of the appropriate size and use the compound operator (instead of a for loop) whenever possible. As usual, test your operations on a variety of inputs.

    a. The scalar product of a scalar and a matrix.

    b. The sum of two matrices.

    c. The product of a row vector and a matrix.

    d. The product of a matrix and a column vector.

    e. The product of two matrices.

    f. The nth power of a matrix.

    g. The sum of the elements in the ith row of a matrix.

    g. The sum of the elements in the jth column of a matrix.

6.3.4 Consider the matrices:

$$A = \begin{bmatrix} 2 & -1 \\ 3 & 6 \end{bmatrix} \qquad B = \begin{bmatrix} 1 & 0 & -2 \\ 9 & 3 & -5 \end{bmatrix}$$

Find the following (as you perform the matrix operations, it helps to think about how ISETL might evaluate the given expressions):

   a. AB

   b. BA

   c. $A^2 - 2A + I$

   d. B^T

   e. $(A^T)^T$

   f. A^{-1}

   g. $trace(A)$

   h. $B^T A$

   i. $B^T A^T$

6.3.5 **The laws of matrix algebra.** You are already familiar with the
       laws of equivalences for Boolean expressions (Chapter 2, page 79)
       and the corresponding rules for set expressions (Chapter 3, page 115).
       Here is a list of equivalent matrix expressions, where A, B, and C
       are arbitrary matrices of the appropriate sizes, Z is the zero matrix,
       I_p is the $p \times p$ identity matrix, r and s are scalars, and 0 is the scalar
       zero.

Commutative law of addition

$$A + B = B + A$$

Associative laws

$$(A + B) + C = A + (B + C)$$
$$(AB)C = A(BC)$$

Distributive laws

$$(r + s)A = rA + sA$$
$$r(A + B) = rA + rB$$
$$A(B + C) = AB + AC$$
$$(A + B)C = AC + BC$$

Existence of additive identity

$$A + Z = Z + A = A$$

Existence of multiplicative identity

$$AI_n = I_n A = A \qquad \text{whenever } A \text{ is } n \times n$$

Existence of additive inverses

$$A + (-1)A = Z$$

Simplification laws

$$
\begin{aligned}
0A &= Z \\
rZ &= Z \\
I_m A &= A \qquad \text{whenever } A \text{ is } m \times n \\
A I_n &= A \qquad \text{whenever } A \text{ is } m \times n
\end{aligned}
$$

a. Prove each of these laws using the definitions of the matrix operations and the properties of the real numbers (e.g., real addition and multiplication are commutative, associative, etc.).

b. The interesting thing to notice is not what is in the list but what is *not* in the list. Although the elements in a matrix behave like real numbers, the matrices themselves do not. For example, matrix multiplication is not commutative. Another example is that you can multiply two *nonzero* matrices together and get the zero matrix. These things don't happen when you're dealing with the real numbers. Determine whether each of the following statements is true or false. If it is false, find a counterexample to the given statement. If it is true, explain why.

i) $AB = BA$, whenever AB and BA are defined.
ii) $(AB = Z) \Longrightarrow ((A = Z) \vee (B = Z))$.
iii) $(A \neq Z) \Longrightarrow (A^{-1} \text{ exists})$.
iv) $(A^2 = Z) \Longrightarrow (A = Z)$.
v) $(A^2 = I) \Longrightarrow ((A = I) \vee (A = -I))$.
vi) $(A - B)(A + B) = (A^2 - B^2)$.
vii) $(AC = BC) \Longrightarrow (A = B)$.
viii) $(CA = CB) \Longrightarrow (A = B)$.

c. Rewrite each of the statements in part b., replacing matrix multiplication with function composition, the inverse of a matrix with the inverse of a function, and so on. For example, in place of the first statement, consider

$$F \circ G = G \circ F \qquad \text{whenever the compositions are defined}$$

Determine whether each statement is true or false, giving a counterexample if the statement is false.

6.3.6 Prove the following properties pertaining to the transpose of a matrix:

a. $(A + B)^T = A^T + B^T$, whenever $A + B$ is defined.

b. $(rA)^T = r(A^T)$.

c. $(A^T)^T = A$.

d. $(AB)^T = B^T A^T$, whenever AB is defined.

e. $\text{trace}(A) = \text{trace}(A^T)$, whenever A is a square matrix.

f. $A = A^T$, whenever A is a diagonal matrix.

g. $(A^{-1})^T = (A^T)^{-1}$, whenever A is invertible.

6.3.7 Prove or disprove the following statements involving invertible matrices.

a. The inverse of an invertible matrix is unique.

b. The inverse of a diagonal matrix is a diagonal matrix.

c. A diagonal matrix is invertible if and only if its trace is not equal to 0.

d. If $A = BDB^{-1}$, where B is invertible and A and D are square matrices, then $A^n = BD^n B^{-1}$ for all $n \geq 1$.

e. If $AC = BC$, where C is invertible, then $A = B$.

f. If

$$A = \begin{bmatrix} a & b \\ c & d \end{bmatrix} \qquad \text{where } ad - bc \neq 0$$

then

$$A^{-1} = \frac{1}{ad - bc} \begin{bmatrix} d & -b \\ -c & a \end{bmatrix}$$

6.3.8 Assume $AB = BA$. Show that the following hold:

a. $(AB)^2 = A^2 B^2$

b. $A^2 B = BA^2$

c. $(A + B)^2 = A^2 + B^2$, whenever $AB = Z$.

d. $(A + B)(A - B) = A^2 - B^2$

6.3.9 Representation of a system of equations using matrices. You can express the *system of equations*

$$\begin{array}{rcrcr} 2x & + & 3y & = & -1 \\ x & + & 4y & = & 3 \end{array}$$

in *matrix form*, $AX = B$, as follows:

$$\begin{bmatrix} 2 & 3 \\ 1 & 4 \end{bmatrix} \begin{bmatrix} x \\ y \end{bmatrix} = \begin{bmatrix} -1 \\ 3 \end{bmatrix}$$

where A is the *coefficient matrix* and X is the *matrix of unknowns*.

    a. Solve the corresponding system of equations by finding the inverse of A.

    b. In general, what can you say about the corresponding system of equations if the coefficient matrix is not invertible?

6.3.10 Matrix representation of permutations. Consider the following permutations of $\{1..4\}$:

$$\chi = [4, 3, 1, 2], \qquad \psi = [1, 2, 4, 3], \qquad \omega = [3, 2, 4, 1]$$

    a. Find the permutation matrices of

       i) χ

       ii) ψ

       iii) ω

       iv) ω^{-1}

       v) χ^{-1}

    b. Use the results from part a. to find the permutation matrices of

       i) $\chi \circ \psi$

       ii) $\psi \circ \chi$

       iii) $\chi \circ \psi \circ \omega$

       iv) ω^3

       v) $(\omega \circ \chi)^{-1}$

6.3.11 Given a function

$$f : \{x_1, \ldots, x_m\} \longrightarrow \{y_1, \ldots, y_n\}$$

the *incidence matrix* of f is the $m \times n$ matrix $F = (f_{ij})$, where

$$f_{ij} = \begin{cases} 1 & \text{if } f(x_i) = y_j \\ 0 & \text{otherwise} \end{cases}$$

Consider

$$h : \{x_1, x_2, x_3\} \longrightarrow \{y_1, y_2\}$$

where

$$h(x_1) = h(x_2) = y_1, \qquad h(x_3) = y_2$$

and

$$g : \{y_1, y_2\} \longrightarrow \{z_1, z_2, z_3\}$$

where

$$g(y_1) = z_1, \qquad g(y_2) = z_3$$

    a. Find the incidence matrices of

i) g

ii) h

iii) $g \circ h$

b. Explain how the incidence matrix of $g \circ h$ is related to the incidence matrices of g and h.

6.3.12 Defining operations for matrices with Boolean entries. In the text, we only consider matrix operations on matrices whose entries are numbers. But, what about operations for matrices whose entries are propositions? Consider, for example, the $m \times n$ matrices $A = (a_{ij})$ and $B = (b_{ij})$, where

$$a_{ij} = (i < j), \qquad b_{ij} = ((i + j) \bmod 2 = 0)$$

These entries are Boolean values. It is possible to define operations on Boolean matrices by replacing the product of two numbers by the Boolean operator **and** ($\wedge$) and the sum by the operator **or** ($\vee$).

a. With these ideas in mind, use the definitions of A and B to find the values of the following expressions:

i) The scalar product

$$(2 < 3) \cdot [\text{“}a\text{”} \, \text{in} \, \text{“}cat\text{”}, 4 \bmod 2 = 1, true]$$

ii) The vector sum

$$[F, T \wedge F, \neg T] + [F \vee T, T \implies F, F]$$

where T denotes the value *true* and F is *false*.

iii) $A + B$, where A and B are 3×3.

iv) BA, where A is 3×2 and B is 2×4.

b. Define an ISETL **func** that

i) accepts two Boolean matrices and returns their sum.

ii) accepts two Boolean matrices and returns their product.

iii) accepts a Boolean scalar and a Boolean matrix and returns their scalar product.

c. Does the set of all $n \times n$ Boolean matrices have an additive identity? A multiplicative identity?

6.3.13 Normally an $m \times n$ matrix must have n items in each row. However, an $m \times n$ *raggedy matrix* is a matrix with m rows in which n is the *maximum* number of items in a row instead of the required number for every row.

a. How would you represent a raggedy matrix in ISETL?

    b. Use ISETL to define the product of a scalar and a raggedy matrix represented in ISETL.

    c. Use ISETL to define the sum of two raggedy matrices represented in ISETL.

    d. Explain how you might find the product of two raggedy matrices with appropriate sizes. What is the size of the product?

6.4 Determinants

The determinant of a square matrix is a number (scalar) associated with the matrix. It can be used to determine if the matrix is invertible, and if the matrix is, it can be used to find the inverse of the matrix. The determinant can also be used to solve a system of n equations with n unknowns. Since the determinant assigns to each square matrix exactly one real number, it is a function, denoted by *det*, whose domain is the set $\mathcal{M}_n$ of all $n \times n$ matrices with entries from the set of real numbers $\mathcal{R}$ and whose range is $\mathcal{R}$. That is,

$$det : \mathcal{M}_n \longrightarrow \mathcal{R}$$

We will continue throughout this section to use *det* for the determinant function regardless of the value of n. This is actually a little sloppy because the definition of *det* depends on the value of n, and hence, there is actually one function for each positive integer n.

    Of course, the first question we need to think about is "Given an $n \times n$ matrix, how do we find this associated real number?" You may already know how to evaluate $det(A)$ for a few small values of n. For instance, when $n = 2$,

$$det \left(\begin{bmatrix} 1 & 2 \\ 3 & 4 \end{bmatrix} \right) = 1 \cdot (1 \cdot 4) + (-1) \cdot (2 \cdot 3) = -2$$

However, instead of learning how to find $det(A)$ for specific values of n, our goal is to develop the definition for any $n > 0$.

    The definition of determinant is a little complex, so, with the aid of ISETL, we'll first examine two basic concepts underlying the definition, both of which involve permutations: the *sign* of a permutation and the *elementary product* determined by a given permutation. For instance, in the example of the 2×2 matrix that we considered above, there are two possible permutations of $\{1, 2\}$, $\sigma = [1, 2]$ and $\tau = [2, 1]$. As you will see, the sign of σ is $+1$ and it determines the elementary product $1 \cdot 4$, while the sign of τ is -1 and it determines the elementary product $2 \cdot 3$. Now, the value of the determinant of a matrix equals the sum of the *signed elementary products*, which in this case is -2.

First, we need to find the "sign" of a given permutation. A permutation, π of $\{1..n\}$, is said to perform an *inversion* whenever

$$\pi(i) > \pi(j) \qquad \text{but} \qquad i < j \qquad \text{where } 1 \le i, j \le n$$

To find the *sign* of π, we must first find the permutation's total number of inversions. For example, consider the permutation τ of $\{1..4\}$ given by

$$\tau = [1, 4, 3, 2]$$

To determine the total number of inversions performed by τ, fix $i = 1$ and consider what happens for $j = 2, 3, 4$. Now,

$$\tau(1) < \tau(2), \qquad \tau(1) < \tau(3), \qquad \tau(1) < \tau(4)$$

so τ has no inversions when $i = 1$. However, when $i = 2$, if you examine what happens for $j = 3$ and 4, you see that

$$\tau(2) > \tau(3), \qquad \tau(2) > \tau(4)$$

so τ has two inversions for this value of i. Finally, when $i = 3$, τ has one more inversion since

$$\tau(3) > \tau(4)$$

Therefore, τ's total number of inversions is $0 + 2 + 1$ or 3. With this information, you can then determine the sign of τ, $sgn(\tau)$, by raising -1 to the power equal to the permutation's total number of inversions. Since τ has 3 inversions,

$$sgn(\tau) = (-1)^3 = -1$$

In this case, we say that τ is an *odd* permutation, since the total number of inversions is odd. If the total were an even number, then we say that the permutation is *even*.

Observe that *sgn* is a function from the set of all *n*-tuples whose entries form the set $\{1..n\}$ to the set $\{-1, 1\}$. With this in mind, let's define an ISETL func that accepts a permutation of $\{1..n\}$, p, represented by an *n*-tuple, and returns its sign.

```
perm_sgn := func(p);
            local n, num_inversions;
            n := #p;
            num_inversions := #{[i,j] :
                i in [1..n-1], j in [i+1..n] | p(i)>p(j)};
            return (-1)**num_inversions;
          end;
```

Notice that the func perm_sgn and the definition of the sign of a permutation are almost identical. Why is i+1 the lower bound on j?

The second concept necessary for defining the determinant of a matrix is the elementary product. Suppose A is an $n \times n$ matrix represented in ISETL and p is an *n*-tuple representing a permutation of $\{1..n\}$. Then, the *elementary product* determined by p is

$$A(1)(p(1)) * A(2)(p(2)) * \cdots * A(n)(p(n))$$

which can be expressed using the compound operator % as

$$\%*[A(i)(p(i)) : i \text{ in } [1..n]];$$

or using mathematics notation as

$$\prod_{i=1}^{n} a_{i\,p(i)}$$

where $A = (a_{ij})$. Where do its terms come from? To calculate the product, ISETL iterates through each of the rows in A by having i iterate through [1..n], so clearly the product contains one item from each row. The column, however, of each term in the product is determined by the permutation p, so not only does the product contain exactly one entry from each row, it also contains exactly one entry from each column. Therefore, if

$$A := [[2, 1, 0, 9],$$
$$[1, 0, 1, 3],$$
$$[4, 2, 7, 0],$$
$$[2, 4, 6, 8]];$$

then the elementary product determined by p = [1, 4, 3, 2] is

$$A(1)(p(1)) * A(2)(p(2)) * A(3)(p(3)) * A(4)(p(4));$$

or

$$A(1)(1) * A(2)(4) * A(3)(3) * A(4)(2);$$

which has the value

$$2 \cdot 3 \cdot 7 \cdot 4$$

Now, p has three inversions so its sign is -1, and the signed elementary product determined by p is

$$(-1) \cdot (2 \cdot 3 \cdot 7 \cdot 4) \text{ or } -168$$

Therefore, combining the idea of the sign of a permutation and the elementary product determined by the permutation, and assuming that perm_sgn is an ISETL **func** that returns the sign of a permutation, we define the *signed elementary product* to be the product of the sign and the elementary product. That is,

```
        perm_sgn(p) * (%*[A(i)(p(i)) :   i in [1..n]]);
```

or assuming $A = (a_{ij})$ and that π is a permutation on $\{1..n\}$,

$$sgn(\pi) \cdot \prod_{i=1}^{n} a_{i\,\pi(i)}$$

Finally, by adding together all possible signed elementary products, you have the determinant of the given $n \times n$ matrix, **A**:

```
%+[[perm_sgn(p) * (%*[A(i)(p(i)) :
                    i in [1..n]]) :   p in permset];
```

where **n** = **#A** and **permset** is the set of all possible permutations of $\{1..n\}$. This can be expressed using $\sum$ and $\prod$ as

$$det(A) = \sum_{\pi \in \mathbf{P}_n} sgn(\pi) \cdot \prod_{i=1}^{n} a_{i\,\pi(i)}$$

where $A = (a_{ij})$ is an $n \times n$ matrix and $\mathbf{P}_n$ is the set of all permutations of $\{1..n\}$. Frequently, $det(A)$ is denoted by $|A|$.

Let's construct **permset** for $n = 3$ and call it **perm3**. Now, **perm3** is a set of 3-**tuples** containing all possible combinations of the integers 1, 2, and 3, without any repetitions. So,

```
perm3 := {[x,y,z] :   x in {1..3}, y in {1..3}-{x},
                      z in {1..3}-{x,y}};
```

Therefore,

```
perm3 = {[1,2,3], [1,3,2], [2,1,3],
         [2,3,1], [3,1,2], [3,2,1]]
```

Then, noting that [1,2,3], [2,3,1], and [3,1,2] are all even permutations while the others are odd, we can use the definition to find the determinant of

$$A = \begin{bmatrix} 0 & 1 & 2 \\ 3 & 6 & 9 \\ 6 & 2 & 1 \end{bmatrix}$$

by finding the sum of the signed elementary products:

$$a_{11} \cdot a_{22} \cdot a_{33} - a_{11} \cdot a_{23} \cdot a_{32} - a_{12} \cdot a_{21} \cdot a_{33} + a_{12} \cdot a_{23} \cdot a_{31} + a_{13} \cdot a_{21} \cdot a_{32} - a_{13} \cdot a_{22} \cdot a_{31}$$

So,

$$det(A) = 0 \cdot 6 \cdot 1 - 0 \cdot 9 \cdot 2 - 1 \cdot 3 \cdot 1 + 1 \cdot 9 \cdot 6 + 2 \cdot 3 \cdot 2 - 2 \cdot 6 \cdot 6 = 33$$

What properties do determinants satisfy? We will discuss a few of them here and leave some others for you to think about in the exercises. One important property is

If B is the matrix obtained from the matrix A by interchanging the rth and sth rows of A, then $det(B) = -det(A)$.

To help convince yourself that this is true, first note that the set of all
elementary products associated with A and the set associated with B must
be equal, since each elementary product contains precisely one entry from
each row and one from each column; the only difference, we claim, is that
the permutations associated with a particular elementary product for A
and B have opposite signs. For instance, if p_A is the permutation that
corresponds to a given elementary product for A, where

$$p_A = [i_1, i_2, \ldots, i_r, \ldots, i_s, \ldots, i_n]$$

then the permutation for the equivalent product with respect to B would
be

$$p_B = [i_1, i_2, \ldots, i_s, \ldots, i_r, \ldots, i_n]$$

However, when you interchange two entries in a permutation, the number
of inversions increases or decreases by an odd number (see Exercise 6.4.3),
so the two permutations, p_A and p_B, must have opposite signs. Since this
is true for each corresponding pair of terms in the two determinants, the
property holds.

One consequence of this fact is

If a matrix has two equal rows, then its determinant is 0.

To see that this is the case, let A be a matrix with two equal rows. Interchange the equal rows. The matrix is unchanged, but the above property
tells you that $det(A) = -det(A)$. The only real number, however, for which
this is true is 0. So, $det(A) = 0$.

Two other fundamental properties of determinants are

If B is the matrix obtained from A by multiplying the entries
in the rth row of A by some real number c, then $det(B) = c \cdot det(A)$.

and

If B is the matrix obtained from A by adding c times the
entries in the pth row to the corresponding entries in the rth
row, then $det(A) = det(B)$ whenever c is any nonzero real
number.

To see that the first property holds, look at the definition of determinant,
noting that if $A = (a_{ij})$ and $B = (b_{ij})$ then

$$b_{ij} = \begin{cases} a_{ij} & \text{if } i \neq r \\ c \cdot a_{ij} & \text{if } i = r \end{cases}$$

Therefore,

$$
\begin{aligned}
det(B) &= \sum_{\pi \in \mathbf{P}_n} sgn(\pi) \cdot b_{1\,\pi(1)} \cdots b_{r\,\pi(r)} \cdots b_{n\,\pi(n)} \\
&= \sum_{\pi \in \mathbf{P}_n} sgn(\pi) \cdot a_{1\,\pi(1)} \cdots c\, a_{r\,\pi(r)} \cdots a_{n\,\pi(n)} \\
&= c \cdot \sum_{\pi \in \mathbf{P}_n} sgn(\pi) \cdot a_{1\,\pi(1)} \cdots a_{r\,\pi(r)} \cdots a_{n\,\pi(n)} \\
&= c \cdot det(A)
\end{aligned}
$$

We leave the proof of the second property for you to think about in Exercise 6.4.4.

You can always evaluate the determinant of a matrix by using the definition. However, by applying the various properties, you can develop alternate ways to evaluate the determinant. We conclude this section by discussing one of these, which follows from applying the above properties along with the following fact:

> The determinant of an $n \times n$ upper triangular matrix A is equal to the product of the entries on its diagonal.

Therefore, in ISETL, the determinant of an upper triangular matrix A is equal to

```
%*[A(i)(i) :  i in [1..n]];
```

or, in mathematical notation,

$$
\prod_{i=1}^{n} a_{ii}
$$

To see that this holds, examine the elementary products associated with an upper triangular matrix. All the elementary products will be equal to zero, *except* the ones that are determined by a permutation where

$$
\pi(1) < \pi(2) < \cdots < \pi(n)
$$

(why?). But, the only time this occurs is when $\pi = [1, 2, \ldots, n]$, and since the sign of π is 1 and its associated elementary product is precisely the product of the entries on the main diagonal, the value of the determinant equals the product of the diagonal entries.

It appears at first glance that this last property only helps you to find the determinant of a very small class of matrices, namely, only those that have all zero entries below the main diagonal. However, any matrix can be *reduced* to an upper triangular matrix by using the *elementary row operations*, which are

1. Interchange row r and row s.

2. Multiply the entries in the rth row by the scalar c.

3. Add a nonzero scalar multiple of the entries in the pth row to the corresponding entries in the rth row.

These operations should look familiar—each one corresponds to a property we discussed previously. If you start with the original matrix and perform the given row operations to reduce it to an upper triangular matrix, then the properties tell you exactly what the impact will be on the value of the determinant. According to them, the first operation changes the sign of the determinant, the second multiplies it by the scalar c, and the third leaves the value of the determinant unchanged.

We can use these operations to reduce a matrix to an upper triangular matrix, keeping track of the changes in the determinant. Then, when the reduction is complete, we can go back and adjust the value of the determinant. For example, consider the following reduction of the matrix A to the matrix B, which is in upper triangular form:

$$A = \begin{bmatrix} 0 & 1 & 2 \\ 3 & 6 & 9 \\ 6 & 2 & 1 \end{bmatrix} \begin{matrix} \rightarrow R_2 \\ \rightarrow R_1 \\ \rightarrow R_3 \end{matrix} \begin{bmatrix} 3 & 6 & 9 \\ 0 & 1 & 2 \\ 6 & 2 & 1 \end{bmatrix} \begin{matrix} \rightarrow \frac{1}{3}R_1 \\ \rightarrow R_2 \\ \rightarrow R_3 \end{matrix} \begin{bmatrix} 1 & 2 & 3 \\ 0 & 1 & 2 \\ 6 & 2 & 1 \end{bmatrix}$$

$$\begin{matrix} \rightarrow R_1 \\ \rightarrow R_2 \\ \rightarrow R_3 - 6R_1 \end{matrix} \begin{bmatrix} 1 & 2 & 3 \\ 0 & 1 & 2 \\ 0 & -10 & -17 \end{bmatrix} \begin{matrix} \rightarrow R_1 \\ \rightarrow R_2 \\ \rightarrow R_3 + 10R_2 \end{matrix} \begin{bmatrix} 1 & 2 & 3 \\ 0 & 1 & 2 \\ 0 & 0 & 3 \end{bmatrix} = B$$

where, for instance, we have used the notation "$\rightarrow R_i$" to indicate that we have replaced the designated row with the values in the ith row.

Of course, there are many different ways to reduce a given matrix, but in order to avoid spending a lot of time going around in circles, it helps if your reduction steps have a general pattern. What we have done in our example is to try to get a 1 in the (i, i) position and then use this 1 to get 0 in the other entries in the ith column that are below the main diagonal. That is, we got a 1 in the $(1,1)$ position by first interchanging rows 1 and 2 and then multiplying the new row 1 by $1/3$. Next, we used the 1 in the $(1,1)$ position to get a 0 in the $(3,1)$ position by adding -6 times the first row to the third row. Finally, we used the 1 in the $(2,2)$ position to get 0 in the $(3,2)$ position by adding 10 times the second row to the third row. Having done all this, what is the value of the determinant of A? The only operations that affect the value of the determinant is to interchange rows or multiply by a scalar. Thus, since in obtaining B from A, we interchanged two rows and multiplied a row by $1/3$, we have

$$det(B) = -\frac{1}{3}det(A)$$

But, B is an upper triangular matrix, so

$$det(B) = 1 \cdot 1 \cdot 3 = 3$$

Therefore,

$$det(A) = -9$$

We'll describe some other approaches to evaluating determinants in the exercises.

Summary of Section 6.4

The goal of this section is to help you develop a firm understanding of the definition of determinant and, hence, to be able to use the definition to help convince yourself that determinants satisfy a number of different properties. We have not spent a lot of time discussing alternative methods for evaluating determinants because we are concerned that this leads to "rote" evaluation without thinking about what a determinant really is. Instead, our hope is that when you think about evaluating a determinant, you have this dynamic mental picture of ISETL iterating through the set of all n-permutations and finding the sum of the associated signed elementary products—we are convinced that if you have this, then the mathematical definition of determinant (which many students find baffling) will seem reasonable to you. The exercises are designed to help you achieve this goal and to introduce you to some applications.

Exercises

6.4.1 Determinants and ISETL.

    a. Construct a **func** called **perm_sgn** that accepts a permutation (of $\{1..n\}$) and returns its sign.

    b. Define **perm4** to be a **func** that calculates the set of all permutations of $\{1..4\}$.

    c. Using parts a and b., define an ISETL **func**, **det4**, that accepts a 4 × 4 matrix and returns its determinant.

    d. Test your **func** by finding the determinants of the following matrices. As usual, "think ISETL!"

        i) `[[1,2,3,4], [-3,0,6,9], [1,-2,10,0], [0,1,1,1]]`

        ii) `[[(i+j) mod 3 :  j in [1..4]]:  i in [1..4]]`

        iii) `[[2,1,0,9], [1,0,1,0], [-4,-2,0,-18], [2,4,6,8]]`

        iv) `[[1.1,0.0,1.2,0.0], [-3.0,0.0,1.0,0.0],`
           `[0.0,0.0,1.0,0.0], [-2.5,-5.0,2.0,0.0]]`

$$\text{v)} \begin{bmatrix} 21 & -2 & 20 & 0 \\ 1 & 10 & 0 & -8 \\ 5 & 0 & -10 & 15 \\ 0 & 4 & 12 & -16 \end{bmatrix}$$

6.4.2 The given properties and ISETL.

a. Define a **func** called **swap** that accepts a square matrix A and two **integer** values, r and s, and returns the matrix obtained from A by interchanging the rth and sth rows. Compare the values of

$$\texttt{det4(A) and det4(swap(A,r,s))}$$

for the matrix

$$\texttt{[[1,2,3,4], [-3,0,6,9], [1,-2,10,0], [0,1,1,1]]}$$

and several values of r and s.

b. Define a **func** called **row_add** that accepts a square matrix A, a **floating-point** number c, and **integers** p and r and returns the matrix obtained from A by adding c times the pth row to the rth row of A. Compare the values of

$$\texttt{det4(A) and det4(row\_add(A,c,p,r))}$$

for the matrix

$$\texttt{[[1,0,0,0], [6,4,0,0], [-9,8,11,0], [19,-26,78,2]]}$$

and several values of c, p, and r.

c. Define a **func** called **row_mult** that accepts a square matrix A, a **floating-point** number c, and an **integer** r and returns the matrix obtained from A by multiplying the entries in the rth row by c. Compare the values of

$$\texttt{det4(A) and det4(row\_mult(A,c,r))}$$

for the matrix represented in part b.

6.4.3 Counting questions.

a. What is the maximum number of inversions a permutation of $\{1..n\}$ can perform? Express your answer in terms of n.

b. Use an ISETL **tuple** former to construct the permutation that has the maximum number of inversions on $\{1..n\}$.

c. What permutation has the minimum number of inversions?

d. How many signed elementary products (not necessarily distinct) are associated with a given $n \times n$ matrix?

e. Show that when you interchange two entries in a permutation the number of inversions increases or decreases by an odd number. *Hint: First examine what happens when you interchange two adjacent entries.*

6.4.4 Suppose A and B are $n \times n$ matrices. Discuss why the following properties hold by considering the definition of determinant:

a. If A has an all zero row, then $det(A) = 0$.

b. If B is obtained from A by adding a nonzero scalar multiple of one row to another row, then $det(B) = det(A)$.

c. For any real number c, $det(c\,A) = c^n det(A)$.

d. $det(A^T) = (det(A))^T$, where A^T is the transpose of A. (Remark: As a result of this important fact, we can replace the word row by the word column in most statements involving properties of determinants.)

6.4.5 Prove the following:

a. $det(I_n) = 1$.

b. If the matrix A contains one row that is a multiple of another, then $det(A) = 0$.

6.4.6 Fact: $det(AB) = det(A)det(B)$ for any $n \times n$ matrices A and B.

a. Prove that this property holds for any 2×2 matrices $A = (a_{ij})$ and $B = (b_{ij})$.

b. Use the given property and the fact that $det(I_n) = 1$ to show that

i) if A is an invertible matrix, then $det(A^{-1}) = (det(A))^{-1}$.

ii) a matrix is invertible if and only if its determinant is not equal to 0.

6.4.7 *Cramer's rule* for solving a system of n equations in n unknowns. A system of n equations in n unknowns,

$$
\begin{aligned}
a_{11}x_1 &+ a_{12}x_2 + \;\ldots\; + a_{1n}x_n &= b_1 \\
a_{21}x_1 &+ a_{22}x_2 + \;\ldots\; + a_{2n}x_n &= b_2 \\
&\;\;\vdots \\
a_{n1}x_1 &+ a_{n2}x_2 + \;\ldots\; + a_{nn}x_n &= b_n
\end{aligned}
$$

can be represented in matrix form by

$$
\begin{bmatrix}
a_{11} & a_{12} & \cdots & a_{1n} \\
a_{21} & a_{22} & \cdots & a_{2n} \\
\vdots & & & \\
a_{n1} & a_{n2} & \cdots & a_{nn}
\end{bmatrix}
\begin{bmatrix}
x_1 \\
x_2 \\
\vdots \\
x_n
\end{bmatrix}
=
\begin{bmatrix}
b_1 \\
b_2 \\
\vdots \\
b_n
\end{bmatrix}
$$

or as

$$AX = B$$

Cramer's rule states that if $det(A) \neq 0$ then the solution to the system is given by

$$x_j = \frac{det(A_j)}{det(A)} \qquad \text{for } 1 \leq j \leq n$$

where A_j is the matrix obtained from A by replacing the jth column of A by B.

   a. Show that Cramer's rule holds for an arbitrary system of 2 equations and 2 unknowns whenever $det(A) \neq 0$.

   b. Use Cramer's rule to solve the following systems of equations (if possible):

      i)
$$\begin{aligned} 2x + 3y &= -1 \\ x + 4y &= 3 \end{aligned}$$

      ii)
$$\begin{aligned} 98x - 53y + 27z &= 11 \\ 107y - 22z &= 0 \\ 36z &= 0 \end{aligned}$$

      iii)
$$\begin{aligned} -26s + t - 78u + 13v &= 0 \\ 16s + 21u - 35v &= 67 \\ 52s - 2t + 156u - 26v &= 17 \\ 10t - 109u - 6v &= 45 \end{aligned}$$

   c. The ISETL version of Cramer's Rule.

      i) Define an ISETL func called replace that accepts an $n \times n$ matrix A, an n-tuple b, and an integer j and returns the matrix obtained from A by replacing the jth column of A with the entries in B.

      ii) Use your ISETL funcs det4 (see Exercise 6.4.1) and replace to define a func cramer4 that accepts a 4 × 4 coefficient matrix A and a 4-tuple B and returns a 4-tuple containing the solution to the associated system of equations whenever $det(\text{A}) \neq 0$.

      iii) Test cramer4 on the following system:

$$\begin{aligned} 6p - 7r + 10s &= 11 \\ -8p + 2q + 5r &= 0 \\ p + q + r + s &= 13 \\ 4q - 8r - 15s &= -12 \end{aligned}$$

      iv) Run cramer4 on the systems given in part b.

6.4.8 Assume $A = (a_{ij})$ is a given 3 × 3 matrix.

a. Show that

$$det(A) \quad = \quad a_{11} \cdot det \begin{bmatrix} a_{22} & a_{23} \\ a_{32} & a_{33} \end{bmatrix} - a_{12} \cdot det \begin{bmatrix} a_{21} & a_{23} \\ a_{31} & a_{33} \end{bmatrix}$$
$$+ a_{13} \cdot det \begin{bmatrix} a_{21} & a_{22} \\ a_{31} & a_{32} \end{bmatrix}$$

b. Find the determinant of A where

$$A = \begin{bmatrix} 13 & 0 & 2 \\ -6 & 4 & 1 \\ 10 & 16 & 5 \end{bmatrix}$$

using the following methods:

i) Apply the definition of determinant.

ii) Reduce A to upper triangular form and find the product of the diagonal entries.

iii) Use the formula given in part a.

iv) Define an ISETL func that accepts a 3 × 3 matrix and returns its determinant. Run your func on A.

c. Generalize the formula given in part a. to find the determinant of a 4 × 4 matrix.

d. The general form of the method given in part a. is as follows: For a given $n \times n$ matrix $A = (a_{ij})$ and a fixed i, where $1 \leq i \leq n$,

$$det(A) = \sum_{j=1}^{n} (-1)^{i+j} a_{ij} \cdot det(M_{ij})$$

where M_{ij} is the $(n-1) \times (n-1)$ matrix obtained by deleting the ith row and jth column of A. Write a recursive ISETL func that accepts an $n \times n$ matrix and uses this formula to return the determinant of the matrix.

Chapter 7

Mathematical Induction

To the Instructor

As with several topics in this book, our presentation of mathematical induction is somewhat different from standard approaches. We are not alone in realizing that understanding this very important method of proof requires, at the very least, that students work a variety of problems that go beyond the usual proof that a given finite series has a certain closed form. In addition to presenting problems arising from a wide range of situations, we provide examples in which various steps in the induction proof are not so obvious. In some, the base case is hard to determine; in others, it is not obvious how to set up the problem formally as a Boolean valued function of the positive integers corresponding to a proposition. There will be situations in which it is difficult to see what, in the problem, corresponds to the positive integer. Of course, we give a number of problems in which it is not so easy to prove the implication from n to $n + 1$.

There has been a fair amount of research on how undergraduates learn and we feel that it tells us that experience with many problems, even from a wide variety of contexts, is not enough to learn a concept as complicated as mathematical induction. It appears that understanding such a concept and being able to use it requires the construction of several mental schemas that are linked together to form one large schema corresponding to *proof by induction*. This must be done by the student over a period of time and the main role of an educational experience is to encourage and assist this construction. In previous chapters, we have set numerous tasks for the student aimed at preparing her or him for this development.

The main schemas that the student must understand in order to become capable of making induction proofs are, first, *proposition valued functions of the positive integers*, which are really the mathematical constructs required to even state the typical assertion that is to be proved by induction. Second is *modus ponens*, by which we mean an awareness of the total process that infers from the assertion of an implication $A \Rightarrow B$ together with the proposition A the conclusion B. Third, the student must understand that implications are objects and can form the range of a function, so that there can be an *implication valued function of the positive integers*.

The instructor should be aware that we are using the terms *proposition valued function* and *implication valued function* in a psychological rather than a mathematical sense. Mathematically, they are both *Boolean valued functions*, that is, their range is the two element set {*true, false*}. Psychologically, however, the first refers to any proposition and the second refers to a proposition that happens to be an implication.

Once these concepts are firmly in place, they can be combined to form the concept of proof by induction. Beginning with a proposition valued function of the positive integers P, one constructs the corresponding implication valued function Q defined by

$$Q(n) = (P(n) \Rightarrow P(n+1)), \qquad n = 1, 2, 3, \ldots$$

The next step is to establish $P(1)$ or, more generally, $P(n_0)$ for some specific value n_0, often referred to as the *base case*. Finally, one shows that $Q(n)$ holds for all $n \geq n_0$. If all of the concepts up to this point have been constructed by the student, he or she will be able to imagine using modus ponens to coordinate $P(n)$ with $Q(n)$ to obtain $P(n+1)$ successively for $n = n_0, n_0 + 1, \ldots$.

It should be fairly clear at this point that many activities in previous chapters were designed with all of this in mind. We will start at the beginning of mathematical induction in this chapter and run through a number of computer activities with ISETL that we feel will help the student to form mental constructs corresponding to what we have just described. There will be a certain amount of repetition of previous work that we feel will serve as both review and reinforcement.

7.1   Preview

Now you are ready for a great adventure. Almost everybody has trouble with making proofs by induction, and as far as undergraduate students in Mathematics courses are concerned, they usually do not succeed in learning how to do it. The first of these reactions will, most likely, be yours as well. You will find induction hard to learn. The second will *not* be your experience. You will, in fact, succeed in learning it. Provided, of course, that you take this chapter seriously and are not misled by its relative shortness of length. The hours you spend will not be short. But, if you spend them, then you, like the many students before you who have used our approach, will learn how to make proofs by induction. You might even come to like it.

We begin this chapter by explaining how the method works, and in particular, how the tasks you will be performing with ISETL will lead you to develop, that is, construct, certain ideas in your mind. In the end, you will simply put these ideas together and, lo and behold, you will know induction! (It's not really magic, it's just fun to write like that.)

Some of the things that are discussed in this chapter have already been covered earlier in the book. You may or may not have grasped them then. If you did, then this is review; if you didn't, then this is your second chance. We will treat everything as if you are seeing it for the first time, only referring to past activities when we wish to make explicit use of them as, for instance, in particular examples.

The first concept that you must acquire in order to understand mathematical induction is *proposition valued function of the positive integers*. By this we mean a function (corresponding to a proposition) whose domain is the set $\mathcal{N}$ of positive integers and range is the set $\{true, false\}$. We will use ISETL funcs that take a positive integer as input and return a Boolean value. It is also useful to think of a tuple in ISETL as an instrument for representing a proposition valued function of the positive integers, that is, a sequence of Boolean values. (Of course, as in Chapter 4, because of the finite nature of ISETL, a tuple can only represent an approximation to a sequence that has been obtained by restricting its domain to a finite subset of the positive integers.) Practice with expressing various statements (that later will become theorems to be proved by induction) in ISETL helps you become comfortable in thinking about this important concept.

The next thing is a dissection of the implication $A \Rightarrow B$. If you know that this implication is true and you know that A holds, then you know that B holds. This little reasoning is a mental process that you go through all the time. You need to be conscious of doing it. We will try to help you raise this consciousness by having you write programs that work with implications as character strings, taking them apart, and combining them to arrive at various conclusions. We call this reasoning process *modus ponens*, which is Latin for *method of the bridge*.

At the same time, you must understand that these implications are objects, just as any proposition is an object when you think of it as a totality. Since they are objects, or data, they can appear as the result of a function, so that we can have *implication valued functions of the positive integers*. Of course, this is a special case of proposition valued functions of the positive integers, but the two constructs will play different roles in induction.

Now, it will be time to start putting things together. You will come to understand the connection between a proposition valued function of the positive integers and its *corresponding* implication valued function by writing an ISETL func that converts any example of the former into the latter. The actual process of making an induction proof can be simulated by an ISETL program. You will write it and apply it to various theorems.

An interesting feature of this *induction simulator* is that if everything is correct, it will run forever! Obviously, the point of the simulator is not to get an answer but to construct something whose action corresponds to the thought process you must go through in making the proof in your mind (and your mind, of course, *can* think of a process that runs forever).

A second mechanism that will help you think about the induction process comes from representing both the proposition valued function and its corresponding implication valued function as `tuples`. You want to prove that every component in one `tuple` (the proposition) has the value *true*, and you do it by proving that its first component and every component of the other `tuple` (the implication) has the value *true*. What happens is that each component of the implication `tuple` (which you know has the value *true*) allows you to add one more *true* value in the next component of the proposition `tuple`. This is mathematical induction. Its value lies in the fact that *in practice* it is often much easier to establish that the implication valued function has the constant value *true* than to prove the same thing for the original proposition valued function.

Once you have begun to develop your concept of mathematical induction, you will have an opportunity, in the last section of the chapter, to work with it in several variations and applications. This should help you to solidify your understanding.

This introduction may appear to you as a little bit general and abstract. If, at this point, you feel the need for some examples to help you make sense of it all, that is exactly what is coming next. In any case, don't worry if you didn't completely understand everything in this section. After you have worked with some of the material that follows, go back and read it again. You will be surprised at how much clearer it seems.

The adventure begins.

7.2 Proposition Valued Functions of the Positive Integers

At this point, you should be sufficiently familiar with ideas surrounding the function concept to understand perfectly well what is meant by the title of this section. We are referring to a function, corresponding to a proposition, whose domain is the set of positive integers and whose range is the set $\{true, false\}$. For example, we could write

$$2^n > n^2 + 2n - 2, \qquad n = 1, 2, \ldots$$

Then, the integer in the domain is symbolized by the n in the expression and the value of the function, $P(n)$, is the truth or falsity of the inequality. Thus,

$$P(1) = \text{truth or falsity of the assertion, } 2^1 > 1^2 + 2 \cdot 1 - 2$$

$P(2) =$ truth or falsity of the assertion, $2^2 > 2^2 + 2 \cdot 2 - 2$

$P(3) =$ truth or falsity of the assertion, $2^3 > 3^2 + 2 \cdot 3 - 2$

and so on. Of course, things don't always have to begin with $n = 1$. In many problems, the values of n can begin at 0 or some other integer, positive or negative. They don't even have to go up by ones. They can increase according to any rule that is appropriate to the problem. The only important thing is that there is, in any given problem, a first value, called the *base case*, and a clear understanding, for every value of n, which value comes next. These variations do not present any special difficulties, so we will not emphasize them. In this chapter, when we don't think there is any danger of confusion, we will often use the phrase *proposition valued function* and let you figure out the domain from the context of the problem.

The first thing you need to be able to do in order to make a proof by induction is to read the statement of a problem and identify it as a proposition valued function. This means that you have to figure out what, in the problem, is the meaning of the integer and you must be able to express precisely what is the meaning of $P(n)$. This is easy in the above example. The domain variable n is just the n that appears in the inequality and $P(n)$ is the truth or falsity of the inequality for that value of n. The *problem*, incidentally, is to find a base value n_0 and show that the function P, restricted to the domain $\{n : n \in \mathcal{N} \mid n \geq n_0\}$, has the constant value *true*. Put more succinctly, the problem is to show that

$$\forall\, n \geq n_0, \quad P(n) = \textit{true}$$

But, we will not be concerned with the actual problem until a little later in this chapter.

Identifying the statement of the theorem to be proved is not always so easy. Here are a few examples for you to think about while we consider something else. See if you can figure out what is the meaning of n and what exactly is the meaning of $P(n)$ in each of the following. Then, after a brief discussion of using ISETL to represent proposition valued functions, we will return to these examples and give *our* version.

1. $\displaystyle\sum_{i=1}^{k}(3i^2 - 3i + 1) = k^3$

2. The sum of the cubes of any three consecutive positive integers is divisible by 9.

3. A gambling casino can represent a given amount of dollars using only \$5 and \$9 chips.

4. The relation $x - y = z - w$ is a *loop invariant* for the while loop in the following ISETL func:

```
what1 := func(x,y);
            local z,w;
            z := x;
            w := y;
            while w>0 do
                z := z-1;
                w := w-1;
            end;
            return z;
        end;
```

5. If $2^n > n^2 + 2n - 2$, then $2^{n+1} > n^2 + 4n + 1$.

While you are thinking about how to express the above five statements as proposition valued functions, let's consider how to represent these functions in ISETL. We will use two ISETL constructs, **funcs** and **tuples**. The **func** just takes the integer as parameter input and, after checking that it is in the proper range, evaluates the proposition (when this is possible—sometimes it is just your mind that imagines a **func** evaluating a proposition). Thus, for the example that we used at the beginning of this section, we could write the following **func**:

```
P1 := func(n);
          if is_integer(n) and n>=1 then
              return 2**n > n**2 + 2*n - 2;
          end;
      end;
```

Whether you actually write and run the **func** or not, you should always think about representing a proposition valued function this way and try to imagine the computer running through the **func** for various values of the parameter in order to evaluate the proposition.

Using a **tuple**, you can represent a proposition valued function by simply listing its values—or at least the ones that you know. Since, in our context, the domain values will always have a definite starting point and then move along one–by–one (this is what is meant by *Discrete Mathematics*), the indices of the **tuple** can represent the domain values and the components can represent the values of the function. Thus, in general, we can write

$$[P(1), P(2), P(3), \ldots]$$

where P is the proposition valued function. Actually, when thinking about this in ISETL, we don't use the three dots as an ellipsis. We make use of our interpretation of a **tuple** as having an unlimited number of components, but with only finitely many of the components having been given a value. When it is a proposition valued function, since the value is either *true* or *false*, we can interpret the situation as saying that the **tuple** represents the entire function, but we only *know* the value of the function for finitely many values of the integer. Thus, in our example, if the value of P1 at n is given by the truth or falsity of the inequality and if we had determined the first eight values of this function (starting at n=1), then this could be represented with a tuple as follows:

[true, false, false, false, false, true, true, true]

Now, let's return to the above five examples and see how we can express them as proposition valued functions.

1. This one is easy. The domain variable is k, which counts the number of terms in the sum, and $P(k)$ is the truth or falsity of the equation

$$\sum_{i=1}^{k}(3i^2 - 3i + 1) = k^3$$

2. Here, there are several possibilities. One is to let the domain variable be the first of the three consecutive numbers. If we call it n, then we can express, in ISETL, the value of the proposition valued function at n as follows:

(n**3 + (n+1)**3 + (n+2)**3) mod 9 = 0

3. This one sometimes gives difficulty. The domain variable is the number of dollars and the value of the function is the truth or falsity of the existence of appropriate numbers of $5 and $9 chips. We can express this mathematically using quantification. The statement that d dollars can be represented looks like this:

$$\exists a, b \in \{0, 1, 2, \ldots, d\} \; \ni \; 5a + 9b = d$$

4. There is a little subtlety in this one. You will have difficulty understanding it unless you think about how the computer might process this loop and have a definite picture in your mind of running through it over and over again. Then, you can see that the domain variable is the number of times that the loop has been processed. It might be reasonable to start at 0. If we call the count n, then $P(n)$ is the assertion that after the loop has been processed n times, the relation

$$x - y = z - w$$

holds. In particular, $P(0)$ is the assertion that this relation holds before the loop has been processed at all—but after the variables z and w have been initialized.

Loop invariants like this can be used to determine exactly what a loop does, and this is useful in verifying programs.

5. We finish with a simple one. The domain variable is n and the value of $P(n)$ is the truth value of the implication

$$(2^n > n^2 + 2n - 2) \Rightarrow (2^{n+1} > n^2 + 4n + 1)$$

Summary of Section 7.2

The purpose of this section was to recall to your mind the idea of a proposition valued function and to concentrate on the case where the domain of the function was the set $\mathcal{N}$ of positive integers, the integers from some point on, or any discrete set that can be counted by the positive integers.

The main activity in this section was to represent such a proposition valued function as an ISETL func and to think of it as a Boolean tuple for which only finitely many of the values are known. The main goal of the exercises is that, when you are finished with them, you will be able to read a statement and express it as a proposition valued function of the integers or some similar discrete domain set.

Exercises

7.2.1 Each of the following statements expresses a proposition valued function. Where possible, write an ISETL func that will evaluate it. Run your func for several representative values of the domain variable.

a. $\displaystyle\sum_{i=1}^{n} i^3 = \left(\frac{n(n+1)}{2}\right)^2$

b. $\displaystyle\frac{1}{1\cdot 3} + \frac{1}{3\cdot 5} + \cdots + \frac{1}{(2n-1)(2n+1)} = \frac{n}{2n+1}$

c. $\displaystyle e^{1-n} + \sum_{i=1}^{n} \frac{1}{\sqrt{i}} > \frac{19}{10}\sqrt{n}$

d. $\displaystyle n - 2 < \frac{n^2 - n}{12}$

e. $\displaystyle\frac{60(n-1)}{2^n} + n! > 4^n$

f. The values are given by

$$F_n = \frac{1}{\sqrt{5}}\left[\left(\frac{1+\sqrt{5}}{2}\right)^{n+1} - \left(\frac{1-\sqrt{5}}{2}\right)^{n+1}\right]$$

where $F = (F_n), n \geq 1$, is the sequence of numbers defined by

$$F_1 = F_2 = 1, \qquad F_{n+2} = F_n + F_{n+1}, \qquad n \geq 1$$

This sequence of numbers is called the *Fibonacci sequence*.

g. There are exactly 2^n binary words of length n. (A *binary word* is a finite sequence of zeros and one. Its length is the number of terms in the sequence.)

h. $B_{n,k+1} + B_{n,k} = B_{n+1,k+1}$

where, for n a positive integer and $k = 1, 2, \ldots, n$, the symbol $B_{n,k}$ denotes the number of binary words of length n containing exactly k ones.

i. $11^{n+2} + 12^{2n+1}$ is divisible by 133.

j. Any integer composed of 3^n identical digits is divisible by 3^n.

k. A casino with only \$7 and \$9 chips can represent a given number of dollars.

l. If the integer p divides the product of a finite set of integers, then it must divide one of them.

7.2.2 For each of the functions in Exercise 7.2.1 for which you wrote a `func`, use your `func` to produce a `tuple` of the first 30 values of the function. In each case, make a guess as to what you think is the value of the remaining components of the tuple.

7.2.3 In each of the following situations, determine a proposition valued function that represents it, identify the domain variable, and specify the meaning of the value of the function for a particular value of the domain variable.

a. $\displaystyle\sum_{i=1}^{n}(i + i^3) = \frac{n(n+1)(n^2+n+2)}{4}$

b. $\begin{aligned}
1^3 &= 1\\
2^3 &= 3+5\\
3^3 &= 7+9+11\\
4^3 &= 13+15+17+19\\
&\vdots
\end{aligned}$

c. If 8 divides $5^{n+1} + 2 \cdot 3^n + 1$, then 8 divides $5^{n+2} + 2 \cdot 3^{n+1} + 1$.

d. $F_n < \left(\frac{7}{4}\right)^n$ for every positive integer n, where $(F_n)_n$ is the Fibonacci sequence defined in Exercise 7.2.1.

e. Every third term of the Fibonacci sequence is even.

f. The expression $a(n) = 2(3^n) - 5$ defines the only function which satisfies the following relations:

$$\begin{aligned}
a(0) &= -3\\
a(1) &= 1\\
a(n) &= 4a(n-1) - 3a(n-2) \qquad \text{for } n \geq 2
\end{aligned}$$

g. The relation $yw + z = x + y^2$ is a loop invariant for the `while` loop in the following `func`:

```
what2 := func(x,y);
           local z,w;
           z := x;
           w := y;
           while w>0 do
                   z := z + y;
                   w := w - 1;
           end while;
           return z;
end;
```

h. The complement of the union of any finite collection of sets is equal to the intersection of the complements of the sets.

i. For each positive integer n, there are more than n prime numbers.

j. The relations

$$
\begin{aligned}
x_1 &= \sqrt{2} \\
x_{n+1} &= \sqrt{1 + x_n} \qquad \text{for } n \geq 1
\end{aligned}
$$

define a sequence of irrational numbers.

k. For any positive integer n and $a_1, a_2, \ldots, a_n$ in the domain of the binary operation $\odot$, the expression

$$a_1 \odot a_2 \odot \ldots \odot a_n$$

is unambiguously defined. Here, $\odot$ is a binary operation that is associative, that is, $a \odot (b \odot c) = (a \odot b) \odot c$ for all a, b, and c in the domain of $\odot$.

7.3   Modus Ponens

Modus ponens, or *the method of the bridge*, is a good analogy for beginning to think about implications. If there is a bridge connecting two vantage points and you can get to the first one, then, in two steps, you can get to the second one. First, you get to the first one, and then you use the bridge. Thus, if the following picture represents a river and two islands where the river banks and the islands are connected by bridges, then if you are on the A side of the river, and you want to get to the island marked B, you can first go to the bank marked A and then use a bridge to get to B. If you want to go from the bank marked A all the way to the bank marked C, then you cannot do this directly, but you must put together two of these "implications," that is, you use modus ponens once to get from A to B and then, having arrived at B, use it again to go from B to C.

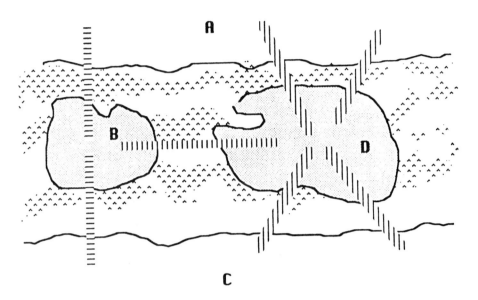

You can see that a bridge is two things. Of course, it is an object—made of stone and metal—but it is also an idea—the idea of going through the process of crossing the bridge. It is important to keep in mind that a bridge refers to both of these things—an object and a process.

Now let's review how all of this looks with logical implications. One big difference between bridges and implications is that you can cross most bridges in either direction, but an implication only goes one way. (If it goes both ways, we call it an *equivalence*.) Otherwise, the two are pretty much the same. An implication $A \Rightarrow B$ is a process that begins with the assertion A and, if that holds, concludes that the assertion B must, *of necessity*, hold as well. But an implication is also an object. We can represent it in ISETL as a string, "A impl B" and work with this string as we did in Chapter 2 (see page 94) with the medical diagnosis problem. We can also represent its action as a func, such as

```
AtoB := func(A,B);
          if (A and not B) then
              return false;
          else return true;
          end if;
     end;
```

Now, AtoB is an identifier whose value is the func that we just defined.

It can be evaluated, passed as a parameter, stored, and so on. Its value represents a function that evaluates the **Boolean** expression **A impl B**.

Let's look at an example that is a little more meaty. Consider the following labeled assertions.

> C : Today is the last class of the semester.
> P : The project is due today.
> W : This is the last week of the semester.
> Y : Yesterday was Thursday.
> F : Today is Friday.
> Q : W and F

Now, it is not hard to understand what is the situation if these assertions and the following list of implications between them all hold.

> C impl P
> Q impl C
> Y impl F

You can then imagine an ISETL program that could accept values for the Boolean variables **Y** and **W** and decide whether or not the project is due today. Again, the important thing here is that one could store each of these implications as **strings**. In this sense, they are objects. An ISETL program could analyze these **strings** and, assuming that each of the implications is true, wherever it finds the hypothesis true, move from hypothesis to conclusion and assert the conclusion. In this sense, the implications represent processes.

Moreover, it is possible to take two or more of these objects, match hypotheses with conclusions and construct new, valid implications. This corresponds to performing several related processes in sequence. For example, since both **C impl P** and **Q impl C** are *true*, we may construct the implication **Q impl P** and assert that it is *true*.

We can use ISETL **funcs** to express many of these things. For example, suppose that **implset** is the **set** consisting of three **strings** of length 8 having the form **"A impl B"** and corresponding to the three implications in the above list of three expressions. Here is a **func** that accepts as parameter a character and determines whether it is possible to link two implications in the **set** so as to draw some conclusion from the assertion given by the input parameter.

```
I := func(hyp);
        if exists imp1, imp2 in implset |
                ((imp1(1) = hyp) and imp1(8) = imp2(1)) then
            return imp2(8);
        end;
    end;
```

Then, for example, the value of `I("Q")` would be `"P"`, whereas the value of `I("Y")` would be `OM`.

The implications do not necessarily have to be represented as **strings**. It would also be possible (and perhaps a little more convenient) to represent them as **tuples** of length 2 or ordered pairs in which the first component is the hypothesis and the second is the conclusion. For instance, the implication `C impl P` can be represented by the pair `["C","P"]`.

We can consider other kinds of situations. Here is one that relates more directly to induction. Suppose that P is a proposition valued function, for instance, assume P is the assertion that for a positive integer n

$$2^n > n^2 + 2n - 2$$

Now, suppose for some reason that we were not able to evaluate P directly for various values of n, but that we did have a **func** that could evaluate the implication valued function Q whose value at n is $P(n) \Rightarrow P(n+1)$. (*Incidentally, this situation is not as unrealistic as it may sound. Of course, the computer doesn't care which expression it has to evaluate, but if you were doing it by hand, then P(n) could be laborious, and you couldn't do it if you didn't know the value of n. But, look at what a little discussion does for the implication:*)

$$\left(2^n > n^2 + 2n - 2\right) \Rightarrow \left(2^{n+1} > (n+1)^2 + 2(n+1) - 2\right)$$

or

$$\left(2^n > n^2 + 2n - 2\right) \Rightarrow \left(2 \cdot 2^n > n^2 + 4n + 1\right)$$

or

$$\left(2^n > n^2 + 2n - 2\right) \Rightarrow \left(2^n > \frac{1}{2}(n^2 + 4n + 1)\right)$$

and for this it is enough to show that

$$n^2 + 2n - 2 \geq \frac{1}{2}(n^2 + 4n + 1)$$

or

$$n^2 \geq 5$$

which is an assertion that you can evaluate with minimal knowledge of n.)

In any case, assuming that we have Q, where $Q(n) = (P(n) \Rightarrow P(n+1))$, here is ISETL code that will tell whether any $P(k)$ is *true*, provided that we know that $P(n_0)$ is true for some $n_0 < k$ and that $Q(n)$ is *true* for $n = n_o, \ldots, k - 1$.

```
P(n0) := true;
for n in [n0..k-1] do
    if (P(n) and Q(n)) then
        P(n+1) := true;
    end;
end;
if P(k) then print "P(k) is true";
else print "P(k) not established";
end;
```

Summary of Section 7.3

The main point of this section is that an implication is both an object and a process. It is essential that you maintain both interpretations and are able to pass back and forth at will. You can think of an implication as a **string** that can be parsed to direct a passing from hypothesis to conclusion—or you can represent it as a **func** that can be treated as data or called to transform data. You may also consider an implication as an ordered pair of propositions.

Exercises

7.3.1 Interpret the following chart of Mathematics course prerequisites as a set of implications. For example, the fact that Calculus III is a *direct* prerequisite for Probability can be interpreted as (registering for) Probability implies (that you have taken) Calculus III. These implications can be stored in ISETL as **tuples**. For example, the pair ["Pr", "CIII"] would be included, where we have used "Pr" to denote Probability and "CIII" to denote Calculus III.

We do not consider here *indirect* prerequisites, so the pair ["Pr", "CII"] would *not* be included.

   a. Write an ISETL **func** that will accept as input any one of these courses and return some direct prerequisite course.

   b. Write an ISETL **func** that will accept as input any one of these courses and return all direct prerequisite courses.

   c. Write an ISETL **func** that will accept as input any one of these courses and return every course that is a prerequisite for a course that is a prerequisite for the input course.

   d. Write an ISETL **func** that will accept as input any one of these courses and return every course that must be taken before the input course can be taken. *(Hint: You can make this problem a lot easier if you use the fact that the longest "chain" has six courses. You will learn a lot more if you ignore this hint!)*

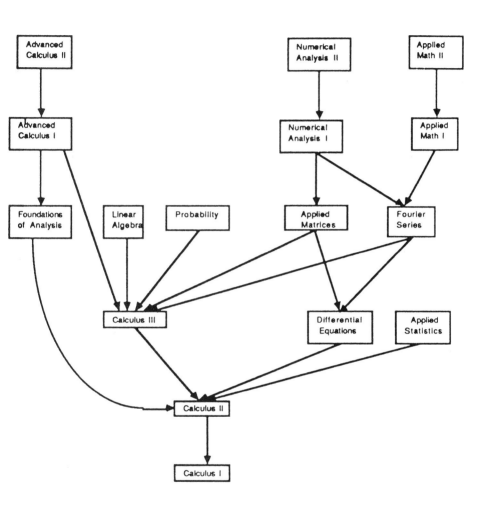

Mathematics   Course   Prerequisites

7.3.2 Write an ISETL **func** that will accept as input a proposition valued function on an appropriate domain and return a **tuple** of 25 values of implications between successive values of this function. Apply your program to each of the functions in Exercise 7.2.1 for which it makes sense.

7.3.3 Construct an interesting "expert system" for planning your day. Consider various categories of information, alternatives, and implications between them, such as the weather, the day of the week, what clothes you wear, what (if any) classes you must attend, contents of your refrigerator, shopping that must be done, means of travel, and so on. Write various ISETL **funcs** that can answer interesting questions about your day.

7.4   Implication Valued Functions of the Positive Integers

An implication is one kind of proposition, and we have already considered proposition valued functions, so you might think that this is a review section. Well, largely it is, and review is often a good idea. But, we will emphasize a different collection of examples and there is one new idea to be considered. Any time you have a proposition valued function of the positive integers, you can construct a *corresponding* implication valued function of the positive integers.

It works like this. Suppose that P is a proposition valued function of the integers. Then, you can define a new function Q, which is an implication valued function, by

$$Q(n) = (P(n) \Rightarrow P(n+1)), \qquad n = 1, 2, \ldots$$

Perhaps you can imagine how to implement this conversion with an ISETL **func**. You'll have a chance to do so in the exercises. In the meantime, let's look at some examples in which we can simplify the resulting expression.

Let P be the example discussed in Section 7.2,

$$P(n) = \left(2^n > n^2 + 2n - 2\right), \qquad n = 1, 2, \ldots$$

Then, the corresponding implication valued function Q is given by

$$Q(n) = \left((2^n > n^2 + 2n - 2) \Rightarrow (2^{n+1} > (n+1)^2 + 2(n+1) - 2)\right)$$

or, after a little simplification,

$$Q(n) = \left((2^n > n^2 + 2n - 2) \Rightarrow (2^{n+1} > n^2 + 4n + 1)\right)$$

which is exactly the same as Example 5 at the end of Section 7.2, in case you hadn't noticed. Recall also that, at the end of Section 7.3, we concluded that this implication will hold provided that $n^2 \geq 5$.

Here's another one. Let P be the function given by the expression in Example 1 of Section 7.2,

$$P(k) = \left(\sum_{i=1}^{k} (3i^2 - 3i + 1) = k^3 \right), \qquad k = 1, 2, \dots$$

Then, $Q(k)$ is

$$\left(\sum_{i=1}^{k} (3i^2 - 3i + 1) = k^3 \right) \Rightarrow \left(\sum_{i=1}^{k+1} (3i^2 - 3i + 1) = (k+1)^3 \right)$$

or

$$\left(\sum_{i=1}^{k} (3i^2 - 3i + 1) = k^3 \right)$$
$$\Rightarrow \left(\sum_{i=1}^{k} (3i^2 - 3i + 1) + 3(k+1)^2 - 3(k+1) + 1 = (k+1)^3 \right)$$

or, using the equality on the left-hand side of the implication to substitute for the summation on the right-hand side, $Q(k)$ is

$$\left(\sum_{i=1}^{k} (3i^2 - 3i + 1) = k^3 \right)$$
$$\Rightarrow \left(k^3 + 3(k+1)^2 - 3(k+1) + 1 = (k+1)^3 \right)$$

or

$$\left(\sum_{i=1}^{k} (3i^2 - 3i + 1) = k^3 \right) \Rightarrow \left(k^3 + 3k^2 + 3k + 1 = (k+1)^3 \right)$$

which is true.

We can use our interpretation of proposition valued functions of the positive integers as **tuples** to give an interpretation of the relation between a proposition valued function and its corresponding implication valued function. Consider the first example discussed above where the value of the function at n is the assertion $2^n > n^2 + 2n - 2$ and suppose that we have used a **tuple** P to represent it in ISETL as shown in the the following figure where the **T** indicates that the value of the function at 7 is *true*. We also show the corresponding implication valued function Q and indicate that its value at 7 is *true* as well ($7^2 \geq 5$).

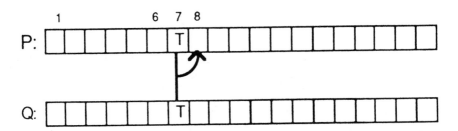

The curved arrow underneath P at component 7 indicates an application of modus ponens. That is, the fact that Q(7) has the value *true* means that the implication P(7) impl P(8) holds, and since according to our figure, P(7) is *true*, we may "cross the bridge" and conclude that P(8) is *true* as well. Thus, we could put a T in the eighth component of P.

In the next figure, we have put in some more information that we know. By simple calculation, we determined that the first component of P has the value *true*, its next four components have the value *false*, and its sixth component has the value *true*. As we noted above, $Q(n)$ holds provided that

$$n^2 \geq 5$$

so, in particular, from the sixth component on, we know that Q(n) has the value *true*. Since this is the first component for which both P and Q have the value *true*, we don't care about the earlier components of Q.

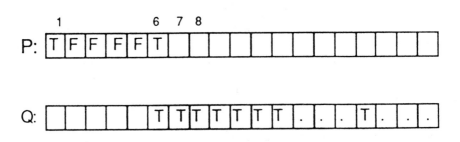

Finally, we start applying modus ponens, as indicated by the curly arrows in the following figure.

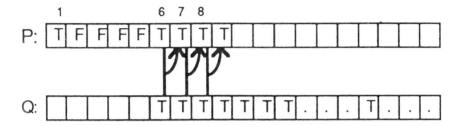

As you can see, it would be possible to apply modus ponens forever and thereby fill in all of the remaining components of P with T indicating that the value is *true*. In this way, we may conclude that P(n) is *true* for all values of n from six on.

We close this section with an important observation about the previous discussion. Ignoring the first few components, we showed that both functions P and Q have the constant value *true*. This was done in two different ways. For Q, we proved it using a direct proof that argued from the algebraic expression for $Q(n)$ to obtain the desired conclusion by ordinary manipulations. For P, we did something very different. We established that $P(6) = true$ and then used the fact that $Q(n) = true$ for $n \geq 6$ to conclude, one component at a time, that $P(n) = true$ as well for $n \geq 6$. The method we used is called (roll the drums) *mathematical induction*, which we explain in the next section.

Summary of Section 7.4

In this section, we turned again to proposition valued functions with particular emphasis on the case in which the proposition is an implication. You can always get an implication valued function of the positive integers by taking any proposition valued function of the positive integers and forming the implication between two successive values of the proposition valued function. This gives the corresponding implication valued function. We also introduced in this section the idea of moving along a **tuple** of **Boolean** values as a metaphor for using modus ponens to apply values of the implication valued function (when the value is *true*) to obtain new values of the original proposition valued function (when the previous value is also *true*). This motion will form the essential idea of mathematical induction in the next section.

Exercises

7.4.1 Write an ISETL **func, impl_fn,** that will accept a proposition valued function of the positive integers and return the corresponding implication valued function. Your **func** can accept its input in the form of either another **func** or a **tuple**, but its output must be a **func**.

Run your **func** on each of the functions constructed in the solution of Exercise 7.2.1 that can be represented in ISETL.

7.4.2 For each of the proposition valued functions determined by the situations in Exercise 7.2.3, write out the corresponding implication valued functions.

7.4.3 For each of the implication valued functions that you obtained in solving Exercise 7.4.2, determine its values for all possible values of the domain variable.

7.4.4 In this problem, you are to construct a system that we call an *induction simulator*. In your system, it should be possible to do each of the following for a given problem of the kind we have been discussing in this chapter.

    a. Determine the precise statement of the proposition valued function for the given problem and write a **func** called **prop_fn** that will evaluate it for any element of its domain.

    b. Write a program that begins by initializing an empty **tuple**, P. This means that we do not know if **prop_fn(n)** holds for any n. Next, have a loop in your program that runs **prop_fn** for n beginning at 1 and halts at the first n for which it returns **true**. This will be a possible n0 for this problem. Using the **func** **impl_fn** that you wrote for Exercise 7.4.1, add the following code to your program.

```
n := n0;
P(n0) := true;
while (((prop_fn(n) /= om) and
              P(n)) and impl_fn(prop_fn)(n)) do
   P(n+1) := true;
   n := n+1;
   print "The value of prop_fn is true for n = ", n;
end;
print "prop_fn is not proved for n = ", n+1;
```

    c. Apply your system to each of the problems in Exercise 7.2.1 that can be represented in ISETL. Explain what happens and how your system would have to be used in order to actually prove one of the statements in the exercise.

7.5   Making Proofs by Induction

Actually, you already know how to do this. Individual problems may give you difficulty and you may not always succeed in finishing one off, but as far as going about the induction proof, what you have done so far in this chapter includes all of the steps that you need to take in applying this method. All we have to do in this section is put it all together and tie a fancy ribbon around it, and you can get on with the task of practicing with this very powerful method.

Making an induction proof amounts to showing that the proposition valued function that expresses the situation has the value *true* for all but finitely many values of the independent variable. That is, if P is this function, then one has to show that, after throwing away finitely many values in the domain, P is a constant function and this constant is the value *true*. In most cases, when the domain is the set of integers, this is the same as showing that there is some base case, n_0, such that

$$P(n) = true \qquad \text{for} \quad n \geq n_0$$

The way we do this is to find P, find n_0, and show that $P(n_0)$ holds. Then, we "prove the implication" by assuming for an arbitrary integer $n \geq n_0$, that $P(n)$ holds and using this, along with the fact that $P(n_0)$ holds, we show that $P(n+1)$ holds. We now describe and exemplify this process in more detail.

There are four steps to an induction proof. The first and third correspond, respectively, to Sections 7.2 and 7.4 of this chapter, the second step is usually not too difficult, and the fourth is the only one that could give you any serious trouble at all, depending on the particular problem. After we explain these steps, we will demonstrate how they are applied in the following three examples, each of which you have met earlier in this chapter.

a. Show that, for n sufficiently large,

$$\sum_{i=1}^{k}(3i^2 - 3i + 1) = k^3$$

b. Show that, for n sufficiently large,

$$2^n > n^2 + 2n - 2$$

c. Show that the relation $x - y = z - w$ is a loop invariant for the **while** loop in the following ISETL **func**:

```
what1 := func(x,y);
          local w,z;
          z := x;
          w := y;
          while w>0 do
              z := z-1;
              w := w-1;
          end;
          return z;
end;
```

Now, let's run through the four steps. Although you should always do **Step 1** first, and **Step 3** has to be done before **Step 4**, the relative order in which you do **Steps 2** and **4** depends on the particular problem since either one of these steps could depend on the other.

Step 1. Express the problem as a proposition valued function of the positive integers. Determine what in the problem corresponds to the domain variable, the positive integer. Express as formally as you can the meaning of the value of $P(n)$, where P is the function and n stands for a value of the domain variable.

Step 2. Determine the base case. Find a value of the independent variable, for example, n_0, such that $P(n_0) = true$ and the corresponding implication valued function (see **Step 4**) is *true* for all values of the independent variable larger than or equal to n_0.

Step 3. Derive the implication valued function Q corresponding to the proposition valued function obtained in **Step 1**. It is important to have a very clear understanding for the specific problem you are considering of the meaning of the expression

$$P(n) \Rightarrow P(n + 1)$$

Step 4. Prove that for $n \geq n_0$ (the base case), $Q(n)$ has the constant value *true*. This means that you must show that $P(n) \Rightarrow P(n+1)$. For this, you may assume that $P(n)$ is *true* and, perhaps using that assumption explicitly, prove that $P(n + 1)$ is also *true*. In addition, you may have to use the fact that $P(n_0) = true$ as well.

Here is how it looks for the above three examples.

a. This one is completely straightforward. The domain variable is just the k in the expression, and the value of $P(k)$ is the truth or falsity of the equation $\sum_{i=1}^{k}(3i^2 - 3i + 1) = k^3$. For **Step 2**, we note that when $k = 1$, the left-hand side of the equation is $3(1)^2 - 3(1) + 1 = 1$ and the right-hand side is $(1)^3 = 1$, so they are equal. Thus,

$k_0 = 1$ is a candidate for the base case. Moving to **Step 3**, recall that we already determined in Section 7.4 that the corresponding implication valued function Q is given by

$$Q(k) = \left(\left(\sum_{i=1}^{k} (3i^2 - 3i + 1) = k^3 \right) \right.$$
$$\left. \Rightarrow \left(\sum_{i=1}^{k+1} (3i^2 - 3i + 1) = (k+1)^3 \right) \right)$$

So, we may pass to **Step 4** and prove the implication. To do this, we assume $P(k)$, that is, we assume that

$$\sum_{i=1}^{k} (3i^2 - 3i + 1) = k^3$$

for some particular k, and we show, for this k, that

$$\sum_{i=1}^{k+1} (3i^2 - 3i + 1) = (k+1)^3$$

To this end, we write, using our assumption,

$$\sum_{i=1}^{k+1} (3i^2 - 3i + 1) = \sum_{i=1}^{k} (3i^2 - 3i + 1) + 3(k+1)^2 - 3(k+1) + 1$$
$$= k^3 + 3(k+1)^2 - 3(k+1) + 1$$
$$= k^3 + 3k^2 + 6k + 3 - 3k - 3 + 1$$
$$= k^3 + 3k^2 + 3k + 1 = (k+1)^3$$

Thus, we have shown that $P(1)=true$ and $Q(k)$ holds for all positive integers k. The method of induction allows us to conclude that $P(k)$ holds for all positive integers k.

b. Again, there is no problem with **Step 1**. The domain variable is the n in the inequality, and the value of $P(n)$ is the truth or falsity of the inequality $2^n > n^2 + 2n - 2$. Turning to **Step 2**, however, things get a little tricky. The inequality holds for $n = 1$, but this is not the base case because, as we saw when we were looking at the `tuples` in Section 7.4, $P(n) = false$ for $n = 2, 3, 4$, and 5 and $P(6) = true$. Thus, this one seems to go back and forth between *true* and *false*. It is better to have a look at **Steps 3** and **4**. For **Step 3**, we already observed in Section 7.4 that

$$Q(n) = \left(2^n > n^2 + 2n - 2 \right) \Rightarrow \left(2^{n+1} > (n+1)^2 + 2(n+1) - 2 \right)$$

and we determined that $Q(n) = true$ provided that $n^2 \geq 5$, which
is the case if $n \geq 3$. Thus, in looking for the first value n_0 for which
$P(n_0) = true$ and $Q(n) = true$ for all $n \geq n_0$, we see that $n_0 = 6$
works, so this will be our base case.

To summarize, we see that $P(6) = true$, and for every $n \geq 6$, we
have $P(n) \Rightarrow P(n + 1)$. So, it is possible to insert the value $true$ in
the sixth component of the **tuple** for P and, using Q, add one more
value of $true$ to the next component, continuing this process forever.
Of course, we don't actually do this, but we *imagine* it being done
and thus conclude that every component of P from the sixth on has
the value $true$.

c. In our previous discussion of this example, we saw that the domain
variable is the number of times that the loop has been processed
and the value of $P(n)$ is the truth or falsity of the assertion that
after n times through the loop the relation $x - y = z - w$ holds.
This is **Step 1**. For **Step 2**, we might think of starting at $n = 0$. It
means that we assert that the relation holds after 0 times through
the loop, that is, at the beginning of the loop. This is definitely the
case since, before starting the loop, we have $z = x$ and $w = y$, so
surely $x - y = z - w$. Moving now to **Step 3**, we recall that $Q(n)$
is the assertion

> if, after n times through the loop, we have $x - y = z - w$,
> then, after $n + 1$ times through the loop, we still have
> $x - y = z - w$

If you think about it for a second, you see that what this really says
is that if $x - y = z - w$ just before the evaluation enters the loop,
then this relation still holds just after we leave the loop. This is why
the term *loop invariant* is used. To see that this is the case, and
thereby complete **Step 4**, we proceed, as usual, by assuming that
$x - y = z - w$ after n times through the loop, and we look at what
happens inside the loop. The only thing that happens is that both
z and w are decreased by 1. So, the quantity $z - w$ is unchanged,
and the relation $x - y = z - w$ is still $true$. Thus, the method of
induction has been successfully applied, and we may conclude that
$x - y = z - w$ no matter how many times the loop is executed, and
so it is a loop invariant.

We conclude with a comment on the loop invariant. One way in which a
loop invariant is used is to determine what calculation the loop is actually
making. Notice that the essential calculation is the value of z. This is
completed and returned exactly when the condition on the **while** ceases
to hold, that is, when $w = 0$. The loop invariant relation, $x - y = z - w$,
continues to hold in this case, and so we have $z = x - y$. Hence, one may
conclude that this loop calculates $x - y$.

Summary of Section 7.5

In this section, we put it all together. The method of induction was explained in full and applied to several examples. Now, you are ready to solve problems on your own.

Exercises

In the induction problems that follow, the problem is often stated as an assertion to be proved for whatever values of the domain variable it can be established. In some cases, you are not actually given the assertion but only some data from which you must formulate the assertion before trying to prove it. There may be a few statements that are not true, so be careful!

7.5.1 For each of the assertions in Exercise 7.2.1, choose an appropriate domain variable and prove the assertion for the largest range of values you can.

7.5.2 Choose an appropriate lower bound for n and prove $3^n > 2^n + 64$.

7.5.3 State and prove a general result of which the following three statements are special cases.

    a. $6^{n+2} + 7^{2n+1}$ is divisible by 43

    b. 73 divides $8^{n+2} + 9^{2n+1}$

    c. $11^{n+2} + 12^{2n+1}$ is divisible by 133

7.5.4 Consider the following **func**.

```
what2 := func(x);
            local y,z;
            z := 0;
            y := 0;
            while y/=x do
                z := z + x;
                y := y + 1;
            end;
            return z;
        end;
```

    a. Show that the relation $z = xy$ is a loop invariant for the **while** loop.

    b. Use the loop invariant to determine what calculation the loop is making.

7.5.5 Let Σ be an *alphabet* (that is, a finite set of characters) and A and B *languages* in Σ^* (that is, sets of finite strings of the characters in Σ), with the property that $A \subset B$. Show that $A^* \subset B^*$. (The superscript $*$ over a set means the set of all strings made up by concatenating finitely many elements of the set.)

7.5.6 Show that if a set has n elements then it has 2^n subsets.

7.5.7 Show that for every integer $n \geq 2$ the number of lines obtained by joining n points in the plane, no three of which are colinear, is $n(n-1)/2$.

7.5.8 Show that for n an integer and θ any real number we have

$$(\cos\theta + \sin\theta)^n = \cos n\theta + \sin n\theta$$

7.5.9 For each positive integer n, let $E(n)$ be an expression in $n + 1$ variables $x_1, \ldots, x_{n+1}$ that satisfies the following two conditions.

$$\begin{aligned} E(2) &= (x_1 \Rightarrow x_2) \\ E(n+1) &= (E(n) \wedge (x_n \Rightarrow x_{n+1})) \end{aligned}$$

Show that

$$E(n) = (x_1 \Rightarrow x_n), \qquad n \geq 1$$

7.5.10 If $A_1, A_2, \ldots, A_n$, and B are any sets, then prove

$$\left(\bigcup_{k=1}^{n} A_k \right) \cap B = \bigcup_{k=1}^{n} (A_k \cap B)$$

7.5.11 Consider the following **func**.

```
what3 := func(x);
          local a,b,c,z;
          z := 0;
          a := 1;
          b := 0;
          c := x;
          while c>0 do
                z := z + a + b;
                b := b + 2*a + 1;
                a := a + 3;
                c := c - 1;
          end;
          return z;
      end;
```

    a. Show that the relation $z = (x - c)^3$ is a loop invariant for the **while** loop.

    b. Use the loop invariant to determine what calculation the loop is making.

7.5.12 Show that the sum of the cubes of three consecutive positive integers is divisible by 9.

7.5.13 Show that for n, a positive integer, and $x > -1$, a real number,

$$(1 + x)^n \geq 1 + nx$$

7.5.14 Show that n straight lines in the plane, no two of which are parallel and no three of which intersect in a single point, divide the plane into $(n^2 + n + 2)/2$ regions.

7.5.15 Show that

$$1^3 + 3^3 + 5^3 + \cdots + (2n - 1)^3 = n^2(2n^2 - 1)$$

7.5.16 Show that

$$1 \cdot 2 \cdot 3 + 2 \cdot 3 \cdot 4 + \cdots + n(n + 1)(n + 2) = \frac{n(n + 1)(n + 2)(n + 3)}{4}$$

7.5.17 Find and prove a formula for

$$\prod_{j=1}^{n} 3^j$$

7.5.18 Consider the following **func**.

```
what4 := func(x,y);
          local w,z;
          z := 1;
          w := y;
          while w>0 do
                  z := z * x;
                  w := w - 1;
          end;
          return z;
     end;
```

    a. Show that the relation $zx^w = x^y$ is a loop invariant for the **while** loop.

b. Use the loop invariant to determine what calculation the loop is making.

7.5.19 Show that

$$\frac{1}{2n} \leq \frac{1 \cdot 3 \cdot 5 \cdots (2n-1)}{2 \cdot 4 \cdot 6 \cdots (2n)}$$

7.5.20 Show that

$$\sum_{i=1}^{2n} \frac{i}{2} = (n + \frac{1}{4})^2$$

7.5.21 State and prove a general result of which the following four formulas are special cases.

$$\frac{1}{1 \cdot 2} + \frac{1}{2 \cdot 3} + \cdots + \frac{1}{n(n+1)} = \frac{n}{n+1}$$

$$\frac{1}{1 \cdot 3} + \frac{1}{3 \cdot 5} + \cdots + \frac{1}{(2n-1)(2n+1)} = \frac{n}{2n+1}$$

$$\frac{1}{1 \cdot 4} + \frac{1}{4 \cdot 7} + \cdots + \frac{1}{(3n-2)(3n+1)} = \frac{n}{3n+1}$$

$$\frac{1}{1 \cdot 5} + \frac{1}{5 \cdot 9} + \cdots + \frac{1}{(4n-3)(4n+1)} = \frac{n}{4n+1}$$

7.5.22 Let x be the sequence defined by

$$x_1 = 1, \qquad x_{n+1} = x_n + 2\sqrt{x_n} + 1, \qquad n \geq 1$$

Show that every term in the sequence is an integer.

7.5.23 Let Σ be an alphabet and A a language in Σ^*, with the property that $A^2 = A$. Show that $A^* = A$. (A^2 is the set of all strings obtained by concatenating two strings, each from A. See Exercise 7.5.5 for explanations of the other terms.)

7.5.24 If (A_i) is a sequence of sets, any two of which are disjoint (we say that the sequence is *pairwise disjoint*), then

$$\#\left(\bigcup_{i=1}^{n} A_i \right) = \sum_{i=1}^{n} \#(A_i)$$

7.5.25 Consider the following func.

```
what3 := func(x);
            local a,b,c,z;
            z := 0;
            a := 1;
            b := 0;
            c := x;
            while c>0 do
                z := z + a + b;
                b := b + 2*a + 1;
                a := a + 3;
                c := c - 1;
            end;
            return z;
        end;
```

a. Show that the relation $b = 3(x - c)^2$ is a loop invariant for the while loop.

b. Use the loop invariant to determine what calculation the loop is making.

7.5.26 Find and prove a general formula for $2 + 4 + 6 + \cdots + 2n$.

7.5.27 Show that $n! + (n + 1)! + (n + 2)! = n!(n + 2)^2$.

7.5.28 For n, a positive integer, and $k = 1, 2, \ldots, n$, let $B_{n,k}$ denote the number of binary words of length n containing exactly k ones. Show that

$$2^n < B_{2n,n} < 4^n, \qquad n > 1$$

7.5.29 Show that for any n there is an algorithm for iterating through the set of all binary words of length n, without repetitions, changing exactly one bit each step.

7.5.30 Show that for $n \geq 1$

$$\frac{1}{\sqrt{1}} + \frac{1}{\sqrt{2}} + \cdots + \frac{1}{\sqrt{n}} \leq 2\sqrt{n} - 1$$

7.5.31 Find and prove a general formula for

$$\frac{1}{2!} + \frac{2}{3!} + \frac{3}{4!} + \cdots + \frac{n}{(n + 1)!}$$

7.5.32 Show that

$$(\cos x + \cos 2x + \cdots + \cos nx) \sin\left(\frac{x}{2}\right) = \cos\left(\left(\frac{x}{2}\right)(n+1)\right) \sin\left(\frac{nx}{2}\right)$$

7.5.33 Find and prove a formula for

$$\prod_{j=5}^{n} 3^j$$

7.5.34 Show that if a is a sequence of positive numbers, then

$$(a_1 a_2 \cdots a_{2^n})^{1/2^n} \leq \frac{a_1 + a_2 + \ldots + a_{2^n}}{2^n}$$

7.5.35 Show that $x + y$ is a factor of the polynomial $x^{2n+1} + y^{2n+1}$.

7.5.36 Consider the following func.

```
what3 := func(x);
            local a,b,c,z;
            z := 0;
            a := 1;
            b := 0;
            c := x;
            while c>0 do
                z := z + a + b;
                b := b + 2*a + 1;
                a := a + 3;
                c := c - 1;
            end;
            return z;
        end;
```

    a. Show that the relation $a = 3(x - c) + 1$ is a loop invariant for the while loop.

    b. Use the loop invariant to determine what calculation the loop is making.

7.5.37 Show that 8 divides $3^n + 7^n + 6$.

7.5.38 Show that $n^2 + 5n + 1$ is even for n sufficiently large.

7.5.39 Show that

$$\frac{1}{1 \cdot 5} + \frac{1}{5 \cdot 9} + \cdots + \frac{1}{(4n - 3)(4n + 1)} = \frac{n}{4n + 1}$$

7.5.40 Show that

$$\sum_{i=1}^{n} \frac{1}{\sqrt{i}} > 1.99\sqrt{n}$$

7.5.41 Show that $2^{n-1}(3^n + 4^n) > 7^n$ for n a sufficiently large integer.

7.5.42 If $(A_i)_i$ is a sequence of sets, then

$$\# \left(\bigcup_{i=1}^{n} A_i \right) = \sum_{i=1}^{n} \#(A_i) - \# \left(\bigcap_{i=1}^{n} A_i \right)$$

7.5.43 If $(A_i)_i$ is a sequence of sets, then

$$\# (A_1 \times A_2 \times \cdots \times A_n) = \prod_{i=1}^{n} \#(A_i)$$

7.5.44 If $C(n,r)$ is the number of combinations of n things taken r at a time, then

$$(x + y)^n = \sum_{r=0}^{n} C(n,r) x^{n-r} y^n$$

7.5.45 If $C(n,r)$ is the number of combinations of n things taken r at a time, then

$$2^n = \sum_{r=0}^{n} C(n,r)$$

7.5.46 The truth table for a Boolean expression in n variables has 2^n lines.

7.5.47 If D is a diagonal matrix and n is any positive integer, then

$$det(D^n) = (det(D))^n$$

7.5.48 If A, B, and D are square matrices of the same size, B is invertible, $A = BDB^{-1}$, and n is any positive integer, then $A^n = BD^n B^{-1}$.

7.5.49 If A and B are square matrices of the same size with $AB = BA$ and n is any positive integer, then $(AB)^n = A^n B^n$.

7.5.50 If A and B are square matrices of the same size with $AB = BA$ and n is any positive integer, then $A^n B = BA^n$.

7.5.51 If A and B are square matrices of the same size with $AB = BA$ and n is any positive integer, then $(A + B)^n = A^n + B^n$.

7.5.52 Show that for any real number c,

$$\sum_{i=1}^{n} c = cn$$

7.5.53 Show that

$$\sum_{i=1}^{n} i = \frac{n(n+1)}{2}$$

7.5.54 Show that

$$\sum_{i=1}^{n} i^2 = \frac{n(n+1)(2n+1)}{6}$$

7.6   Variations on the Induction Theme

7.6.1   Double Induction

Sometimes an assertion depends not on just *one* integer but on *two* (or even more) integer variables. Consider, for example, the following *double loop* calculation:

```
count := 0;
for i in [1..k] do
    for j in [1..n] do
        count := count + 2*i + 3*j;
    end;
end;
```

which adds up weighted sums of the indices in a $k \times n$ array. In mathematical terms, this is a *double summation* and is expressed as follows:

$$\sum_{j=1}^{k} \sum_{i=1}^{n} (2i + 3j)$$

It is possible, using various mathematical techniques, to figure out that the value of **count** when the loop is finished, that is, the value of the double summation, is given by the following expression:

$$\frac{nk}{2}(2n + 3k + 5)$$

The derivation of formulas like this can be quite complicated, and it is good to have a means of checking your accuracy. Mathematical induction can be used for making such a check. Alternatively, you might do a very loose derivation with lots of guessing. This can save time, and using induction to prove the formula once you have it should catch any errors you might have made.

Let us consider, then, how one might prove a formula such as

$$\sum_{j=1}^{k}\sum_{i=1}^{n}(2i+3j)=\frac{nk}{2}(2n+3k+5)$$

Considering this equation to be an assertion, you can see that it depends on two integer variables, k and n. Thus, we have a proposition valued function of two variables. If we call this function P, then $P(3,7)$, for example, is the assertion that the above equality is true when 3 is substituted for k and 7 is substituted for n.

Our overall strategy for dealing with this situation is an application of the general method we used in Chapter 4, Section 4.3.8, to work with functions of two or more variables. Recall the **func** *curry* that we defined on page 202, which takes a function of two variables and returns a function that gives, for each value of the first variable, a function of the second variable. If we think of a mathematical operation *curry* that can do this for any function of two variables, then we may consider *curry*(P). Now, for example, *curry*$(P)(4)$ is the truth value of the assertion that if 4 is substituted for k in our equation then it holds for every value of n. Specifically, *curry*$(P)(4)$ says that

$$\sum_{j=1}^{4}\sum_{i=1}^{n}(2i+3j)=2n(2n+17) \qquad \text{is true for every value of } n$$

Now, here is how we prove such an assertion, using double induction. We will break it down into six steps.

Step 1. We begin with the overall strategy of considering *curry*(P) to be a proposition valued function of the positive integers. Its value, for an integer k, is a certain proposition valued function, to wit, the function that assigns to any positive integer n the truth value of the assertion that $P(k)(n)$ is *true*. We use induction to show that this function, *curry*$(P)(k)$, has the constant value *true*.

Step 2. First, we show that *curry*$(P)(1)$ is *true*. That is, we must show that

$$\sum_{j=1}^{1}\sum_{i=1}^{n}(2i+3j)=n(n+4) \qquad \text{holds for all positive integers } n$$

We do this by *induction on n*.

Step 3. Taking $n=1$, we must show

$$\sum_{j=1}^{1}\sum_{i=1}^{1}(2i+3j)=5$$

which is obvious since there is only one term.

Step 4. Next, to prove the implication, we assume that the relation

$$\sum_{j=1}^{1}\sum_{i=1}^{n}(2i+3j) = n(n+4)$$

holds for n and prove it for $n+1$. This is a single summation of a kind you have done several times, so we leave it as an exercise for the reader.

Step 5. At this point, we have completed the proof in **Step 2**, that is, we have shown that $curry(P)(1)$ is *true*. Next, we show that

$$curry(P)(k) \Rightarrow curry(P)(k+1)$$

This means that we assume that $curry(P)(k)$ holds for every n, and we must show that $curry(P)(k+1)$ holds for every n.

Step 6. The proof is completed with the following calculations, where we use the assumption of **Step 5** in going from the second to the third line.

$$
\begin{aligned}
curry(P)(k+1)(n) &= \sum_{j=1}^{k+1}\sum_{i=1}^{n}(2i+3j) \\
&= \sum_{j=1}^{k}\sum_{i=1}^{n}(2i+3j) + \sum_{i=1}^{n}(2i+3(k+1)) \\
&= \frac{nk}{2}(2n+3k+5) + 3n(k+1) + \sum_{i=1}^{n}2i \\
&= \frac{nk}{2}(2n+3k+5) + 3n(k+1) + n(n+1) \\
&= \frac{n(k+1)}{2}(2n+3k+8)
\end{aligned}
$$

which is the desired expression when k is replaced by $k+1$.

7.6.2   Strong Induction

There are some problems in which it is not possible to prove the implication $P(n) \Rightarrow P(n+1)$ because the assumption that $P(n)$ has the value *true* simply does not provide enough information to establish the value of $P(n+1)$. In many situations, $P(n+1)$ does not have a convenient relation with $P(n)$, but it *is* related to $P(k)$ for some value (or even several values) of k less than n. Consider, for example, the following assertion.

> An integer is equal to a product of primes.

We can express this statement as a proposition valued function P with domain equal to the set of integers. If $n \leq 0$, then the assertion is not very interesting and, since primes are not less than 2, $P(1) = \textit{false}$. Thus, the best we can hope to prove is that $P(n)$ has the value *true* for $n \geq 2$. That is, we want to show that

> Every integer greater than or equal to 2 is equal to a product of primes.

To begin the proof, we take $n_0 = 2$ and observe that since 2 itself is a prime, then $P(2)$ holds.

Next, however, if we try to prove that the implication $P(n) \Rightarrow P(n+1)$ is *true*, we observe that assuming n to be a product of primes does not seem to help us much with factoring $n + 1$. To deal with $n + 1$, we could begin by arguing that if $n + 1$ is a prime, then we are finished. If it is not a prime, then it can be written as the product $k \cdot j$ of two integers, each of which is larger than 1 and less than $n + 1$. Unfortunately, neither of them will be n (since an integer greater than 1 never divides its successor).

The way out is to observe that if we think of solving our problem as representing P as a **tuple** P and filling up its components, one-by-one with the value *true*, then by the time we get to the $(n+1)$st component, we have already filled in, not only the nth component, but *every component from n_0 to n*. In particular, for this problem, we *do* know that both $P(k)$ and $P(j)$ have the value *true*, so we can complete our proof by arguing that $n + 1 = k \cdot j$ and both k and j are products of primes, and therefore, *their* product, $n + 1$, is a product of primes as well.

Our problem is solved, and we can, by reflecting on what has been done, establish a general strategy. The point is that in establishing $P(n + 1)$ we used the assumption that $P(k)$ and $P(j)$ had the value *true*. All we knew about k and j was that they were both between n_0 and n. Thus, we would have to assume that $P(i) = \textit{true}$ for all $i = n_0, n_0 + 1, \ldots, n$. If this is enough to establish that $P(n + 1) = \textit{true}$, then, by thinking about filling up the components of a **tuple** with the value *true*, you can see that $P(n)$ will have the value *true* for every $n \geq n_0$. We summarize this strategy as the *principle of strong induction*.

Step 1. Express the problem as a proposition valued function of the integers P.

Step 2. Determine n_0 and show that $P(n_0) = \textit{true}$.

Step 3. Assume that $P(k) = \textit{true}$ for all values of k that satisfy $n_0 \leq k \leq n$.

Step 4. Use this assumption to show that $P(n + 1) = \textit{true}$.

We can compare the usual induction method with the principle of strong induction. The first two steps in the above list are the same for both. The proof is completed in standard induction by establishing the following implication for every value of $n \geq n_0$:

$$P(n) \Rightarrow P(n+1)$$

In strong induction, the last two steps amount to establishing the following implication for every value of $n \geq n_0$.

$$(P(n_0) \wedge P(n_0 + 1) \wedge \cdots \wedge P(n)) \Rightarrow P(n+1)$$

or

$$(\forall i \in \{n_0, n_0 + 1, \ldots, n\}, \; P(i)) \Rightarrow P(n+1)$$

7.6.3   Recursively Defined Sequences

An important theme that has run through our entire discussion of mathematical induction has been the idea of filling up a `tuple` with components having the value *true* or *false* depending on the values of a proposition valued function of the positive integers. We can extend this idea to filling up the components of a `tuple` with any values whatsoever. This can be done, as with induction, by determining the value of a given component based on knowledge of the value of the previous component, some of the previous components, or even all of them. The result is that a sequence has been constructed. When it is done in this way, we say that the sequence has been *recursively defined*.

This is a different way of defining a sequence than giving a description of a process that, for a given positive integer n, determines the value of the sequence for that integer, which is called the *closed form* definition of a sequence.

In the beginning of this chapter, in Exercise 7.2.1, part f., we introduced the Fibonacci sequence by defining it both ways. Here is what we wrote:

The values are given by

$$F_n = \frac{1}{\sqrt{5}} \left[\left(\frac{1 + \sqrt{5}}{2} \right)^{n+1} - \left(\frac{1 - \sqrt{5}}{2} \right)^{n+1} \right]$$

where $F = (F_n), n \geq 1$, is the sequence of numbers defined by

$$F_1 = F_2 = 1, \qquad F_{n+2} = F_n + F_{n+1} \text{ for } n \geq 1$$

The first part of the statement defines the sequence in closed form and the second part gives the recursive definition.

A number of questions arise immediately. One important question concerns methods for deriving the closed form from the recursive definition. There is no general method that will solve every problem, but there are techniques for solving large classes of problems; unfortunately, these are beyond the scope of this book.

Another problem that leads to a simple application of mathematical induction is to show that the two definitions define the same sequence. (This assumes, of course, that you have both definitions.) To prove it by induction, you begin with the sequence given by a recursive definition and then show that the expression in the closed form satisfies the recursive definition. One can then argue that, first, the two sequences agree at the base case, and second, assuming that they agree on the appropriate previous cases, it follows that they agree at the "next case." Hence, they agree everywhere.

Let's see how this works for the Fibonacci sequence. Given the recursive definition for $F = (F_n), n \geq 1$,

$$F_1 = F_2 = 1, \qquad F_{n+2} = F_n + F_{n+1} \text{ for } n \geq 1$$

let us write $G = (G_n), n \geq 1$, for the closed form so that

$$G_n = \frac{1}{\sqrt{5}} \left[\left(\frac{1 + \sqrt{5}}{2} \right)^{n+1} - \left(\frac{1 - \sqrt{5}}{2} \right)^{n+1} \right]$$

Now it is easy to check by substituting 1 and 2 for n so that $G_1 = G_2 = 1$. It remains to assume that $G_{n-1} = F_{n-1}, G_n = F_n$, and to show that $G_{n+1} = F_{n+1}$. We have

$G_{n-1} + G_n$

$$= \frac{1}{\sqrt{5}} \left[\left(\frac{1 + \sqrt{5}}{2} \right)^n + \left(\frac{1 + \sqrt{5}}{2} \right)^{n+1} \right.$$

$$\left. - \left(\frac{1 - \sqrt{5}}{2} \right)^n - \left(\frac{1 - \sqrt{5}}{2} \right)^{n+1} \right]$$

$$= \frac{1}{\sqrt{5}} \left[\left(\frac{1 + \sqrt{5}}{2} \right)^n \cdot \left(\frac{3 + \sqrt{5}}{2} \right) - \left(\frac{1 - \sqrt{5}}{2} \right)^n \cdot \left(\frac{3 - \sqrt{5}}{2} \right) \right]$$

$$= \frac{1}{\sqrt{5}} \left[\left(\frac{1 + \sqrt{5}}{2} \right)^n \cdot \left(\frac{1 + \sqrt{5}}{2} \right)^2 - \left(\frac{1 - \sqrt{5}}{2} \right)^n \cdot \left(\frac{1 - \sqrt{5}}{2} \right)^2 \right]$$

$$= \frac{1}{\sqrt{5}} \left[\left(\frac{1 + \sqrt{5}}{2} \right)^{n+2} - \left(\frac{1 - \sqrt{5}}{2} \right)^{n+2} \right]$$

$$= \quad G_{n+1}$$

so that the recursion relation is satisfied for $n+1$ and the proof is complete.

There are some situations in which it is not difficult to derive the recurrence relation by elementary means. For example, consider an amoeba that splits into two identical parts every day. Under "ideal" conditions, how many amoeba will there be after n days?

If a is a function of the positive integers where $a(n)$ is the number of amoeba after n days, then

$$a(1) = 2, \qquad a(2) = 4, \qquad a(3) = 8, \ldots$$

It is easy to guess that $a(n) = 2^n$. It is also easy to check this by induction. It is true for $n = 1$ and if $a(n) = 2^n$, then after one more day the number doubles and we obtain

$$a(n + 1) = 2 \cdot a(n) = 2 \cdot 2^n = 2^{n+1}$$

which is correct.

As a final example, consider the following **func** that accepts two nonempty **tuples** of integers, each of which is in increasing order, and merges them into a single **tuple** in increasing order.

```
merge := func(s,t);
    $ s,t are nonempty tuples of integers in ascending order
        if s = [ ] then return t; end;
        if t = [ ] then return s; end;
        if s(1) >= t(1) then
            return [t(1)] + merge(s, t(2..));
        else
        return [s(1)] + merge(s(2..), t]);
        end;
    end;
```

The problem is to find a formula for the maximum number of comparisons that could be made in terms of the cardinalities, n and k of **s** and **t**, respectively. If c is a function of two variables that gives this number, then it is reasonable to argue that, looking at the recursive program, it appears that each comparison involves exactly one integer that is finding its proper place, and this integer is never compared again. Hence, $c(n, k) = n + k$ is a reasonable guess.

If each tuple has one element, then only one comparison is needed, but $n + k = 1 + 1 = 2$, so the guess is wrong. If **s** has 2 elements and **t** has one, then only one comparison is needed if $\mathbf{s(1)} \ >= \ \mathbf{t(1)}$, but two will be needed otherwise. Arguing symmetrically, we conclude that

$$c(2, 1) = c(1, 2) = 2$$

This suggests that $c(n, k) = n + k - 1$, and after checking a few more examples, we can try to prove it. Obviously, double induction is required for.

If $k = 1$, then the worst that could happen is that we must compare the single element of **t** with every element of **s**. Hence,

$$c(n, 1) = n = n + 1 - 1$$

so the formula holds for all n when $k = 1$.

Now assume, for a given value of k, that the formula holds for all n. We will show that it holds for all n when this k is replaced by $k + 1$. This is done by induction on n. That is, assuming

(1) $$c(n, k) = n + k - 1 \qquad \text{for } n = 1, 2, \ldots$$

we must show by induction on n that

$$c(n, k + 1) = n + k \qquad \text{for } n = 1, 2, \ldots$$

When $n = 1$ again, we can argue that the worst case is that the single element of **s** must be compared with all $k + 1$ elements of **t**, so $c(1, k+1) = k + 1$ as desired.

Finally, we add to our assumptions that $c(n, k + 1) = n + k$, and we take a look at the **func** when **#(s)** has the value $n + 1$ and **#(t)** has the value $k + 1$. Neither **tuple** is empty, so we skip the first two conditions and make a single comparison that must be counted. If **s(1) >= t(1)**, the test passes and we apply **merge** to the **tuple s** and the **tuple** obtained from **t** by removing its first component. Hence, the total number of comparisons is, by our assumptions in (1),

$$1 + c(n + 1, k) = 1 + n + 1 + k - 1 = n + k + 1$$

which is the proposed value of $c(n + 1, k + 1)$.

The last alternative is that **s(1) < t(1)**, so the test fails and we apply **merge** to the **tuple** obtained from **s** by deleting its first component and the **tuple t**. Thus, we still must count the one comparison, and we get, using our last assumption,

$$1 + c(n, k + 1) = 1 + n + k$$

which completes the proof.

Summary of Section 7.6

In this section, we put together a number of things that have been discussed in this text. Double induction combines our study of induction from the proposition valued function point of view with our considerations of functions of several variables in Chapter 4. To understand strong induction, you must, of course, have a good concept of mathematical induction and, in addition, be able to work with functions whose values are functions (Chapter 4) along with conjunctions of a list of propositions (Chapter 2) and universal quantification (Chapter 5). Finally, in the last subsection, we gave a very brief introduction to *recurrence relations*.

Exercises

7.6.1 For n a positive integer and $k = 1, 2, \ldots, n$, let $B_{n,k}$ denote the number of binary words of length n containing exactly k ones. Show that

$$B_{n,k+1} = \frac{n-k}{k+1} B_{n,k}$$

7.6.2 For n a positive integer and $k = 1, 2, \ldots, n$, let $B_{n,k}$ denote the number of binary words of length n containing exactly k ones. Show that

$$\sum_{i=0}^{n} B_{k+i,k} = B_{k+n+1,k+1}$$

7.6.3 For n a positive integer and $k = 1, 2, \ldots, n$, let $B_{n,k}$ denote the number of binary words of length n containing exactly k ones. Show that

$$\sum_{i=0}^{n} B_{k+i,i} = B_{k+n+1,n}$$

7.6.4 Find an algebraic formula for the following double summation and show that your formula is correct for all positive integer values of n and k.

$$\sum_{j=1}^{k} \sum_{i=1}^{n} (i-j)$$

7.6.5 Find an algebraic formula for the following double summation and show that your formula is correct for all positive integer values of n and k.

$$\sum_{j=1}^{k} \sum_{i=1}^{n} (i+j)$$

7.6.6 Let F be the Fibonacci sequence defined in Exercise 7.2.1, part f. Show that $F_n < \left(\frac{13}{8}\right)^n$ for every positive integer n.

7.6.7 Let F be the Fibonacci sequence defined in Exercise 7.2.1, part f. Show that

$$F_{n+1}^2 - F_n F_{n+2} = (-1)^n \text{ for } n \geq 1$$

7.6.8 Let F be the Fibonacci sequence defined in Exercise 7.2.1, part f. Show that

$$F_{n+1}^2 - F_n^2 = F_{n-1} F_{n+2} \text{ for } n \geq 1$$

7.6.9 Let F be the Fibonacci sequence defined in Exercise 7.2.1, part f. Show that

$$F_n = 1 + \sum_{k=1}^{n-2} F_k \text{ for } n > 2$$

7.6.10 Let F be the Fibonacci sequence defined in Exercise 7.2.1, part f. Let $V = (V_n), n \geq 1$, be a sequence of numbers defined by

$$V_1 = 2, \qquad V_2 = 3, \qquad V_{n+2} = V_{n+1} + V_n + 1 \text{ for } n \geq 1$$

Find and prove a relation between V and F.

7.6.11 Let $L = (L_n), n \geq 1$, be a sequence of numbers defined by

$$L_1 = 1, \qquad L_2 = 3, \qquad L_{n+2} = L_{n+1} + L_n \text{ for } n \geq 1$$

Show that for all positive integers n

$$\sum_{i=1}^{n} L_i = L_{n+2} - 3$$

7.6.12 Let a be a sequence of numbers defined by

$$a_0 = 12, \qquad a_1 = 29, \qquad a_n = 5a_{n-1} - 6a_{n-2} \text{ for } n \geq 2$$

Show that

$$a_n = 5(3^n) + 7(2^n) \text{ for } n \geq 0$$

7.6.13 Let b be the sequence defined by

$$b_1 = 1, \qquad b_n = \sqrt{3b_{n-1} + 1} \text{ for } n \geq 2$$

Show that $b_n < \frac{7}{2}$ for n sufficiently large.

7.6.14 Let c be the sequence defined by

$$c_0 = c_1 = 1, \qquad c_n = 3c_{n-1} - 2c_{n-2} \text{ for } n \geq 2$$

Find and prove a general formula for c_n.

7.6.15 Show that $a(n) = 5(2^n) + 1$ is the only function that satisfies the following relations:

$$a(0) = 6, \qquad a(1) = 11, \qquad a(n) = 3a(n-1) - 2a(n-2) \text{ for } n \geq 2$$

7.6.16 Let a be a sequence of positive numbers. Let x be the sequence defined by

$$x_1 = a_1, \qquad x_n = \frac{1}{x_{n-1}} + a_n \text{ for } n > 1$$

and let y be the sequence defined by

$$y_1 = a_1, \qquad y_2 = a_1 a_2 + 1, \qquad y_n = a_n y_{n-1} + y_{n-2} \text{ for } n > 2$$

Show that

$$y_n = \prod_{i=1}^{n} x_i \text{ for } n \geq 1$$

7.6.17 Let a be a sequence of positive numbers. Let x be the sequence defined by

$$x_1 = a_1, \qquad x_n = \frac{1}{x_{n-1}} + a_n \text{ for } n > 1$$

and let y be the sequence defined by

$$y_1 = a_1, \qquad y_2 = a_1 a_2 + 1, \qquad y_n = a_n y_{n-1} + y_{n-2} \text{ for } n > 2$$

Show that

$$x_n = \frac{y_n}{y_{n-1}} \text{ for } n > 1$$

7.6.18 The *bubble sort* algorithm for sorting a `tuple` of items for which the relation < is defined proceeds as follows. Starting with the last two elements in the `tuple` and moving toward the beginning of the `tuple`, compare successive elements and interchange them if necessary. After this first pass, the first element of the `tuple` will be correct, and the process can be repeated ignoring the first element. Continuing in this way, the `tuple` is eventually sorted.

Express this algorithm in ISETL using a double `for` loop, derive the total number of comparisons that are needed as a function of the cardinality of the `tuple`, and prove it by induction.

7.6.19 Suppose that a bank compounds interest daily at a fixed rate of 0.02%. Derive a formula that will give the total amount after n days if the starting amount was \$100. Use mathematical induction to prove that your formula is correct.

7.6.20 Consider a rectangular grid with dimensions $a \times b$. Suppose that you want to travel from the lower left-hand corner to the upper right-hand corner and are only allowed to take steps to the right and up. Derive a formula, in terms of the dimensions, that gives the total number of different paths you could take. Use induction to prove that your formula is correct.

7.6.21 For n a positive integer and $k = 1, 2, \ldots, n$, let $B_{n,k}$ denote the number of binary words of length n containing exactly k ones. State and prove a general formula of which the following three relations are special cases.

$$B_{n,2} = \frac{n(n-1)}{2}$$

$$B_{n,3} = \frac{n(n-1)(n-2)}{6}$$

$$B_{n,4} = \frac{n(n-1)(n-2)(n-3)}{24}$$

7.6.22 Prove that if a `tuple` of 2^n numbers is in ascending order, then it can be determined if a given number is in the `tuple` using at most $n + 1$ comparisons.

7.6.23 The *merge sort* algorithm orders the elements of a `tuple` by breaking it at the middle into two `tuples` (of lengths that differ by at most 1), calling itself to sort them separately, and using `merge` (page 356) to put them together. Write an ISETL `func` that will implement this algorithm, derive a formula for the maximum number of comparisons it requires, and prove your formula by induction.

7.6.24 Prove that if $C(n, r)$ is the number of combinations of n things taken r at a time and $0 \le r \le n$, then

$$C(n, r) = \frac{n!}{r!(n-r)!}$$

7.6.25 Prove that if $P(n, r)$ is the number of permutations of n things taken r at a time and $0 \le r \le n$, then

$$P(n, r) = \frac{n!}{(n-r)!}$$

7.6.26 No book on Discrete Mathematics could possibly end without discussing the *Towers of Hanoi*. Most of you know about the monks who were given three pegs (abstract towers) and a bunch of disks with holes at the center so that they fit on the pegs. The disks all have different diameters and they were arranged on one peg with the largest at the bottom and the pile going up with disks of decreasing diameter. The goal given to the monks was to get the entire pile of disks, in the same order, from the original peg onto one of the other pegs. They could use the third peg as an intermediary, but they had to follow two rules in moving pegs.

Rule 1. Only one disk at a time can be moved.

Rule 2. No disk may be placed on top of a smaller disk.

Your job, in this problem, is to write an ISETL func that will print out a succession of legal moves that will transfer the disks, derive a formula for the total number of moves in terms of the number of disks, and prove your formula by induction.

We offer a hint in the form of a suggestion of how to think about an algorithm for moving the disks. It consists of the following 3 steps.

1. Move all but the bottom disk to the peg other than the one to which you want to move the whole pile.

2. Move the bottom disk of the orginal pile (which is now on a peg by itself) onto the peg to which you want to move the whole pile.

3. Move the remaining disks from the peg they are on to the peg on which the original bottom disk was placed.

It is probably not true that the world will end as soon as you solve this problem—but you should be careful about running your program.

Chapter 8

Relations and Graphs

To the Instructor

This chapter introduces the student to relations and graphs. We emphasize the similarities and differences between functions and relations and examine various representations of a relation by a **map**, and a Boolean matrix, as well as graphically by an arrow diagram, graph, digraph, or Hasse diagram when possible. One of our goals is to help the student to be able to go back and forth between the various representations.

In the second section, we discuss properties of relations, such as what it means for a relation to be reflexive, symmetric, antisymmetric, or transitive, and we introduce equivalence relations and partial order relations. We give a lot of examples and try to aid the students in developing an intuitive understanding of each of the properties. Then, in the exercises, we ask them to write ISETL **funcs** that accept a relation represented by a **map** and test whether or not the input satisfies a particular property. The definition of their **funcs** is very similar to the standard mathematical definitions, and in writing the **funcs**, they develop (discover) the definitions on their own rather than memorize their statements.

We conclude this volume with a brief introduction to graph theory. We discuss representation of vertices in ISETL by **atoms**, a shortest path algorithm for digraphs, and necessary conditions for a graph to have an Eulerian circuit.

8.1 Preview

This chapter is about *relations* and *graphs*. Actually, you've had a lot of experience with relations, since every function corresponds to a relation. Not every relation, however, is a function. One difference between them is that the domain of a function from a set A to a set B must be all of A, while the domain of a relation from A to B can be any subset of A. The major difference, however, between the two concepts is that while a function is a process that transforms an object in its domain to exactly one item in its range, a relation may transform an object in its domain to *many* different items in its range. Since a function transforms each input to a single value, it is sometimes referred to as a *single-valued* map or *smap*, while a relation, on the other hand, is said to be a *multivalued* map or *map*.

A relation from one finite set to another can be represented graphically, by a Boolean valued matrix, or in ISETL by a `set` of `2-tuples` that we will call a `map`. We'll look carefully at each of these kinds of representations and discuss how to go back and forth from one to the other.

Just as a function can satisfy properties, such as being one-to-one or onto, a relation can satisfy certain properties, too. We'll discuss the general idea behind some of these properties, illustrate them by considering the graphs of relations that satisfy them, and then let you develop your own working definitions by writing ISETL property testers.

The chapter concludes with a general introduction to graphs, where we are thinking about the term graph to mean a collection of points (vertices) and lines connecting the points (edges). A lot of interesting problems arise when you think about traveling around a graph. For instance, you might want to find the length of the shortest path from one vertex to another, or you might be interested in determining whether or not it is possible to start at one vertex of a given graph, travel every edge exactly once, and return to the initial vertex before returning home. These types of questions are part of an area in Mathematics called graph theory. Our discussion will serve as a pointer to future studies and as a fun way to say "good-bye for now."

8.2　Relations and Their Representations

The *Cartesian product* of sets A and B, denoted by $A \times B$, is the set of all ordered pairs or `2-tuples`, whose first component is an element of A and whose second component is an element of B. That is,

$$A \times B = \{[x, y] : x \in A, y \in B\}$$

Any subset of $A \times B$ is said to be a *relation from A to B*. So, R is a relation from A to B, whenever

$$R \subseteq A \times B$$

As you can see, a relation is consists of three things: the set A, which is called the *co-range* of the relation, the set B is called the *range*, and a particular subset of their Cartesian product, namely, R. If R is a relation from a set A to itself, then we say that R is a *relation on A*.

Since a relation is any subset of the Cartesian product of two sets, you can define a relation by arbitrarily choosing a collection of pairs. For example, the following are relations from $\{1..4\}$ to $\{$ "a", "b", "c" $\}$.

$$R1 = \{[1, \text{``}a\text{''}], [2, \text{``}b\text{''}], [3, \text{``}a\text{''}], [4, \text{``}b\text{''}]\}$$
$$R2 = \{[1, \text{``}a\text{''}], [1, \text{``}b\text{''}], [2, \text{``}c\text{''}], [4, \text{``}c\text{''}]\}$$
$$R3 = \emptyset$$

The set $\{1..4\}$ is the co-range, while the set $\{$ "a", "b", "c" $\}$ is the range of each of these relations. As with a function represented by an **smap**, the *domain* of each relation is the set of all first components of each of the ordered pairs in the relation. So, domain($R1$) = $\{1..4\}$ and domain($R2$) = $\{1, 2, 4\}$, while domain($R3$) = $\emptyset$. An important thing to notice is that the domain of a relation is not necessarily equal to its co-range. We'll discuss this more later on, but for now you should just be aware that this may be the case. One more bit of terminology, again, as with a function represented by an **smap**, the *image* of a relation is the set of second components of the ordered pairs in the relation. So, image($R1$) = $\{$ "a", "b" $\}$, and image($R2$) = $\{$ "a", "b", "c" $\}$, while image($R3$) = $\emptyset$.

We can represent a relation in ISETL as a **set** of **2-tuples**, called a **map**. For example, the relation $R2$ from the set $\{1..4\}$ to the set of characters $\{$ "a", "b", "c" $\}$ can be represented by

```
R2 := {[1, "a"], [1, "b"], [2, "c"], [4, "c"]};
```

A word of caution—taken by itself, the **tuple** representation of a relation does not tell you what the underlying sets are, just as the **smap** representation of a function does not indicate the range of the function. You know the domain of **R2**, namely,

```
domain(R2);
```

and you know the image, which is given by

```
image(R2);
```

but you cannot determine the co-range and range by considering **R2** by itself. Consequently, if we are a little sloppy and do not specify what the co-range and range of a relation represented by a **map** are, you will have to assume that they equal the domain and image, respectively.

Frequently, a relation is constructed by choosing all those items in the Cartesian product that satisfy a given condition. For example, if S is the set of all students at your school and C is the set of all courses listed in your catalogue, then you can make up a relation F from S to C by saying that the student x is related to the course y, that is, $[x, y] \in F$, when "x has completed y," so

$$F = \{[x, y] : x \in S, y \in C \mid x \text{ has completed } y\}$$

Therefore, if a member of S, "$SueSmith$", has completed courses "$M161$" and "$CS131$" in C, then

$$["SueSmith", "M161"] \text{ and } ["SueSmith", "CS131"]$$

are both members of F.

You can also use an ISETL set former to construct a relation, such as

```
G := {[r,s] :  r,s in {1..10} | even(r) and (r < s)};
```

Then, G is a relation on the set $\{1..10\}$. Write out the items in G. Then find the co-range, range, domain, and image of G.

One example of a relation with which you are familiar is an smap representing a function. For example, in Chapter 4, Section 4.2.4, we set

```
words := {"well", "by", "now", "you", "must", "begin", "to",
          "see", "that", "what"};
```

and considered the smap

```
next := {["well","what"], ["by","must"], ["now","see"],
         ["must","now"], ["begin","by"], ["to","well"],
         ["see","that"], ["that","to"], ["what","you"]};
```

The smap next represents a function whose domain is the set of all items in words, with the exception of the string "you", that is, words - {"you"}. Its range is words, and its image is everything in words except "begin", that is, words - {"begin"}. However, next also represents a relation on words.

Although an smap defines a relation, not every relation is an smap. An smap representing a function from A to B is not an arbitrary collection of elements from $A \times B$; it must satisfy the requirements that its domain equals A and that every item in the domain appears as a first component of some member of the smap exactly once. With a relation from A to B, however, the domain may be any subset of A and each item in the domain may appear as the first component of an element in its associated map many times. So, we use the terms smap and map to distinguish between the representation in ISETL of a function and a relation, noting that every smap is a map, but not vice versa.

As another example, what if instead of alphabetizing the items in words, you specify that x in words is related to y in words if "y follows x in lexicographic ordering." You could represent this relation, follows, using an ISETL set former by

```
follows := {[x, y] :  x, y in words | y > x};
```

Then,

```
["now", "see"], ["now", "that"], ["now", "to"],
["now", "well"], ["now", "what"]
```

are elements in `follows`. So, `follows` is a relation on `words` that clearly is not a function. To determine the `set` of all items in `words` that are related to `"now"`, you can enter

`follows{"now"};`

which returns

`{"see", "that", "to", "well", "what"};`

On the other hand,

`follows{"what"};`

returns `{}`, indicating that `"what"` is not in the domain of the relation `follows`.

Just as you can sometimes form a new function by finding the composition of two existing functions, it may be possible to construct a new relation by composing two relations. If the functions $S : A \longrightarrow B$ and $T : B \longrightarrow C$ are represented by `smaps`, then finding the `tuples` in the representation of their composition is fairly straightforward—since $image(S) \subseteq domain(T)$, for each item $[x, y]$ in S, you find the `tuple` in T that has y as its first component, $[y, z]$, form a new `tuple` $[x, z]$, and add it to the `smap` representing the composition, $T \circ S$. The difference between composing functions and composing relations is that if S is a relation from A to B and T is a relation from B to C, then corresponding to each $[x, y]$ in S there may be many distinct `tuples` in T that have y as their first component, for example, $[y, z_1], \ldots, [y, z_n]$. In this case, you must form the `tuples` $[x, z_1], \ldots, [x, z_n]$ and add each one to the `map` representing the composition. On the other hand, for a given $[x, y]$ in S, there may not be any `tuples` in T having y as their first component, in which case the `tuple` $[x, y]$ does not contribute anything to the composition. Consider, for example, the relation

`char_num := {[y,z]: y in words, z in {1..10}`
`          | #y>=3 and z<=#y};`

Then, for instance,

`["that",3], ["that",4]`

are both in `char_num`. Now, the relation `follows`, which we defined above, is a relation from `words` to `words`, while `char_num` is a relation from `words` to $\{1..10\}$, and thus their composition,

`char_num .comp follows`

is a relation from `words` to $\{1..10\}$, where `comp` is an ISETL `func` that accepts two relations represented by `maps` and returns their composition. But what `tuples` are in the `map` representing the composition? To determine

this, take an element in follows, for example, ["by","that"]. Then, since
["that",3], and ["that",4] are both in char_num, you include ["by",3]
and ["by",4] in the map representing the composition. On the other hand,
if you choose ["now","to"] from follows, then no new items are added
to the composition, since "to" is not in the domain of char_num. Continue
in this manner, examining each member of follows, searching for all the
linking elements in char_num (if any), forming the new tuples, and adding
them to the map representing the composition of char_num and follows.
You can express this process in ISETL using a set former, such as

{[x,z] : x in words, z in {1..10} | (exists y in words |
                [x,y] in follows and [y,z] in char_num)};

Another way to construct a new relation from an old one is to finds its
inverse. Every relation has an *inverse*, where if R is a relation from A to B
then its inverse, R^{-1}, is a relation from B to A. For example, the inverse
of next is a relation from {1..10} to words, represented by

inv_next := {[w(2), w(1)] : w in follows};

The *identity* relation I_A on a set A is the relation where every element
in A is related to itself and only itself. Thus,

$$I_A = \{[x, x] : x \in A\}$$

If you compose next with its inverse to form inv_next .comp next, you
get the identity relation on the domain of next. Conversely, next .comp
inv_next evaluates to the identity relation on the image of next. In Chapter 4, Section 4.3.5, we noted that the only time you can construct the
inverse of a function and get another function is when the given function
is both one-to-one and onto. But, now you can see that every function
does have an inverse. The difference is that the inverse is not necessarily a
function, but a relation.

Up to this point, we've represented a relation as an ISETL map, that is,
by a set of 2-tuples. Let's look at some different graphic representations
of a relation. If R is a relation from A to B where A and B are finite sets,
then one way to represent R graphically is with an *arrow diagram*—that
is, draw diagrams for sets A and B and then indicate that $[x, y] \in R$ by
drawing an arrow starting at $x \in A$ and terminating at $y \in B$. Consider
again the relation $R2$ from {1..4} to {"a", "b", "c"} given by

$$R2 = \{[1, \text{``}a\text{''}], [1, \text{``}b\text{''}], [2, \text{``}c\text{''}], [4, \text{``}c\text{''}]\}$$

The arrow diagram for $R2$ is shown in Figure 8.1. A member of A is the
initial end of an arrow whenever it is in the domain of the relation, and
isolated points in the diagram for A correspond to items in the co-range of
R that are not in the domain. Similarly, an element of B is the terminal

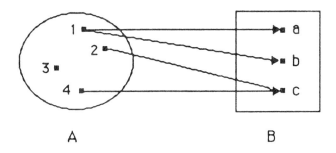

Figure 8.1: Arrow diagram representing $R2$

end of an arrow whenever if it is in the image of R, and isolated points in the diagram for B correspond to items in the range of R that are not in the image.

Although all of our examples thus far have involved finite sets, you can have, of course, a relation from A to B where A and B are not finite. For example, if $\mathcal{R}$ is the set of all real numbers, then you might be interested in the relation K on $\mathcal{R}$ where

$$K = \{[a, b] : a, b \in \mathcal{R} \mid a^2 + b^2 \le 4\}$$

This relation can be represented graphically as in Figure 8.2, where the coordinates of each point in the shaded area correspond to an item in the relation.

Sometimes a relation on a finite set, that is, from a finite set to itself, can be represented by a *graph*. Here, we are thinking of a graph in a general sense as a collection of points or *vertices* and a collection of lines or *edges*, where an edge is determined by two (not necessarily distinct) vertices. For example, suppose $Cities = \{a, b, c, d, e, f, g\}$ is a collection of names of cities, and T is a relation on the set $Cities$ where two cities are related whenever they are connected by a road. Assume T is represented by the graph shown in Figure 8.3. How do you read the graph? That is, how can you determine from the graph the representation of the relation as a map? Well, the graph has one vertex for each member of $Cities$ and one edge connecting each related pair of cities. For example, since there is an edge between vertex a and vertex d, this indicates that a is related to d and d to a, so both $[a, d]$ and $[d, a]$ are in T. What about the edge that connects a with itself? This type of edge is called a *loop* and indicates that there is a road starting at city a and returning to a, that is, $[a, a] \in T$. Continuing to interpret the given graph in this manner, we see that

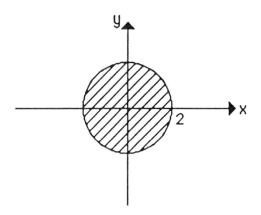

Figure 8.2: Graphic representation of K

$$T = \{[a,a], \; [a,c], \; [a,d], \; [b,c], \; [c,a], \; [c,b],$$
$$[c,d], \; [d,c], \; [d,a], \; [e,f], \; [f,e]\}$$

Notice that the *isolated* vertex, g, is not connected by an edge to itself or to any other vertex. Thinking in terms of cities and roads, this implies that there is no road between g and any of the other members of *cities*, including itself. So, g is in the co-range and the range of T, but not in the domain or the image of T and, hence, does not appear in the map representation of T. The graph representing T is actually composed of three *connected subgraphs*: one consisting of the collection of vertices a, b, c, and d and their associated edges, another is formed by vertices e and f and their connecting edge, and the third subgraph is the isolated vertex g. Each connected subgraph represents the collection of cities (vertices) that can be reached from one another and the roads (edges) connecting them.

The edges of a graph do not have any direction; consequently, each edge that is not a loop adds two items to the map representing a relation on a finite set, while a loop adds one. But, some relations do not have this symmetric property. To represent these types of relations, we specify the direction of an edge by placing an arrow on the edge in the appropriate direction. Then, an edge *from* vertex a *to* vertex b corresponds to $[a,b]$ being in the relation. We call this a *directed graph* or *digraph*, and it can be used to represent any relation on a finite set. As an example, look once more at the diagram on page 333. This is a digraph that represents a relation, let's call it prereq, on the set of all Mathematics courses, where [p,q] in prereq means that course p has course q as a prerequisite. The

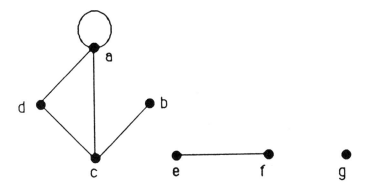

Figure 8.3: Graph representing T

vertices in the digraph consist of all the possible courses, and [p,q] in
prereq when there is a directed edge starting at p and ending at q. Notice,
however, that an edge starting at p and ending at q implies *only* that [p,q]
is a member of prereq, and not that [q,p] is in prereq.

With the digraph representation of a relation, each vertex that is the
initial end of an edge is in the domain, while each one that is a terminal end
is in the image. Isolated vertices are in neither the domain nor the image.
It is interesting to observe that the digraph of the inverse of a relation can
be obtained from the digraph of a relation by simply changing the direction
of the edges.

Another way to represent a relation R from $A = \{x_1, \ldots, x_m\}$ to $B = \{y_1, \ldots, y_n\}$ is by an $m \times n$ *Boolean matrix* that we will denote by $M(R)$.
A matrix is said to be a Boolean matrix whenever each entry in the matrix
has the value *true* or the value *false*. The general idea is that if x_i is related
to y_j then $M(R)$ has the value *true* in the (ij)-position, otherwise, it has
the value *false*. That is, if $M(R) = (m_{ij})$, then

$$m_{ij} = \begin{cases} true & \text{if } [x_i, y_j] \in R \\ false & \text{otherwise} \end{cases}$$

Let's look again at the relation $R2$ from $\{1..4\}$ to $\{``a", ``b", ``c"\}$ defined
by

$$R2 = \{[1, ``a"], [1, ``b"], [2, ``c"], [4, ``c"]\}$$

Then, $M(R2)$ is the 4×3 matrix

$$
M(R2) = \begin{array}{c} \\ 1 \\ 2 \\ 3 \\ 4 \end{array}
\begin{array}{ccc}
\text{``}a\text{''} & \text{``}b\text{''} & \text{``}c\text{''} \\
\left[\begin{array}{ccc}
T & T & F \\
F & F & T \\
F & F & F \\
F & F & T
\end{array} \right]
\end{array}
$$

where we have used T for *true* and F for *false*, and we have labeled the rows and columns with the members of A and B, respectively. Observe that the $(1, 2)$-entry is *true* since $[1, \text{``}b\text{''}] \in R2$, while the *false* in the $(1, 3)$-position indicates that $[1, \text{``}c\text{''}] \notin R2$, and so on. Also notice that all the entries in the third row of $M(R2)$ have the value *false* since 3 is in the co-range but not in the domain of $R2$. What happens to $M(R)$ if there exists an element in the range that is not in the image?

This is not the first time you've seen a relation represented by a matrix. In Chapter 6, Section 6.3.2, we represented a permutation π of $\{1..n\}$ by an $n \times n$ matrix, $P = (p_{ij})$, where

$$
p_{ij} = \left\{ \begin{array}{ll} 1 & \text{if } \pi(i) = j \\ 0 & \text{otherwise} \end{array} \right.
$$

which we called a permutation matrix. Now, every permutation of $\{1..n\}$ can be represented by a relation on $\{1..n\}$. For instance, if $\pi = [3, 2, 1]$ is a permutation of $\{1..3\}$, then you can construct the corresponding relation, P on $\{1..3\}$, by including $[i, j]$ in P, whenever $\pi(i) = j$. So, in this case $P = \{[1, 3], [2, 2], [3, 1]\}$.

A logical question to ask is, Given a permutation, what is the relationship between its permutation matrix and the Boolean matrix of its corresponding relation? We claim that they are the same, where 1 corresponds to *true* and 0 to *false*. To help convince yourself that this is the case, find the permutation matrix representing $\pi = [3, 2, 1]$ and the Boolean matrix representing P. Compare them. Notice that if $\pi(i) = j$ then the (ij)-entry in its permutation matrix is 1. Furthermore, if $\pi(i) = j$, then $[i, j]$ is a member of the **map**, P, representing π, so the (ij)-entry in π's Boolean matrix is *true*. Conversely, if $\pi(i) \neq j$, then the (ij)-entry in the permutation matrix is 0 and $[i, j]$ is not a member of the associated **map**, so the (ij)-entry in its Boolean matrix is *false*. Consequently, the Boolean matrix representing a permutation can be obtained from the permutation matrix by replacing 0 with *false* and 1 with *true*, and vice versa.

We also discussed in Chapter 6, Section 6.3.2, the fact that if σ and τ are permutations of $\{1..n\}$ with matrix representations S and T, respectively, then $\sigma \circ \tau$ is represented by the matrix product TS. That is, the matrix representing the composition of the permutations equals the product of the matrices corresponding to the permutations, in the *reverse* order. This is also true for relations. As an example, suppose R is a relation from $\{1..4\}$

to $\{$ "x", "y", "z" $\}$ given by

$$R = \{[1, \text{"}x\text{"}], [1, \text{"}y\text{"}], [2, \text{"}y\text{"}], [4, \text{"}x\text{"}], [3, \text{"}z\text{"}]\}$$

and Q is a relation from $\{$ "x", "y", "z" $\}$ to $\{$ "A", "B" $\}$ given by

$$Q = \{[\text{"}y\text{"}, \text{"}A\text{"}], [\text{"}y\text{"}, \text{"}B\text{"}], [\text{"}z\text{"}, \text{"}A\text{"}]\}$$

Then, the map representation of the relation $Q \circ R$ from $\{1..4\}$ to $\{$ "A", "B" $\}$ is

$$Q \circ R = \{[1, \text{"}A\text{"}], [1, \text{"}B\text{"}], [2, \text{"}A\text{"}], [2, \text{"}B\text{"}], [3, \text{"}A\text{"}]\}$$

On the other hand,

$$M(R) = \begin{array}{c} \\ 1 \\ 2 \\ 3 \\ 4 \end{array} \begin{array}{ccc} \text{"}x\text{"} & \text{"}y\text{"} & \text{"}z\text{"} \\ \left[\begin{array}{ccc} T & T & F \\ F & T & F \\ F & F & T \\ T & F & F \end{array}\right] \end{array} \qquad M(Q) = \begin{array}{c} \\ \text{"}x\text{"} \\ \text{"}y\text{"} \\ \text{"}z\text{"} \end{array} \begin{array}{cc} \text{"}A\text{"} & \text{"}B\text{"} \\ \left[\begin{array}{cc} F & F \\ T & T \\ T & F \end{array}\right] \end{array}$$

and

$$M(Q \circ R) = \begin{array}{c} \\ 1 \\ 2 \\ 3 \\ 4 \end{array} \begin{array}{cc} \text{"}A\text{"} & \text{"}B\text{"} \\ \left[\begin{array}{cc} T & T \\ T & T \\ T & F \\ F & F \end{array}\right] \end{array}$$

We claim that

$$M(Q \circ R) = M(R) \cdot M(Q)$$

The first thing we need to decide is how to multiply Boolean matrices. You may have already thought about how you might do this, if you completed Exercise 6.3.12. The idea is to find the product of the matrices in the usual fashion, but replace the operation of real addition by the Boolean operator or ($\vee$) and the operation of multiplication by and ($\wedge$). For instance, to find the $(1, 1)$-entry in the product

$$\begin{aligned} (M(R) \cdot M(Q))(1, 1) &= (M(R)(1, 1) \wedge M(Q)(1, 1)) \\ &\quad \vee (M(R)(1, 2) \wedge M(Q)(2, 1)) \\ &\quad \vee (M(R)(1, 3) \wedge M(Q)(3, 1)) \\ &= (T \wedge F) \vee (T \wedge T) \vee (F \wedge T) \\ &= F \vee T \vee F \\ (M(R) \cdot M(Q))(1, 1) &= T \end{aligned}$$

Now, $M(Q \circ R)(1, 1)$ also has the value *true*, so

$$M(Q \circ R)(1, 1) = M(R) \cdot M(Q)(1, 1)$$

as desired. In fact if you repeat the process to find the remaining entries in the product you will see that each one equals the corresponding entry in the matrix representing the composition. Why is this the case? Let's examine more closely how the values of the entries in the $(1,1)$-position of $M(R) \cdot M(Q)$ and $M(Q \circ R)$ are determined, and why they are the same. Looking at the first row of $M(R)$, we see that its second entry, $M(R)(1,2)$, has the value *true* since $[1, "y"] \in R$. Similarly, looking at the first column of $M(Q)$, we see that its second entry, $M(Q)(2,1)$, also has the value *true* since $["y", "A"] \in Q$. Thus, $M(R)(1,2) \wedge M(Q)(2,1)$ has the value *true*, and consequently the $(1,1)$-entry in $M(R) \cdot M(Q)$ has the value *true*. On the other hand, the $(1,1)$-entry in $M(Q \circ R)$ must also have the value *true*, since $[1, "y"] \in R$ and $["y", "A"] \in Q$ implies that $[1, "A"] \in Q \circ R$. In fact, this will occur whenever there exists a "link" between an element in R and one in Q. So the product of the Boolean matrices representing R and Q equals the Boolean matrix representing $Q \circ R$.

Summary of Section 8.2

In this section, we introduced the concept of a relation and discussed how to represent a relation by an ISETL map, graphically, and by a Boolean matrix. The exercises provide you with the opportunity to practice going back and forth between the various representations.

We compared relations represented by maps to functions represented by smaps and extended the function definitions for domain, range, and image to relation. We introduced the identity relation and discussed how to construct new relations from old ones by finding their inverses and their composition.

Exercises

8.2.1 Consider the following list of relations:

Example 1. $N = \{[1,2], [2,3], [2,4], [3,4], [3,3], [4,4]\}$ on $\{1..4\}$

Example 2. $D = \{[x, y] : x, y \in \{3, 6, 9, 12\} \mid y \bmod x = 0\}$

Example 3. $T = \{[t1, t2] : t1, t2 \in ATT \mid t1, t2$ have the same area code$\}$, where ATT is the set of all 10-digit telephone numbers.

Example 4. $S1 = \{[A, B] : A, B \in pow(F) \mid A \subseteq B\}$, where $F = \{"fe", "fi", "fo"\}$.

Example 5. $S2 = \{[C, D] : C, D \in pow(\{5, 7..9\}) \mid C \cap D \neq \emptyset\}$

Example 6. $R = \{[str, n] : str \in S, n \in \{1..5\} \mid \#str \leq n\}$, where S is a set of ISETL strings.

Example 7. The relation *SymbolTable* from *ids* to U is constructed by $[X, Y] \in SymbolTable$ exactly when the value of X is Y, where *ids* is the set of all possible ISETL identifiers and U is the set of all possible values of ISETL objects (including OM).

Example 8. $R = \{[x, y] : x, y \in \mathcal{R} \mid x < 2y + 1\}$, where $\mathcal{R}$ is the set of all real numbers.

Example 9. $C = \{[n, m] : n, m \in \mathcal{R} \mid n^2 + m^2 = 9\}$, where $\mathcal{R}$ is the set of all real numbers.

Example 10. $E = \{\}$ on $\{1..5\}$

Example 11. The relation on $\{1..5\}$ in Figure 8.4.

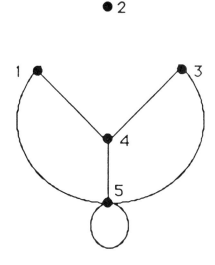

Figure 8.4: Example 11

Example 12. The relation on $\{a, b, c, d, e, f, g,\}$ in Figure 8.5.

Example 13.

$$M(K) = \begin{array}{c} \\ -1 \\ 1 \\ 3 \\ 5 \\ 7 \end{array} \begin{array}{cccc} 0.1 & 0.2 & 0.3 & 0.4 \\ \left[\begin{array}{cccc} T & T & F & T \\ F & T & F & F \\ F & F & T & F \\ T & F & F & F \\ F & F & F & F \end{array} \right] \end{array}$$

For each example, find

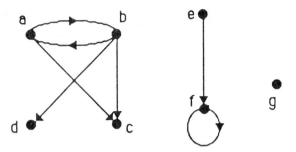

Figure 8.5: Example 12

a. the co-range of the relation.

b. the domain of the relation.

c. the range of the relation.

d. the image of the relation.

e. the inverse of the relation.

8.2.2 Consider the examples listed in Exercise 8.2.1.

   a. Represent the following examples by a digraph:

      i) Example 1.

      ii) Example 2.

      iii) Example 4.

      iv) Example 5.

      v) Example 10.

      vi) Example 11.

      vii) Example 13.

   b. Give a graphic representation in the real plane of the following relations on $\mathcal{R}$:

      i) Example 8.

      ii) Example 9.

8.2.3 Consider the examples listed in Exercise 8.2.1. Represent the following examples by a Boolean matrix:

   a. Example 1.

    b. Example 2.

    c. Example 4.

    d. Example 5.

    e. Example 10.

    f. Example 11.

    g. Example 12.

8.2.4 Consider the following sets:

```
chars := {"a", "b", "c", "d"};
sets := {{"a", "b", "c"}, {"b", "c", "d"}, {"a"}, {}};
```

Suppose E represents a relation from chars to sets, and N represents one from sets to {1..4}, where

```
E := {[x,S] : x in chars, S in sets | x in S};
N := {[S,n] : S in sets, n in {1..4} | #S >= n};
```

    a. Find the value of the following ISETL expressions.

       i) E{"a"};

      ii) E{"d"};

     iii) N{{"a"}};

     iv) N{{"c","b","a"}};

    b. Represent the following relations by arrow diagrams, where comp is a func that denotes composition.

       i) E

      ii) N

     iii) N .comp E

     iv) The inverse of E

      v) The inverse of (N .comp E)

    c. Find the Boolean matrix representation of each of the relations given in part b.

8.2.5 Write the following ISETL funcs:

    a. A func called comp that accepts two relations represented as maps and returns their composition.

    b. A func that accepts a relation represented as a map and returns its inverse.

8.2.6 Suppose R is a relation. Express each of the following in terms of domain(R), image(R), and R:

  a. domain(R^{-1})

  b. image(R^{-1})

  c. $(R^{-1})^{-1}$

8.2.7 Let F be a relation on $A = \{w_1, \ldots, w_n\}$. For each of the following assertions, describe the impact on

  1. the Boolean matrix representing F and

  2. the digraph representing F

  if it is known that the assertion holds for F.

  a. Every element in A is related to itself.

  b. w_2 is not related to any element in A, including itself.

  c. w_i, $1 \leq i \leq n$, is related to every element in A, except itself.

  d. $F\{w_1\} = \{w_2, w_5, w_3\}$

  e. $F\{w_3\} = \{\}$

  f. $F = \emptyset$

  g. $F = A \times A$

  h. $F = I_A$, where I_A is the identity relation on A.

  i. $F = F^{-1}$

8.2.8 Counting problems. Suppose A and B are finite sets.

  a. Express the number of distinct relations from A to B in terms of the cardinalities of A and B.

  b. Express the number of elements in a relation R from A to B in terms of the cardinalities of domain(R) and image(R).

8.2.9 The following involve the digraph of the Mathematics course prerequisites that appears on page 333 in reference to Exercise 7.3.1.

  a. Represent the information contained in the digraph by an ISETL map, called **prereq**. Note: [x,y] in **prereq** means that course **x** has course **y** as a prerequisite.

  b. Using the relation **prereq**, write an ISETL **func** that accepts as input any course in the domain of **prereq** and returns

   i) some prerequisite course.

   ii) the set containing all the prerequisite courses.

   iii) every course that is a prerequisite for a course that is a prerequisite for the input course.

   iv) every course that must be taken before the input course can be taken.

    c. Compare your answers to part b. to your responses to Exercise 7.3.1.

8.2.10 Let R be a relation on a finite set A. Prove that

$$M(R^n) = (M(R))^n \qquad \text{for } n = 1, 2, \ldots$$

8.2.11 If relations R_1 and R_2 on $S = \{x_1, \ldots, x_n\}$ are represented by Boolean matrices $M(R_1)$ and $M(R_2)$, respectively, then

$$M(R_1 \cup R_2) = M(R_1) + M(R_2)$$

That is, the Boolean matrix representing the union of two relations equals the sum of the matrices representing the relations.

Note that you add two Boolean matrices by applying the Boolean operator **or** ($\vee$) to corresponding entries.

    a. Give an example that illustrates this statement.

    b. Explain why the statement is always true.

8.2.12 *n-ary relations.* In this section, we define a relation to be a subset of the Cartesian product of two sets. This is called a *binary relation*, while a subset of the Cartesian product of n sets is referred to as an *n-ary relation*. A data base (see Chapter 4, Section 4.4.1) can be stored in an n-ary relation, that is usually called a *table*. For example, suppose

    − *Senator* is a set of 2-**tuples** whose components are **strings** corresponding to the first and last names of all the senators of the United States.

    − *State* is the set of all **strings** representing the abbreviations of the states.

    − *Party* = {"D", "R"}.

    − *Votes* is the set of positive integers.

Then, the relation *ElectionInfo* on *Senator* × *State* × *Party* × *Votes* can be represented in tabular form, such as in Table 8.1, or by a collection of 4-**tuples**,

```
{[["Ann","Klein"], "NY", "D", 206456],
 [["Joe","Shmoo"], "PA", "R", 51456], ...}
```

    a. The *projection* operation abstracts subtables from a table by deleting some of the table's columns. Write an ISETL **func** **Proj** that accepts a table and a **tuple** containing a list of columns to be retained and returns the projection of the table. For example,

Senator	State	Party	Votes
Ann Klein	NY	D	206,456
Joe Shmoo	PA	R	51,456
⋮	⋮	⋮	⋮

Table 8.1: The tabular representation of *ElectionInfo*

$$Proj(ElectionInfo, [1,2]);$$

returns

$$\{[["Ann","Klein"],"NY"], [["Joe","Shmoo"], "PA"], \dots\}$$

Test your func on an expanded version of *ElectionInfo*.

b. Explain why it is possible that $\#(Proj(T, C)) \leq \#T$ for a table T and list of columns C.

c. The *join* operation joins two tables to form a new table. This is done by combining lines of the tables that agree on shared columns.

Write an ISETL func called join that accepts two tables and returns their join. You may assume that they share the first column. A more general function could be given a correspondence between columns in the two tables.

For example,

$$join(ElectionInfo, PersonalInfo);$$

where *PersonalInfo* is represented in tabular form by Table 8.2, returns

$$\{[["Ann","Klein"], "NY", "D", 206456, 42, false],$$
$$[["Joe","Shmoo"], "PA", "R", 51456, 78, true], \dots\};$$

Senator	Age	Married
Ann Klein	42	false
Joe Shmoo	78	true
⋮	⋮	⋮

Table 8.2: The tabular representation of *PersonalInfo*

8.3   Properties of Relations

In this section, we restrict our discussion to relations whose domain and image are subsets of the same set, that is, to relations defined *on* a set, such as a relation defined on the set of all people in the world, or on the set of all real numbers. Then, just as a function can satisfy various properties, such as being one-to-one or onto, a relation can satisfy certain properties, too. Our goal is for you to develop an intuitive understanding of some of these properties by thinking about examples that satisfy a particular property and by looking at the effect this has on the representation of the relation by a digraph or a Boolean matrix. In the exercises, we ask you to define ISETL **funcs** that test whether or not a relation satisfies a specified property. In the process of writing and testing your **funcs**, you will develop precise, meaningful mathematical statements corresponding to the definition of each of the properties under consideration.

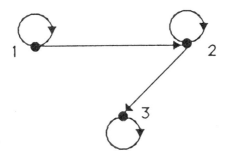

Figure 8.6: Digraph representing reflexive relation L

One of the simplest properties that a relation R on S can satisfy is the requirement that the domain of R equals its co-range S with every element in S related to itself. Whenever this is true, we say that the relation is *reflexive*. For example, if I is a set of positive integers, then the relation R on I defined by

$$[n, m] \in R \text{ exactly when } n \text{ is a factor of } m, \text{ where } n, m \in I$$

is reflexive since every $x \in I$ is a factor of itself.

If a relation on a finite set is reflexive, what impact does this have on its digraph? Well, since every member of the co-range is related to itself, the co-range, domain, range, and image all coincide. Hence, the digraph will have no isolated vertices, and it will have a loop at every vertex. For instance, the graph in Figure 8.6 represents the reflexive relation

$$L = \{[1, 2], [1, 1], [2, 2], [2, 3], [3, 3]\}$$

on the set $\{1..3\}$.

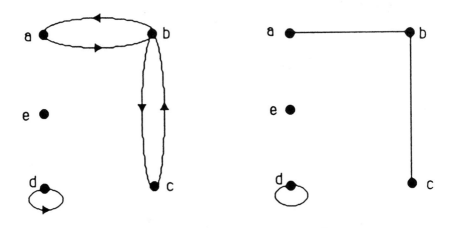

Figure 8.7: Representations of symmetric relation Q

A relation R on S is said to be *symmetric* if it has the property that, for all $x, y \in S$, if x is related to y, then y is related to x. For example, if S is the set of all students at your school, then the relation

$$DM = \{[x, y] : x, y \in S \mid x \text{ lives in the same dorm as } y\}$$

is symmetric since whenever $p \in S$ "lives in the same dorm as" $q \in S$, then q "lives in the same dorm as" p.

In the digraph representing a symmetric relation, between every pair of vertices there are two edges connecting the vertices (which indicates that the vertices are related) or none (which indicates that they are not). As a result, a relation that is symmetric can be represented by a *graph* as well as a *digraph*. For example, the symmetric relation on the set $\{a, b, c, d\}$,

$$Q = \{[a, b], [b, a], [b, c], [c, b], [d, d]\}$$

is represented by both the graph and digraph given in Figure 8.7.

On the other hand, a relation is *antisymmetric* if for all $x, y \in S$, whenever x is related to y and y is related to x, it follows that $x = y$. For example, think about the relation $\leq$ on the set of all real numbers. For any $p, q \in \mathcal{R}$, if $p \leq q$ and $q \leq p$, then $p = q$. The relation defined by $\subseteq$ on a collection of subsets of a given set is also antisymmetric, since if $A \subseteq B$ and $B \subseteq A$, then it must be the case that $A = B$.

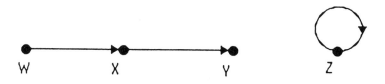

Figure 8.8: Digraph representing antisymmetric relation

In the digraph representing an antisymmetric relation, between every pair of vertices there is at most one edge. For example, the digraph shown in Figure 8.8 represents an antisymmetric relation on $\{w, x, y, z\}$.

The last property we want to discuss is the idea of a *transitive* relation. A relation on S is transitive if for all $x, y, z \in S$, whenever x is related to y and y is related to z, then x is related to z. For instance, both the relations $\leq$ on the set of real numbers and $\subseteq$ on a collection of sets are transitive. As another example, consider the relation defined by "is a sister of" on the set of all women in the world. If a "is a sister of" b and b "is a sister of" c, then a "is a sister of" c, so this relation is transitive. If a relation on a finite set is transitive, then its digraph representation must have an edge from u to w, whenever there are edges from u to v and v to w, for vertices u, v, and w. The relation on $\{$"a", "b", "c", "d"$\}$ represented in Figure 8.9 satisfies the transitive property.

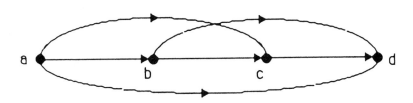

Figure 8.9: Digraph representing transitive relation

Each of the relations represented in Figures 8.6, 8.7, and 8.9 satisfies exactly one of the given properties. Our goal, up to this point, has been for you to be able to recognize which relations satisfy which properties. Now,

let's reverse the process and, starting with an arbitrary relation, extend
the relation to a reflexive, symmetric, or transitive relation by adding the
least number ordered pairs to the map representing the original relation,
by adding the least number of edges to the relation's digraph representation, or by changing the smallest number of *falses* to *trues* in the Boolean
matrix representation. The new relation that extends the given relation is
called the *reflexive closure*, the *symmetric closure*, or the *transitive closure*
depending on the particular property under consideration. Let's examine
how you might construct each of these closures.

To construct the reflexive closure, R_r, of a given relation R on S you
need to extend R so that every member of S is related to itself. Consider,
for example, the relation F on $\{1..4\}$ represented by the map

$$F = \{[1,2],[2,3],[4,4]\} \tag{8.1}$$

Initialize F_r to be F and then for each x in $\{1..4\}$ add $[x,x]$ to F_r. So,

$$F_r = \{[1,2],[2,3],[4,4],[1,1],[2,2],[3,3]\}$$

If a relation is represented by a digraph, such as the representation of
F in Figure 8.10, then adding a loop at every vertex gives you the digraph

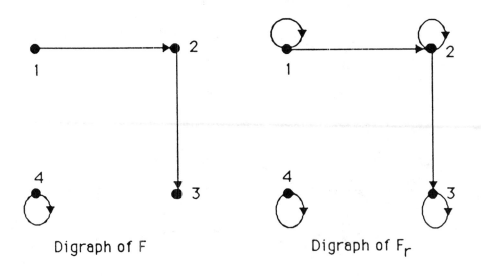

Digraph of F   Digraph of F_r

Figure 8.10: Digraph representing F and F_r

representing its reflexive closure, F_r, which is also shown in Figure 8.10.

If the relation is represented by a Boolean matrix, you can construct the Boolean matrix for the reflexive closure by inserting *true* in every diagonal entry of the matrix. So, the matrices representing F and F_r are

$$M(F) = \begin{bmatrix} F & T & F & F \\ F & F & T & F \\ F & F & F & F \\ F & F & F & T \end{bmatrix} \qquad M(F_r) = \begin{bmatrix} T & T & F & F \\ F & T & T & F \\ F & F & T & F \\ F & F & F & T \end{bmatrix}$$

Finding the symmetric closure, R_s, of a relation R on S is also fairly straight-forward. The basic idea is to initialize the symmetric closure to equal the underlying relation and then for each $[x, y]$ in the relation include $[y, x]$ in R_s. For example, the symmetric closure of F on $\{1..4\}$ defined in equation 8.1 is

$$F_s = \{[1, 2], [2, 3], [4, 4], [2, 1], [3, 2]\}$$

To find the digraph representation of the closure, for each edge in the original digraph, you add an edge with reverse direction. The symmetric closure of a relation can also be represented by a graph, as in Figure 8.11. Inserting

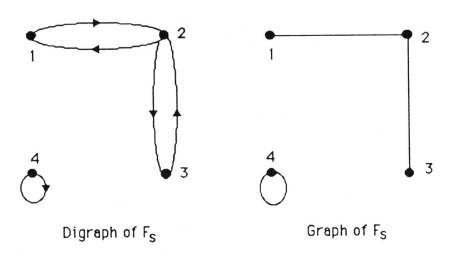

Digraph of F_s Graph of F_s

Figure 8.11: Graphic representations of F_s

trues in the Boolean matrix representing a relation, so that corresponding rows and columns have the same entries, gives the matrix representation of the symmetric closure. So, in our example,

$$M(F_s) = \begin{bmatrix} F & T & F & F \\ T & F & T & F \\ F & T & F & F \\ F & F & F & T \end{bmatrix}$$

Notice that $M(F_s) = (M(F_s))^T$, and so the matrix representing a symmetric relation is symmetric. (See Exercise 6.3.2.)

Finally, the general idea behind the construction of the transitive closure of R on S, or R_t, is to add $[x, y]$ to R_t whenever there exists a "chain" of elements in R that link x and y together. That is, if R is represented by a map, then whenever there exists $z_1, z_2, \ldots, z_k$ in S, such that $[x, z_1], [z_1, z_2], \ldots, [z_k, y]$ are all in R, you include $[x, y]$ in R_t. Consider the relation H on $\{1..3\}$, where

$$H = \{[1, 2], [2, 3], [3, 1]\} \qquad (8.2)$$

To construct H_t, you include $[1, 2]$, $[2, 3]$, and $[3, 1]$ in H_t since $H \subseteq H_t$, and include $[1, 3]$ since $[1, 2]$ and $[2, 3]$ are in H, and add $[1, 1]$ since $[1, 2]$, $[2, 3]$, and $[3, 1]$ are in H, and so on. Continuing the process, you have

$$H_t = \{[1, 1], [1, 2], [1, 3], [2, 1], [2, 2], [2, 3], [3, 1], [3, 2], [3, 3]\} \qquad (8.3)$$

On the other hand, if R is represented by a digraph, the idea is to add an edge from vertex v to w whenever there is a chain of edges connecting v to w. What is the digraph representation of the transitive closure of relation H given in equation 8.2?

Determining the Boolean representation of the transitive closure from the Boolean matrix corresponding to the underlying relation can be a little tricky. It turns out, however, that for any relation R on S, where $\#S = n$,

$$M(R_t) = M(R) + M(R)^2 + \cdots + M(R)^n \qquad (8.4)$$

Before showing that this is true, let's see how it works for the transitive closure of the relation, H, defined in equation 8.2. By looking at the map representation for H_t in equation 8.3, you can see that the Boolean representation of H_t is the 3×3 matrix that has *true* for the value of every entry. Moreover,

$$M(H) + M(H)^2 + M(H)^3$$

$$= \begin{bmatrix} F & T & F \\ F & F & T \\ T & F & F \end{bmatrix} + \begin{bmatrix} F & F & T \\ T & F & F \\ F & T & F \end{bmatrix} + \begin{bmatrix} T & F & F \\ F & T & F \\ F & F & T \end{bmatrix}$$

$$= \begin{bmatrix} T & T & T \\ T & T & T \\ T & T & T \end{bmatrix}$$

$$= M(H_t)$$

which supports our claim.

But why is this *always* the case? Let's examine the construction of R_t again. We know that $R \subseteq R_t$ and that $R^2 \subseteq R_t$, since whenever $[x, z_1]$ and $[z_1, y]$ are in R, $[x, y] \in R^2$ and $[x, y] \in R_t$. In fact, if R is a relation on S, where $\#S = n$,

$$R^k \subseteq R_t \qquad \text{for all } 1 \le k \le n$$

since whenever $[x, z_1], [z_1, z_2], \ldots, [z_{k-1}, y]$ are all in R, $[x, y] \in R^k$ and $[x, y] \in R_t$. Therefore,

$$R \cup R^2 \cup \cdots \cup R^n \subseteq R_t$$

Furthermore,

$$R_t \subseteq R \cup R^2 \cup \cdots \cup R^n$$

since each $[x, y] \in R_t$ corresponds to a sequence of k elements in R, where $1 \le k \le n$ and hence is a member of R^k. Therefore,

$$R_t = R \cup R^2 \cup \cdots \cup R^n$$

But then, using this fact and recalling that the matrix representing the composition of two relations equals the product of the matrices of the given relations and that the matrix corresponding to the union of two relations equals the sum of the relations' respective matrices (see Exercise 8.2.10), we have

$$\begin{aligned} M(R_t) &= M(R \cup R^2 \cup \cdots \cup R^n) \\ &= M(R) + M(R^2) + \cdots + M(R^n) \\ &= M(R) + M(R)^2 + \cdots + M(R)^n \end{aligned}$$

which is precisely what we claimed in equation 8.4.

One very important class of relations includes the relations that are reflexive, symmetric, and transitive. Relations that satisfy all three of these properties are called *equivalence relations*. For instance, the relation DM, defined by the condition "live in the same dorm" on the set of all students at your school, is an equivalence relation, where we are assuming that every student in your school is assigned to one and only one dorm. That is,

- DM is reflexive, since all students in the school live in a dorm, and every student "lives in the same dorm as" her or himself.

- DM is symmetric, since whenever a student x "lives in the same dorm as" a student y, then y must "live in the same dorm as" x.

- DM is transitive, since whenever a student x "lives in the same dorm as" a student y and y "lives in the same dorm as" a student z, then it must be the case that x "lives in the same dorm as" z.

What about the set of all students that are related to a particular student p? If we use a set former to construct DM as we did before, such as

$$DM = \{[x,y] : x, y \in S \mid y \text{ lives in the same dorm as } x\}$$

where S is the set of all students at your school, then the set of all students that "live in the same dorm as" p is given by $DM\{p\}$. The set $DM\{p\}$ is called the *equivalence class* of p.

Now, suppose q is another student at your school. How does this student's equivalence class, $DM\{q\}$, relate to $DM\{p\}$? We claim that because DM is an equivalence relation, that is, reflexive, symmetric, and transitive, there are two possibilities: either they are exactly the same or they have no elements in common. In fact, for all $x, y \in S$,

$$DM\{x\} = DM\{y\} \qquad \text{if and only if} \qquad [x,y] \in DM \quad (8.5)$$
$$DM\{x\} \cap DM\{y\} = \emptyset \qquad \text{if and only if} \qquad [x,y] \notin DM \quad (8.6)$$

Of course, both statements hold for any relation that is an equivalence relation.

To show that statement 8.5 holds, we need to prove two things:

$$DM\{x\} = DM\{y\} \quad \Longrightarrow \quad [x,y] \in DM \qquad (8.7)$$
$$[x,y] \in DM \quad \Longrightarrow \quad DM\{x\} = DM\{y\} \qquad (8.8)$$

To prove equation 8.7, assume that $DM\{x\} = DM\{y\}$ and show that $[x,y] \in DM$. Now, $x \in DM\{x\}$ since DM is reflexive. But by assumption, $DM\{x\} = DM\{y\}$, so $x \in DM\{y\}$, and therefore, $[y,x] \in DM$. But, then $[x,y] \in DM$ since DM is symmetric, and we're done.

To prove statement 8.8, assume that $[x,y] \in DM$ and show that the following are true:

$$DM\{x\} \subseteq DM\{y\} \qquad (8.9)$$
$$DM\{y\} \subseteq DM\{x\} \qquad (8.10)$$

Let's first show that $DM\{x\} \subseteq DM\{y\}$ whenever $[x, y] \in DM$. Let t be an arbitrary element of $DM\{x\}$. Then, $[x, t] \in DM$. Now, by assumption, $[x, y] \in DM$ and DM is symmetric, so $[y, x] \in DM$. Thus, $[y, t] \in DM$ since DM is transitive, and therefore, $t \in DM\{y\}$. Consequently, $DM\{x\} \subseteq DM\{y\}$. The proof that $DM\{y\} \subseteq DM\{x\}$ whenever $[x, y] \in DM$ is totally analogous. Therefore, for all $x, y \in S$,

$$DM\{x\} = DM\{y\} \qquad \text{if and only if} \qquad [x, y] \in DM$$

whenever DM is an equivalence relation. We leave the proof of statement 8.6 for the exercises (Exercise 8.3.14).

What about the set of all equivalence classes associated with an equivalence relation R on S? How do they relate to the set S? Since an equivalence relation is reflexive, every element of S is a member of an equivalence class, namely the one it determines. So,

$$S \subseteq \bigcup_{x \in S} R\{x\}$$

Furthermore, since every equivalence class is a subset of S,

$$\bigcup_{x \in S} R\{x\} \subseteq S$$

Therefore,

$$S = \bigcup_{x \in S} R\{x\}$$

We know that any two equivalence classes of an equivalence relation on a set S are either identical or they are disjoint. We also know that the union of all the equivalence classes equals S. Whenever we have a collection of nonempty subsets of a set that satisfies these two properties, we say that the collection forms a *partition* of the given set. Therefore, the collection of all equivalence classes of an equivalence relation on a set S forms a partition of S. Conversely, every partition of a set defines an equivalence relation on the set. For example, consider the partition

$$P = \{\{a, b\}, \{c\}\}$$

of the set $\{a, b, c\}$. To construct the associated relation, think of each of the subsets of the partition as corresponding to an equivalence class of the new relation, R. Therefore, since $\{a, b\} \in P$, you add $[a, a], [a, b], [b, b], [b, a]$ to R, and since $\{c\} \in P$, you add $[c, c]$ to R. Thus,

$$R = \{[a, a], [a, b], [b, b], [b, a], [c, c]\} \qquad (8.11)$$

is the relation on S determined by the partition P, whose set of equivalence classes equals P.

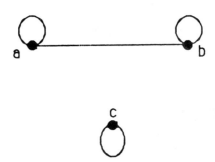

Figure 8.12: Graph representing R

Let's look at the graphic representation of an equivalence relation. The graph representing the relation given in equation 8.11 is shown in Figure 8.12. Observe that since an equivalence relation is symmetric, it can be represented by a graph, instead of a digraph. Furthermore, there is a loop at each vertex in the graph since an equivalence relation is reflexive. Also, there is a one-to-one correspondence between the equivalence classes of the relation and the connected subgraphs of the graph. As a result, the graph representing the equivalence relation DM would have a vertex for each student in your school. Each vertex would have a loop and be connected by an edge to all the other vertices that represent students who live in the same dorm. Each collection of students who live in the same dorm constitute an equivalence class and would form a connected subgraph of the graph.

Another important kind of relation is a relation that is reflexive, antisymmetric, and transitive. A relation that satisfies these three properties is called a *partial order relation*. For example, the relation $\subseteq$ on a collection of subsets of a given set, U, is a partial order relation since

- "$\subseteq$" is reflexive, since for any subset A of U,

$$A \subseteq A$$

- "$\subseteq$" is antisymmetric, since for any subsets A and B of U,

$$A \subseteq B \wedge B \subseteq A \Longrightarrow A = B$$

- "$\subseteq$" is transitive, since for any subsets A, B, and C of U,

$$A \subseteq B \wedge B \subseteq C \Longrightarrow A \subseteq C$$

Another relation that we examined earlier that is also a partial order relation is the relation "is a factor of" on a given set of nonnegative integers. We have already discussed the fact that "is a factor of" is reflexive. Spend a few moments convincing yourself that it is also antisymmetric and transitive.

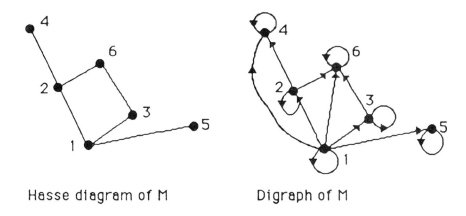

Hasse diagram of M   Digraph of M

Figure 8.13: Graphic representations of *Fac*

A partial order on a finite set can be represented by a digraph, but an alternate graphic representation of a partial order relation is a *Hasse diagram*. A Hasse diagram, representing a relation R on S, consists of a collection of vertices and undirected edges, which are interpreted as follows.

1. For each vertex corresponding to $x \in S$, $[x, x] \in R$.

2. If the vertex corresponding to $y \in S$ is located *above* the vertex corresponding to $x \in S$ and they are connected by an edge, then $[x, y] \in R$.

3. If the vertex corresponding to $y \in S$ is located above the vertex corresponding to $x \in S$ and the two vertices are connected by a sequence of edges, then $[x, y] \in R$.

Therefore, unlike a graph, the *location* of the vertices in a Hasse diagram is important. Furthermore, the edges indicating that the relation is transitive and the loops indicating that it is reflexive are omitted. As an example, consider the partial order relation "is a factor of" on $\{1..6\}$, or

$$Fac = \{[n, m] : n, m \in \{1..6\} \mid m \bmod n = 0\}$$

Then,

$$Fac \ = \ \{[1,1],[1,2],[1,3],[1,4],[1,5],[1,6],[2,2],$$
$$[2,4],[2,6],[3,3],[3,6],[4,4],[5,5],[6,6]\}$$

Figure 8.13, shows the digraph and the Hasse diagram of *Fac*, while Figure 8.14 gives the Hasse diagram of the relation $\subseteq$ on the collection of all subsets of $\{1..3\}$, or

$$Sub = \{[A,B] : A, B \in \text{pow}(\{1..3\}) \mid A \subseteq B\}$$

How many edges, including loops, would you have to add to the Hasse

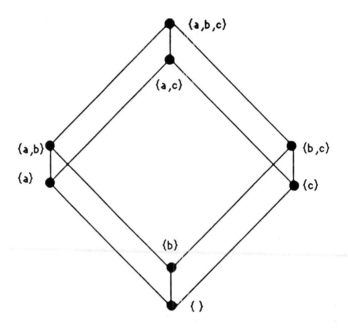

Figure 8.14: Hasse diagram representing *Sub*

diagram of *Sub* if you wanted to represent it by a digraph?

Summary of Section 8.3

This section introduces a number of properties of a relation from a set to itself. We discussed ways of determining when a relation satisfies each of the properties and how to extend a relation to a relation that satisfies a given property.

We also discussed two important classes of relations, namely, equivalence relations and partial order relations. Finally, we introduced Hasse diagrams as another way of representing a partial order.

Exercises

8.3.1 Properties of relations and ISETL `funcs`.

   a. Assume R is a relation on a set S that is represented by a map. Define ISETL `funcs` that accept R and S and return *true* if R satisfies the given property and *false* otherwise.

      i) R is reflexive.

      ii) R is symmetric.

      iii) R is transitive.

      iv) R is antisymmetric.

   b. Use your `funcs` from part a. to define `funcs` that accept a relation R and a set S and return *true* if R is the specified type of relation and *false* otherwise.

      i) R is an equivalence relation.

      ii) R is a partial ordering.

   c. Use your `func` in part b. to define a `func` that accepts a relation R and a set S, tests if R is an equivalence relation on S, and if it is, returns a set containing all the equivalence classes of R.

8.3.2 Determine whether the following examples listed in Exercise 8.2.1 are reflexive, symmetric, transitive, and/or antisymmetric:

   a. Example 1.

   b. Example 2.

   c. Example 3.

   d. Example 4.

   e. Example 5.

   f. Example 8.

   g. Example 9.

   h. Example 10.

   i. Example 11.

   j. Example 12.

8.3.3 Give an example of a relation R on $\{1..10\}$ that satisfies each of the following statements:

    a. R is reflexive, but neither symmetric nor transitive.

    b. R is symmetric, but neither reflexive nor transitive.

    c. R is transitive, but neither reflexive nor symmetric.

    d. R is reflexive and symmetric, but not transitive.

    e. R is reflexive and transitive, but not symmetric.

    f. R is symmetric and transitive, but not reflexive.

8.3.4 Assume R is a relation on A, and I_A is the identity relation. Explain why

    a. R is reflexive if and only if $I_A \subseteq R$.

    b. R is symmetric if and only if $R = R^{-1}$.

    c. R is transitive if and only if $R \circ R = R$.

8.3.5 Assume that R is a relation on A where $A = \text{domain}(R)$. Show that whenever R is symmetric and transitive, then R is also reflexive.

8.3.6 Assume R is a relation on $S = \{x_1, \ldots, x_n\}$ that is represented by a Boolean matrix. Say as much as you can about the matrix representing R if

    a. R is reflexive.

    b. R is symmetric.

    c. R is antisymmetric.

    d. R is transitive.

8.3.7 Suppose R is a relation on S represented by a **map**. Write an ISETL **func** that accepts R and S and returns the relation's

    a. reflexive closure.

    b. symmetric closure.

    c. transitive closure.

8.3.8 Find the reflexive, symmetric, and transitive closures of the following relations defined in Exercise 8.2.1:

    a. Example 1.

    b. Example 10.

    c. Example 11.

    d. Example 12.

8.3.9 Suppose A is an $n \times n$ Boolean matrix. Then, *Warshall's algorithm* states that

$$\sum_{k=1}^{n} A^k = A(I + A(I + A(I + A(\cdots A(I + A(I + A))) \cdots)$$

where I is the $n \times n$ multiplicative identity Boolean matrix. (Notice the similarity between Warshall's algorithm and Horner's method, which we discussed in Chapter 1, Section 1.3.)

    a. Use Warshall's algorithm to find the transitive closure of the relation given in equation 8.2.

    b. Write an ISETL program that accepts an $n \times n$ Boolean matrix A and returns $\sum_{k=1}^{n} A^k$ by applying Warshall's algorithm.

8.3.10 Write an ISETL **func** that accepts a partition of a set and returns the associated equivalence relation represented by a **map**.

8.3.11 Maximal and minimal elements of a partial order relation. If R is a partial order relation on S, then

    1. $M \in S$ is a *maximum* of R if $\forall x \in \text{domain}(R), [x, M] \in R$.

    2. $m \in S$ is a *minimum* of R if $\forall y \in \text{domain}(R), [y, m] \in R$.

    a. Find the maximal and minimal elements of the following relations:

        i) $M = \{[n, m] : n, m \in \{1..6\} \mid m \bmod n = 0\}$

        11) $T = \{[A, B] : A, B \in pow(\{1..3\}) \mid A \subseteq B\}$

    b. What can you say about the Hasse diagram of a partial order relation that has a

        i) maximal element.

        ii) minimal element.

    c. Compare the concept of a maximal element of a partial order relation to the **func d_max** that we discussed in Chapter 1, Section 1.5.1.

8.3.12 For each of the following relations,

    1. Determine whether the relation is an equivalence relation, a partial order, or neither.

    2. If the relation is an equivalence relation on a finite set, represent it by a graph.

    3. If the relation is an equivalence relation, find or describe all of its equivalence classes.

4. If the relation is a partial order relation on a finite set, represent it by a Hasse diagram.

a. The relation "is a sibling of" on the set of all people.

b. $R = \{[g, h] : g, h \in \{1, 2, 3, 5, 6, 10, 15, 30\} \mid h \bmod g = 0\}$.

c. $T = \{[a, b] : a, b \in \mathcal{R} \mid |x| < |y|\}$.

d. The relation "is a subset of" on $\{\emptyset, \{1\}, \{1, 2\}, \{2, 3\}\}$.

e. The relation "is similar to" on the set of all triangles.

f. The relation R on the set M_n of all $n \times n$ matrices with real entries where for $A, B \in M_n$, $[A, B] \in R$ means that there exists $P \in M_n$ invertible such that $A = P^{-1}BP$.

g. $X = \{[a, b] : a, b \in \{1..10\} \mid (b - a) \bmod 3 = 0\}$.

h. $\{[[a, b], [c, d] : a, c \in I, b, d \in I - \{0\} \mid ad = bc\}$, where I is the set of all integers.

8.3.13 A relation R on S is

    − *irreflexive* if $\forall x \in S$, $[x, x] \notin R$.

    − *asymmetric* if for $x, y \in S$, $[x, y] \in R$ implies that $[y, x] \notin R$.

    − a *total order* if R is a partial order and $\forall x, y \in S$, either $[x, y] \in R$ or $[y, x] \in R$.

a. Give an example of an irreflexive relation.

b. What can you say about the digraph and Boolean matrix representations of a relation on a finite set that is irreflexive.

c. Give an example of an asymmetric relation.

d. What can you say about the digraph and Boolean matrix representations of a relation on a finite set that is asymmetric.

e. Show that every asymmetric relation is antisymmetric. Give an example to show that the converse of this statement is not true.

f. Give an example of a total order on a finite set. Represent your example by a Hasse diagram.

8.3.14 Assume R is an equivalence relation on S. Prove that $\forall x, y \in S$,

$$R\{x\} \cap R\{y\} = \emptyset \qquad \text{if and only if} \qquad [x, y] \notin R$$

8.4 Digraphs

In general, a *digraph*, $D = \{V, E\}$, consists of a nonempty finite set of *vertices*, V, and a set of *edges*, E, where each edge is associated with an ordered pair of vertices. Consequently, an edge can be represented by a 2-tuple. In the discussion that follows, we will assume

- D does not have any *parallel* edges, that is, distinct edges that have the *same* initial and terminal vertices.

- D does not have any *isolated* vertices, that is, vertices that are neither initial nor terminal endpoints of an edge.

Under these assumptions, D can be represented by a relation on $V \times V$, namely,

$$\{[x, y] : x, y \in V \mid (\exists e \in E \mid e = [x, y])\} \subseteq V \times V$$

One way to represent a digraph is with a picture. For example, if a, b, c, and d represent vertices, then the digraph

$$D = \{[a, b], [b, b], [b, c], [d, b], [c, d], [d, a], [a, d]\} \qquad (8.12)$$

is represented by *both* of the diagrams in Figure 8.15.

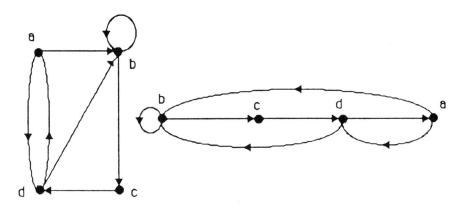

Figure 8.15: Representations of digraph D

Notice that the two representations of D are "structurally the same," since there is a one-to-one, onto correspondence between their vertices and between their edges and the associated vertices.

Finally, we can represent a digraph in ISETL by a map. But, first we need to examine how to represent vertices in ISETL. When we used a digraph to represent a relation R on a set S, we associated each vertex in the digraph with an element in S. Vertices by themselves, however, are abstract points that have no particular properties other than their own identity.   Therefore, we represent them by an ISETL object called an atom. We can create atoms by using the special predefined nonconstant name newat. Each time you call newat it creates a new atom that is distinct from all the previously generated atoms. Atoms can be members of sets or tuples, but the only operations on them are tests for set and tuple membership, equality and inequality, and type. These ideas are illustrated in the following ISETL session. As you read through the code, think about x and y as being ISETL identifiers representing two vertices in a digraph.

```
> x := newat; y := newat;
> x; y;
!1!;
!2!;
> x = y; x /= x; is_atom(x);
false;
false;
true;
> V:= {x,y}; e1:= [x,y]; e2:= [x,x]; D:= {e1,e2};
> V; D; x in e1; e1 notin D;
{!2!,!1!};
{[!1!,!1!], [!1!,!2!]};
true;
false;
> !quit;
```

With these ideas in mind, you can represent the digraph given in equation 8.12 by the map, D, where

```
> V := [ newat :  i in [1..4]];
> D := {[V(1),V(2)], [V(2),V(2)], [V(2),V(3)],
>> [V(4),V(2)], [V(3),V(4)],
>> [V(4),V(1)], [V(1),V(4)]};
```

Then, domain(D) is the set of all vertices that are initial points of edges in D, while image(D) is the set of all vertices that are terminal points. On the other hand, D{v} returns the set of all vertices that can be reached from v in one step by traveling along some edge, where D{v} is empty whenever v notin domain(D).

You can think about walking around a digraph by traveling on the directed edges. Whenever you walk from one vertex to another on a sequence of successive edges, the sequence of edges that you traveled on is called a *path* from the initial vertex to the terminal vertex. The *length* of the path is the number of edges traveled while going from one point to the other. For example, in the digraph represented in Figure 8.15, the **tuple** of 2-tuples,

$$[[c,d],[d,a],[a,d],[d,a],[a,b],[b,b]]$$

represents a path from c to b whose length is 6. This path can also be represented by the **tuple**

$$[c,d,a,d,a,b,b]$$

which indicates that you traveled from c to d, and then from d to a, and so on.

An interesting question is, for a given digraph, what is the length of the shortest path from one specified vertex to another? For instance, if you look again at Figure 8.15, you can see that the shortest path from c to b has length 2. The digraph represented in Figure 8.15 is not very complicated, so it is easy to eyeball the length of the shortest path from one vertex to another. Obviously, this is not always the case. So, let's think about how we might define a **func** that accepts a digraph represented by a **map**, D, two vertices, **f** and **t**, and returns the length of the shortest path, if a path exists, from **f** to **t**. The basic idea is starting at **f**, find the **set** of all the vertices that can be reached from **f** by traveling along one edge. We'll call this **set reached**. Include in **reached** all the vertices that can be reached by traveling along two edges, and so on, until you reach **t**. So, noting that for each **v in reached** the set of vertices that can be reached in one step from **v** is D{v}, we might define our **func** as follows

```
min_length := func(D, f, t);
                local length, reached, next_ver;
                length := 0;
                reached := {f};
                while t notin reached do
                    next_v := %+{D{v} :  v in reached};
                    reached := reached + next_v;
                    length := length + 1;
                end;
                return length;
            end;
```

The problem with our **func** as it now stands is that if there is no path from **t** to **f**, then **t** will never be in the **set reached**, and the **while** loop will execute forever. So, we need to modify our **func** so that we not only keep track of all the vertices we have reached so far (in **reached**) but also keep track of the **set** of *new* vertices—that is, vertices we have not seen before—that can be seen in one step from each of the new vertices that we saw in the last step. We will put these new vertices in the **set new_ver**. If at any point **new_ver** becomes empty, indicating that there are no new vertices to visit, we'll stop the search. So, another version of min_length is

```
min_length := func(D, f, t);
                local length, reached, new_ver;
                length := 0;
                reached := {f};
                new_ver := {f};
                while (t notin new_ver) and (new_ver /= {}) do
                    new_ver := %+{D{v} :   v in new_ver} -
                        reached;
                    reached := reached + new_ver;
                    length := length + 1;
                end;
                if t in new_ver then return length;
                else return om; end;
        end;
```

Up to this point, we have restricted our discussion to directed graphs. A *graph*, $G = \{V, E\}$, on the other hand, is a set of vertices, V, and a set of edges, E, where each edge is associated with an unordered pair of vertices, called the *endpoints* of the edge. Consequently, an edge of a graph can be represented by a **set**, that has cardinality 2 whenever the endpoints of the edge are distinct and cardinality 1 if the edge is a loop. So, a graph that has no multiple edges or isolated vertices can be represented by a **set** of **sets**.

We end this section with a famous problem involving graphs and paths. Many years ago, the people in the town of Königsberg, in what was then called Prussia, used to spend their Sunday afternoons strolling across the bridges over the river Pregel. The figure we used when discussing *modus ponens* in Chapter 7, on page 329, illustrates the location of the various bridges. Their goal was to start at any location, stroll across each bridge exactly once, and return to the place where they started. No one was successful.

In 1735, Leonard Euler used the graph shown in Figure 8.16 to *model* the real-life situation and then proved why it was impossible to find such a path. This was the beginning of graph theory. His argument went something like this: We can associate with each vertex in a graph a nonnegative integer equal to the total number of edges terminating at the given vertex. This is called the *degree* of the vertex, where it is assumed that a loop contributes 2 to the degree. So, for instance, in Figure 8.16, the degree of B is 3, and the degree of D is 5. Euler showed that in order for the desired path to exist, the degree of each vertex in the graph has to be *even*. Why must this be the case? If such a path exists, then, for each vertex, all of the edges terminating at the vertex can be put into pairs, consisting of an

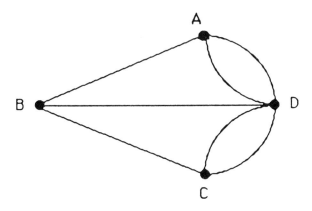

Figure 8.16: Model of Königsberg Problem

"arrival" edge and a "departure" edge, where, for the initial vertex in the
path (which also is the path's terminal vertex), the first and last edges are
paired. Therefore, if a graph has a path that starts and ends at the same
vertex and traverses each edge exactly once—we call a path with these
properties an *Eulerian circuit*—then the degree of each of its vertices must
be even. Since every vertex of the graph shown in Figure 8.16 has odd
degree, the people of Königsberg could never start at any point, cross each
bridge exactly once, and return to where they started.

Summary of Section 8.4

This section offers a brief glimpse at some topics that are important in
graph theory. We looked at various representations of digraphs. We also
discussed two path related problems: finding the length of the shortest
path from one vertex to another in a digraph and showing that if a graph
has an Eulerian circuit then the degree of each vertex must be even.

We discussed how you might represent vertices in ISETL using **atoms**.

Exercises

8.4.1 Represent the following digraphs as **maps** using **atoms** to represent
the vertices.

a. $D1 = \{[v_1, v_1], [v_2, v_2], [v_1, v_2], [v_1, v_1]\}$

b. $D2 = \{[a, b], [c, d], [b, b], [d, c], [e, e]\}$

c. The digraph representation of the relation *Fac* in Figure 8.13.

d. The digraph in Figure 8.17.

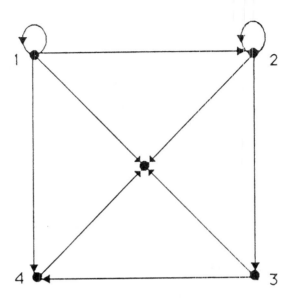

Figure 8.17: Exercise 8.4.1d

8.4.2 Write an ISETL **func** that accepts a positive integer n and returns a map representing a digraph that has n vertices, no loops, and an edge from each vertex to every other vertex.

8.4.3 Write an ISETL **func** that accepts a **map** representing a digraph, a vertex v, and a positive integer n and returns the digraph formed by adjoining n new vertices and n edges from v to each of the new vertices of the original digraph.

8.4.4 Two edges of a digraph are said to be *parallel* if they are determined by the same pair of vertices and they have the same direction.

    a. Give an example of a digraph that has parallel edges. Represent your example with a picture, labeling each edge as well as each vertex.

    b. Describe how you might represent a digraph that has parallel edges in ISETL.

c. Discuss the following statement: Every relation on a finite set can be represented by a digraph, but every digraph does not correspond to a relation.

8.4.5 Paths in digraphs.

a. Write an ISETL func that accepts a path represented by a tuple of 2-tuples, such as $[[a, b], [b, c]]$, and returns its tuple representation, $[a, b, c]$.

b. Write an ISETL func that accepts a digraph, represented by a map, and a path, represented by a tuple, and returns *true* if the path is a Eulerian circuit and *false* otherwise.

c. Run the func min_length on the digraphs given in Exercise 8.4.1 with a variety of inputs.

8.4.6 Two vertices in a graph are *adjacent* if they are endpoints of a common edge. Give an example of a graph, such that "is adjacent to" is an equivalence relation on its set of vertices.

8.4.7 Graphs in ISETL.

a. Describe how you can represent a graph by a set of sets in ISETL, where the graph has no isolated vertices or parallel edges but may have loops.

b. Write an ISETL func that accepts a graph and a vertex and returns the degree of the vertex.

c. Write an ISETL func that tests whether or not a given edge is a loop.

d. Write an ISETL func that accepts a graph and two vertices and tests whether or not the vertices are adjacent. (See Exercise 8.4.6.)

8.4.8 A graph is *connected* if there exists a path between any pair of distinct vertices. Prove that if a graph is connected and every vertex has even degree, then the graph has an Eulerian circuit.

8.4.9 The degree of a vertex of a graph.

a. Can you construct a graph with 4 vertices so that the degree of three of the vertices is 2 and the degree of other is 3?

b. Write an ISETL func that accepts a graph that can be represented by a set of sets (see Exercise 8.4.7) and a vertex and returns the degree of the vertex. Note that a loop contributes 2 to the degree.

c. Write an ISETL **func** that accepts a graph represented by a **set** of **sets** and returns the sum of the degrees of all the vertices in the graph.

d. Show that the sum of the degrees of all of the vertices in a graph is even.

e. Show that in any graph there is an even number of vertices with odd degree.

Index

:=
see assignment statement

$
see comments

=, /=, 23, 46
see equal

<, <=, >, >=, 23, 46
see relational operators

-, 23, 46
see difference
see floating-point operations
see integer operations

%
see compound operator

..
see integers, useful sets, tuples
see slice

+, 23, 46
see concatenation
see floating-point operations
see integer operations
see union

#
see cardinality
see length

/, 23, 46
see floating-point operations

*, 23, 46
see floating-point operations
see integer operations
see intersection
see replication

**, 23, 46
see exponentiation

abs, 23, 46

abstract algebra, 259–260
addition principle, 270
extended, 270
additive identity
matrix, 301
additive inverse
matrix, 301
modular arithmetic, 37
adjacent vertices, 403
alphabet, 279, 343, 346
and
see conjunction
antisymmetric relation, 382
arb, 118, 134, 139–140
arrangement
see permutation
array
one-dimensional, 282
two-dimensional, 283
arrow diagram, 368
assignment statement (:=), 10
associative binary operations, 30
associative law
Boolean expression, 80
matrices, 301
sets, 121
asymmetric relation, 396
atom, 56, 398
attribute, 215

base 2 representation, 15
base case, 320, 323
base conversion, 15, 17
Bierce, Ambrose, 84

Only order after checking to see that there is no local copy of ISETL available to you. ISETL and its documentation may be freely copied, so don't spend your money unless it is necessary. (Instructors: Please provide copies to your students.)

- 256K machines are a tight fit. I'd recommend 512K.

- Check those items desired. All versions except the 1-sided Mac disk include both executable and source code. If you want a version for a machine not listed, please write for further information.

- Include payment in U.S. dollars drawn on a U.S. bank — no purchase orders, please.

- Comments and suggestions may be mailed to the address below or to the following electronic addresses:
  Bitnet: gary@clutx   Internet: gary@clutx.clarkson.edu

Order	Item	Media	Cost
☐	Documentation	Printed	$10
☐	MS-DOS (2.0+)	DS/DD 5 1/4″ floppy	$10
☐	Macintosh	3 1/2″ microfloppy	$10
☐	Mac *1-sided*	3 1/2″ microfloppy	$10
☐	Unix	1600 bpi tape, TAR format	$15
☐	Vax/VMS	1600 bpi tape, Backup format	$15

Total　　　　　　　　　　　　　　　　　　　　　　_____

Please letter neatly　　　　　　　　　　　　*Mail to:*

Name _____

Gary Levin
Dept of Math and CS
Clarkson University
Address _____
Potsdam, NY 13676

_____

City _____ State _____ Zip _____

e-mail address _____

Phone number _____

Minimum memory _____

DEPARTMENT OF MATHEMATICS
BRONFMAN SCIENCE CENTER
WILLIAMS COLLEGE
WILLIAMSTOWN, MA 01267